THE SCIENCE OF LOST WORLDS

BSPS | OPEN

BSPS Open SERIES

ISSN 2564-3169 (Print) ISSN 2564-3177 (Online)

BSPS Open publishes landmark, cutting edge works that represent the full breadth and diversity of the philosophy of science. Diamond Open Access, all books in this series are available freely to readers everywhere.

Published in collaboration with the British Society for the Philosophy of Science.

No. 1 *The Material Theory of Induction* | John D. Norton

No. 2 *The Large-Scale Structure of Inductive Inference* | John D. Norton

No. 3 *The Material Mind: Reduction and Emergence* | Max Kistler

No. 4 *The Multiverse: A Philisophical Introduction* | Jeremy Butterfield

No. 5 *The Science of Lost Worlds* | Adrian Mitchell Currie

UNIVERSITY OF CALGARY
Press

The Science of Lost Worlds

ADRIAN MITCHELL CURRIE

BSPS | OPEN

BSPS Open SERIES
ISSN 2564-3169 (Print) ISSN 2564-3177 (Online)

University of Calgary Press
2500 University Drive NW
Calgary, Alberta
Canada T2N 1N4
press.ucalgary.ca

LIBRARY AND ARCHIVES CANADA CATALOGUING IN PUBLICATION

Title: The science of lost worlds / Adrian Mitchell Currie.
Names: Currie, Adrian, author
Description: Series statement: BSPS open series, ISSN 2564-3169 (print), ISSN (ebook) 2564-3177; no. 5 | Includes bibliographical references and index.
Identifiers: Canadiana (print) 20260173878 | Canadiana (ebook) 20260173908 | ISBN 9781773857091 (hardcover) | ISBN 9781773857107 (softcover) | ISBN 9781773857121 (PDF) | ISBN 9781773857138 (Open Access PDF) | ISBN 9781773857114 (EPUB)
Subjects: LCSH: Paleontology—Philosophy. | LCSH: Art and science.
Classification: LCC QE711.3 .C87 2026 | DDC 560.1—dc23

The University of Calgary Press acknowledges the support of the Government of Alberta through the Alberta Media Fund for our publications. We acknowledge the financial support of the Government of Canada. We acknowledge the financial support of the Canada Council for the Arts for our publishing program.

Alberta Canada Canada Council for the Arts Conseil des Arts du Canada CERTIFIED CANADIAN PUBLISHER

The manufacturer's authorized representative in the EU for product safety is Mare Nostrum Group, B.V., Doelen 72, 4831 GR Breda, The Netherlands. Email: gpsr@mare-nostrum.co.uk

Copy editing by Glen Hoffmann
Cover art: "Wonderful Changes" by Jim Robins
Cover design, page design, and typesetting by Melina Cusano

Contents

List of Illustrations

All illustrations by Jim Robins

List of Figures

List of Tables

Acknowledgements

This book had a rather inauspicious beginning. It was July 2021, around 3 a.m., in a Christchurch hotel, three days into a fourteen-day COVID quarantine. I'd travelled to Aotearoa (New Zealand) from my home in Exeter because my mother, Yvonne, was dying of cancer and, well, I needed to be there. I wish I could say Prof. Ichthy appeared to me in a dream or something suitably dramatic, but rather the title of this book, and the simple thought that what we can know about the past depends on what exists in the present, emerged from a heady mix of jetlag, boredom, and grief.

The journey from that initial title and initial thought to this book took over four years, and its final form has been intimately shaped by far too many discussions and inspirations to list or remember systematically (so a million apologies if I miss you out here). But I'll give it a bash.

Between 2021 and mid-2023, I workshopped what turned out to be chapter 2 with a few hints of chapters 3 to 5 through various presentations at the University of Notre Dame, the London School of Economics, Brown University, and the University of Exeter, as well as meetings of the British Society for the Philosophy of Science and the Philosophy of Science Association (and probably other places missed by my subpar record-keeping). The audiences at those talks were crucial for developing the initial conceptual framework (especially the "loss/erasure" distinction), and hence, for getting the project up and running.

Writing began in earnest over the summer of 2023. This was partly motivated by my failing to get the project funded and a subsequent fear of (as has happened to me in the past) the whole idea foundering due to becoming too emotionally and intellectually entangled with an unsuccessful grant bid. Further motivation came when I decided to incorporate drafts into my undergraduate teaching that autumn. By the end of 2023, the students in the University of Exeter's philosophy module, *The Deep Past, History, and*

Humanity, read through the first drafts of chapters 1 to 5. The following year's students made it to chapter 8, and—I'm planning as I write this—the 2025 batch will read the whole thing. Attaching research to teaching is a great motivation for actually getting the work done. Exposing draft work to engaged students is enormously beneficial for developing the ideas and for articulating them with clarity. I certainly threw my students into the deep end, and I'm enormously grateful to them for their time, focus, and grace. I'm also grateful to other undergraduate groups with whom I've shared some of the research for this book: University of Exeter's undergraduate philosophy society, Katy Shorey's HPS group at Northeastern University, and Luca Corti's class at the University of Padua.

In the summer of 2024, I ran the Philosophy of the Paleosciences Retreat: a few days in the utterly gorgeous nature reserve and farm at Kingcombe Meadows in Dorset. This more or less involved bribing the participants with beautiful scenery and various palaeontology-related escapades (including visiting the Etches Collection Museum of Jurassic Marine Life, several museums in Lyme Regis, and fossil hunting in Charmouth) in exchange for their reading and discussing chapters 1 through 9. Every participant provided enormously helpful feedback and encouragement, and I'm grateful to them: Alisa Bokulich, Angela Cassidy, Max Dresow, Denise Hossom, Chris Manias, Andra Meneganzin, George Newman, Frankziska Reinhard, Ross Pain, Ben Please, Beth Porter, Tom Roberts, Judyth Sassoon, Derek Turner, Katherine Valde, Kirsten Walsh, Joe Wilson, and Yoshinari Yoshida. I'm especially grateful to Kirsten, George, Tom, and Judyth for their hard work in helping me organize everything, keeping everything running smoothly, and dealing with me stressing out about it. Further thanks to Chee Wong both for help running the retreat and more generally for keeping Egenis (the Centre for the Study of the Life Sciences at the University of Exeter) and our immediate research environment from falling into administrative and logistical tatters.

Leading up to the retreat, many of the book's chapters were read by members of Egenis. In addition to those named above, I'd like to thank Tyler Brunet, John Dupre, Ariane Hanemaayer, Geoff Hughes, Zamir Kadodia, Joel Krueger, Celso Neto, Brian Rappert, Adam Toon, Maria Volkova, and Hugh Williamson.

As the book neared completion, I received incisive feedback on various elements from colleagues. Tyler Brunet and Robert Northcott read chapter 8, Mike Stuart read chapters 6 and 7, and Chris Manias read chapter 10. Ideas from later chapters were presented to the Society for Philosophy of Science in Practice, the International Society for the History, Philosophy, and Social Studies of Biology, the UK Integrated History and Philosophy of Science meeting, and Chris Manias' excellent Poppaleo series. Two anonymous referees from the University of Calgary Press (UCP) provided useful feedback, as did David Teira on behalf of the British Society for the Philosophy of Science, and Brian Scrivener from UCP. I'd also like to thank Glen Hoffmann for his flexible and diligent copy-editing of this, at times, idiosyncratic text.

I've been extremely lucky to have had patient feedback from some of the scientists whose work I draw on throughout: many thanks to Alex Liu, Charlotte Kenchington, Nick Butterfield, and Emily Mitchell for their help (and apologies for the stuff I still managed to get wrong). Extra thanks to Charlotte for allowing me to use her haunting rangeomorph image. Similar thanks to Chris Manias for keeping the book's historical elements at least not-too-embarrassing, and to Emily Hauser for her insights into the Prof. Ichthy sections.

Many of my research activities, including the retreat, were only possible due to the Leverhulme Trust's Philip Leverhulme award, which I was extremely lucky to receive. The flexibility of this funding not only gave me precious time to complete the book, but it also enabled me to follow my heart for this project by incorporating more artistic elements, one visual and one auditory.

On the visual side, it was a delight to work with Jim Robins to produce the illustrations of Prof. Ichthy and their colleagues. Jim, besides being an extremely talented artist and designer, is also wonderfully down to earth and sensible—necessary qualities, considering the sketchiness and downright silliness of some of the ideas I threw at him. As with writing the Ichthy vignettes themselves, developing the illustrations formed an active part of the philosophical work of this book. Misty ShadowFang further deserves thanks for allowing her likeness to be used, and I'd also like to thank Lucy Osler, Blaire Andres, Charlie Masquelier, and Jo O'Hara for feedback on the illustrations as they developed.

On the auditory side, I've been working with The Bookshop Band on a series of songs inspired by some of the book's ideas. In 2019, Kirsten Walsh and I met Beth Porter and Ben Please of The Bookshop Band and immediately saw that their highly conversational method of songwriting could potentially work well in a philosophical context. To our great surprise, they were keen to explore collaborations, but (as with many of these things) it wasn't until many years later that the funding—again through Leverhulme's largesse—was available. Beth and Ben attended the summer retreat (and its associated workshop) and spent several months' artist's residency at Egenis. As with Jim's illustrations, thinking through how some of the ideas in this book could be in discussion with music and lyrics became an active part of the research. I'm grateful to them for diving into such an unusual project with such camaraderie, creativity, and enthusiasm. Speaking of the auditory element, I'd also like to thank the team at Sound Gallery Studios for producing the audiobook version, especially Duncan Chave. Boo Jackson deserves special credit for elevating the Ichthy sections with her wonderful performance.

As I hope is obvious, whatever success and skill I might have in philosophy is due to the generous scaffolding provided by my colleagues, and I'm enormously grateful to them. Egenis is an incredible place to do philosophy of science, and my colleagues across philosophy, sociology, anthropology, and STS (Science and Technology Studies) are endless sources of inspiration. Geographically further afield, the philosophy of science and biology generally and the philosophy of the historical sciences (or geosciences, or palaeosciences, or whatever we're calling it) communities have been very kind to me and have proven over and again how supportive, collaborative, and rigorous philosophy of science can be.

Among all those I've mentioned and those that have slipped my mind (sorry!), Kirsten Walsh deserves special mention. Not only is she a loving, supportive partner with whom I'm extremely lucky to build a life, she is also an incisive, creative, and clear-headed philosopher whose intellectual generosity is to blame for the best bits of this book.

My mother lived for just three days after I left quarantine. I'm grateful for those days, and the time I got to spend with my father, Currance, after her passing (and I'm so sorry that my sister Nicole was unable to be there). At her online wake (such were the ways of COVID), it was striking how many people all over the world had been touched by her gregarious, caring ways. In that spirit, I dedicate this book to my mother.

She is lost, but only from a particular perspective; so many lives were enriched by her, she'll not be erased.

Me tangi, kāpā ko te mate i te marama.

1

Prof. Ichthy's Inaugural Lecture

ILLUSTRATION 1.1 Wonderful Changes

The newly minted Prof. Ichthy makes their awkward way to the podium as the auditorium fills with coughs, half-whispered conversations and the susurration of chair-shuffling.[1] The introduction had been (as usual) too long, and the applause too short, leaving an ungainly space in which Ichthy bustles about, fiddles with the overhead projector, gets the slides in place, fondly pats a glass container in which a strange-looking skull rests, and finally straightens their notes with a near ceremonial flourish.

A silence spreads across the audience (punctuated by an out-of-place "whoop!" from one of Ichthy's relatives), as Ichthy clears their throat and begins, rather inauspiciously, not with an introduction but with an instruction: "slide."

Ichthy's graduate student, waiting by the OHP,[2] carefully removes the paper obscuring the first transparency, and the strange and remarkable creature on which Ichthy's career was built is projected onto the screen behind them.

On the screen are two images. One is a fossilized specimen, its head closely resembling the skull on the podium. The specimen is remarkably preserved, more like a mineralized mummy than the usual collection of disarticulated bones. Image two is an illustration Ichthy commissioned to promote their discovery. It depicts a small, mammalian-looking creature. It is clearly a vertebrate, with bilateral symmetry and a rather squat, quadrupedal stance. It is also clearly a mammal, covered in sleek fur. Perhaps more surprising are its webbed feet, and more striking still is its bill. Those versed in anatomy (or who read Ichthy's work in preparation for the lecture) note other non-mammalian features in the illustration: a lack of clear mammaries and a rather reptile-like stance. Again, clearing their throat, Ichthy begins the lecture proper:

"I put it to you that were it not for this remarkable specimen, we would never have guessed nor imagined the existence of this strange extinct species we now call "Broadfoot." No doubt, other fossils have told us of the long-ago 'Age of Mammals.' We know there was a time in the past when Earth's dominant animals were these primitive-seeming, furry, warm-blooded, milk-providing creatures. We know they conquered the land, conquered the air,

1 Alert readers might notice some discrepancies between Currie's text and Robins' visual depictions of Ichthy's lecture. As neither could be present, both have resorted to somewhat unreliable second-hand testimony, and this inevitably leads to differences in interpretation.

2 "Overhead projector" for those too young or forgetful.

and conquered the sea. And we know a lot about them in part because of their fossils."

Ichthy waves one flipper towards the skull on the stage and another at the image behind them.

"But not just their fossils. We know a lot about the ancient mammals because of their surviving cousins. Mammals are smaller nowadays, much diminished by the standards of their grand ancestors: they are reduced to secretive, arboreal, or burrowing creatures. The mammals with whom we share the planet only hint at mammals' past. And the fossil record has provided many mammalian surprises."

Ichthy turns, gazing at the skull, their long snout pointing like an arrow.

"But a mammal like Broadfoot? A web-footed, egg-laying, aquatic mammal with a bill? And moreover, one that, our research suggests, was both venomous and used electricity to detect its prey? Had you told me of such a creature, I would have commended you on your imagination and maybe—*maybe*—would have conceded its existence to be within the realm of possibility. But were it not for the incontrovertible evidence of this remarkable fossilized specimen, I would never have thought it actual. Indeed, I bet, it wouldn't have been imagined at all . . ."

This is a book about the limits of imagination and how the sciences of the past extend and challenge those limits. How we imagine the past is highly constrained by what the present is like. Yet what survives from the past, its traces, often reveal histories that buck our expectations. In the future I've imagined for Prof. Ichthy, there are no platypuses, nor sufficiently similar descendants, nor analogues of platypuses, for the animal's existence to have occurred to them. But the remnants of the past, in this instance that fortuitous fossil, reveal a history differing dramatically from what Ichthy and their contemporaries could have imagined.

I'll call these pasts that diverge from the present *lost worlds*, and the central claim of this book is that without analyzing lost worlds and how scientists tackle them, you won't understand scientific knowledge of the deep past. What is a "lost world," then?

In chapter 2, I'll distinguish between *loss* and *erasure*. One way into the distinction leans on two kinds of evidence that Ichthy highlighted as relevant to Broadfoot. First, fossil evidence: that remarkable specimen. Second, evidence from platypus-like animals contemporaneous with the scientist. To an extent, "erasure" tracks that first kind of evidence, referring to information and its decay over time: fossils might be destroyed, or we might lack the theories or technology required to "decode" them. "Loss" refers to the second kind of evidence, and thus the ongoing existence of things: if platypuses were to go extinct, there would be none for us to examine. The professor is lucky that Broadfoot's fossil wasn't *erased*, because Broadfoot has been *lost* to extinction. Lost worlds, then, are worlds that contain different things, obey differing dynamics—they "work differently"—to the present.

Sometimes the past is hard to know because it has left incomplete records. Sometimes the past is hard to know because the past is weird. "Erasure" is the former, "loss" the latter. The next chapter is devoted to fleshing these concepts out (kindly keep objections in your back pocket until then!), so hopefully, that's enough for now.

Loss is important. Our knowledge of the past continually confronts the possibility of, and sometimes uncovers, loss: histories that work differently, are governed by different rules, and are populated by different creatures and processes than in our own time. Platypuses are unique and do not exist contemporaneously with Ichthy, and this appears to limit the professor's capacity to understand platypuses. Loss generates failures of uniformity, and this both challenges our capacity to acquire knowledge concerning the past and increases that knowledge's significance.[3] Uniformity, the past being like the present, is never guaranteed. Failures of uniformity raise challenges: If our knowledge of the past is mediated by our knowledge of the present, then how can we understand pasts that are significantly different from the present? That is, how can we have a science of lost worlds? An account of how loss is (sometimes) overcome, I'll argue, transforms how we conceive of knowledge about the past, what the past is like, and why knowing about it matters.

In the book's first half (chapters 2 through 5), I provide an account of the strategies scientists adopt in tackling lost or putatively lost worlds. I'll emphasize how historical scientists cleverly characterize their past targets—take perspectives on them—which lets them build various connections between the unknown and the known. Iteratively navigating between these connections enables them to gradually build up a rich picture of the past. Thus, the major

3 This focus on uniformity doesn't force us to tackle the problem of induction, that is, attempt to provide some a priori or a posteriori justification for taking the future to be like the past on pain of our grounds for empirical knowledge giving way.

conceptual innovations I want to introduce to the epistemology of the historical sciences are, first, the notion of *strategic perspectivalism* and, second, *iteration*. As we'll see, these notions bring together a set of ideas implied but insufficiently emphasized by the literature thus far, but they also explain how historical scientists can reconstruct the past even in circumstances where the past is unfamiliar and weird—or lost.

Throughout the book's second half (chapters 6 to 10), I draw philosophical lessons from my analysis. These lessons concern the modal knowledge scientists provide, the nature of scientific imagination, the "peculiar" history they uncover, and the relationship between the past and the present, on the one hand, and the past and the future, on the other. Let's advertise these Big Claims as a bit of motivation for the reader.

Big Claim 1: *The sciences of the deep past are sciences of deep possibility* (chapter 6). Reconstructing lost worlds requires rich modal knowledge about the deep past. I'll argue that this modal knowledge is best understood as objective, rather than epistemic. Under the right conditions, historical scientists generate bona fide knowledge of not only how the past was, but how the past could have been. Further, I'll characterize this modal knowledge in "artifactualist" terms. Scientists targeting the deep past adopt a broad range of strategies that require generating and examining a wide variety of artifacts. These artifacts possess various dispositional properties, and knowledge of those dispositions grounds inferences about the dispositional properties of entities and processes in the past. So, I argue that in terms of method, the practice of historical science intertwines the present with the past; in terms of dividends, historical science is as much, if not more, about what is possible than what was actual.

Big Claim 2: *The imagination involved in investigating lost worlds is artifactualist* (chapter 7). I'll provide an account of scientific imagination that is firmly rooted in the materiality scientists confront and construct. Scientists think, imagine, and speculate through and with various artifacts. The scientific imagination turns on the embodied coupling of scientific agents and these various artifacts. The properties of available artifacts, then, significantly constrain what scientists can imagine. This somewhat vindicates Ichthy's pessimism about their capacity to have imagined Broadfoot without the fossil discovery.

Big Claim 3: *The Paleobiological past is peculiar* (chapter 8). Realizing that the past consists of lost worlds—worlds populated with differing critters and processes operating by differing dynamics; that is, pasts that play by different rules—grounds a view of the metaphysics of history that also makes sense of some aspects of the present. Through history, things are lost, but the past also contains innovation and novelty: the generation of *new stuff*. Life's past contains both extinction and speciation, for example. This is because the past doesn't simply involve one thing happening after another, but nor does it behave with unerring uniformity. Rather, pockets of regularity emerge under particular conditions, and fall away again as those conditions change. This is a form of contingency I call *peculiarity*. Peculiarity underwrites a deeply historical conception of the kind of messy, pluralistic, patchworky metaphysics popular among many scientifically-inspired philosophers. Thus, I'll argue that geological and biological history should be understood in (to crib from Cartwright) "dappled" terms, and that these features of history explain why the natural world is disunified (insofar as it is).

Big Claim 4: *Lost worlds and the unprecedented future are epistemically asymmetric* (chapter 9). You might hope that the strategies historical scientists adopt in trying to understand surprising pasts would apply to apprehending unprecedented futures. I'll argue this is true to some extent—many of the basic strategies useful for uncovering the past are applicable to the future (and knowledge, especially modal knowledge, concerning the past is invaluable insofar as it can be turned toward the future). However, there is an epistemic asymmetry between the past and future that is baked into our status as temporally-limited epistemic agents, as knowers moving through time in a single direction. For past worlds, when we're lucky, we have *traces*, that is, downstream remnants of past events. These records of the past are critical for uncovering history's surprises (thus far exemplified by Ichthy's fossil platypus). However, for future worlds, it is unclear what the analogy of "trace" might be: there are (perhaps) no records of the future. As traces are crucial for uncovering lost worlds, not having "traces of the future" undermines our capacity to acquire knowledge of unprecedented worlds. Given the many surprises the deep past has given us, this asymmetry suggests that expecting to get unprecedented worlds right is wildly optimistic—and perhaps dangerous.

Beyond these Big Claims, I owe the reader a story about why they should care about the general topic of this book. Why is palaeobiological knowledge, or historical knowledge generally, or the nature of the past, something that's worth spending an entire book exploring? Let's see what Ichthy has to say on the topic.

Ichthy regards the audience once more, their maw swinging back from the skull. "As you've come all this way, I should say a thing or two about what I think is valuable about the kind of work I do—and why sitting through this lecture is worth it."

Ichthy glances with mild embarrassment at the gaggle of graduate students in the audience, as they adopt a more philosophical tone than perhaps they're comfortable with.

"Our perspective is typically rooted in the relatively recent here and now. The way the world appears to be now sets a bunch of expectations about how the world usually is. The now, I suppose, is taken to be the default. Current fauna and flora are dominated by gymnosperms, like conifers and ginkgo, by marine reptiles, such as yours truly, and on land by those lumbering sauropods. And that's what we expect life to be like. But discoveries like Broadfoot tell us that the here and now is not permanent, that the past is weird, that history doesn't end, or at least hasn't yet. Once there was a mammalian world, now there is our own, and we've no reason to think that our world is any more stable than theirs was. I don't think this is just a lesson about biology."

Another pause from Ichthy, as they venture even further outside of their comfort zone.

"We live in a particular set of cultural, political, and economic conditions, and these structure how we see the world. Just as we project the living world as default, we take our current social worlds as default. Most interestingly, they structure and constrain how we might imagine the world to be different. I wouldn't have imagined Broadfoot without its specimen, and it is hard to imagine the ways things could be far beyond our immediate experience. History, and the study of history, gives us some of the tools we need to break out of that perspective. So, the kind of work I do is part of that project. Behind all the apparently dry science, work like mine helps us imagine different worlds."

An account of historical knowledge that emphasizes loss and how the peculiarities of history provide rich insight into possibility offers lessons into how we should think about ourselves, and how we might partly break, or at least loosen, the shackles of our temporally-bound perspective. In addition to disrupting entrenched perspectives, knowing the past matters for understanding how we got to where we currently are and for getting some sense of what might happen in the future. In the book's final discussion, I'll turn to the nature of wonder and how knowledge of lost worlds is a powerful source of wonder. Insofar as wonder is transformative, insofar as it can take us outside of ourselves, and insofar as knowing about history is a source of wonder, historical knowledge is itself transformative. If historical knowledge is valuable, then understanding how we *get* historical knowledge is valuable, and that is what this book is about. So, that's my pitch for sitting through the book and Ichthy's lecture.

In the remainder of this introduction, I'll provide two kinds of context, one philosophical and the other historical. I'll start with history, pointing to the role lost worlds played in early Earth science: 19th-century (proto) palaeontologists and geologists claimed revolutionary significance for their work by emphasizing the differences between their present and the pasts they were uncovering. I'll then switch to philosophy, sketching where we are at vis-à-vis the epistemology of the palaeosciences. The sketch will be drawn to highlight how philosophers have focused on trace records, and their incompleteness, as the main source of, and challenge to, historical knowledge. Thus—if my discussion of loss holds water—that work has been too narrow and needs a course correction. I'll finish off, as is traditional for introductions, with a more systematic sketch of the book's structure.

1. Awful (and Marvellous!) Changes

Prof. Ichthy signals the next slide with a meaningful look at a graduate student, who, with a start, switches to the next image. Projected behind the podium is a series of extinct mammals, from the long-necked Camelopardalis to the enormous, aquatic cetacea, all connected by tree-like lines of descent. Broadfoot is prominently displayed amongst the menagerie. Ichthy continues:

"Broadfoot is part of an, until very recently, unknown past, an unknown past unconceived of by our scientific forebearers. After all, who would have imagined a world dominated by the ancestors of the tiny, scampering creatures that are all that remains of Mammalia? That these apparently insignificant animals are the last vestiges of such a remarkable inheritance? I think it no exaggeration to say that the discovery of this mammalian past demonstrated to us that Earth's past has a History."

Ichthy pauses, readjusting their spectacles before attempting some professorial cheekiness:

"I hope you heard The Capitals there: not just 'history,' time passing, but **History**. The past was different from how the present is, in profound, surprising ways. And this matters, I think—the deep past wasn't just one thing after another, but neither did it occur according to some set of determined, destined laws. The deep past unfolds in beautiful, complex, and ephemeral patterns, and it is the job of folks like me to grapple with those patterns."

The notion of loss I'll develop in the next chapter has interesting resonances with the emphases that 18th- and 19th-century proto-historical-scientists placed on their remarkable discoveries. Their arguments for the significance of their work—and the centre around which the spectacle of early geology turned—were the idea that the past differed surprisingly from the present. By way of historical motivation for the next chapter's conceptual analyses, then, a rather broad historical discussion is in order. Before that, I'll introduce a few more themes of the book via a dive into the inspiration behind Prof. Ichthy.

1.1 Prof. Ichthy Versus Professor Ichthyosaurus

Prof. Ichthy is inspired by Henry De la Beche's caricature *Awful Changes* (see Figure 1.1). This was drawn sometime around 1830 and reprinted many times (William Buckland apparently used an enormous print during his lectures). The cartoon shows "Professor Ichthyosaurus"—spectacles, podium, and all—pontificating on a strange skull before a rapt (and, to my eye, somewhat condescendingly amused) crowd of marine reptiles. Professor Ichthyosaurus remarks that the skull's primitive features clearly distinguish it as among the "lower orders" of animals.

There are three major differences between Prof. Ichthy and Professor Ichthyosaurus. First, the time period mocked by the lecture style. Ichthy appears to be stylistically and technologically rooted sometime in the 1970s, '80s, or '90s (although Robins' visual interpretations hark back to both the earlier 20th century and the more distant Victorian era), while Ichthyosaurus is a firmly 19th-century beast. Second, the taxa of the focal specimen differ: Prof. Ichthy is describing an extinct platypus, Professor Ichthyosaurus an extinct *Homo sapiens*. The third is more relevant and will take some unpacking.

Since Martin Rudwick's (1975) interpretation, it has been widely accepted that De la Beche's cartoon pokes fun at some of the ideas in the first volume of Lyell's *Principles of Geology* (1830). There, Lyell famously introduced ideas that were later dubbed (by Whewell) *uniformitarianism*. Uniformitarianism (at least in Lyell's original formulation) can be approximately understood as the claim that the historical processes that produced geological features are still operating today at more or less the same rates. If so, the thinking

FIGURE 1.1 Henry De la Beche's Awful Changes, the Inspiration for Prof. Ichthy. Science Museum Group Collection. © The Board of Trustees of the Science Museum

goes, geologists ought to examine contemporary processes to understand geological features as the products of those processes operating incrementally over significant timescales (see, for instance, Secord 2014; Patton 2014). There is disagreement about how exactly to characterize uniformitarianism in its historical context. Vic Baker (1998), for instance, has associated it with a particular (now outmoded) logic of induction, while David Sepkoski (2020) has emphasized the historical importance of non-uniformitarian thinking, especially as it relates to extinction.

Regarding received ideas about uniformitarianism, we shouldn't accept the historical idealization that Lyell's opponents (dubbed "catastrophists") denied moderate forms of actualism, nor that they clung to somewhat earlier ideas about geological formations occurring through biblically sudden floods and other events. Rather, they thought that some geological phenomena just couldn't be explained in gradualist terms. As Rudwick puts it, "The observed deformations of strata in many mountains . . . seemed to De la Beche to be inexplicable in terms of very gradual and relatively gentle forces, no matter how long the periods of time during which those forces were inferred to have operated" (Rudwick 1975, 545). Happily, the topic at hand doesn't turn on exactly how we characterize uniformitarianism and catastrophism.

De la Beche's problem with Lyell wasn't only about uniformitarianism's explanatory and investigative reach, but concerned a broader debate about the nature of geological history itself. When Lyell's contemporaries examined the fossil record, they discerned a clear pattern of progress through various ages: from fish, to amphibians, reptiles, and so on, to our own species. Further, 18th-century theorists often took the Earth to be steadily cooling (this was the basis of many attempts to estimate the Earth's age) and explained the structure of the continents as one of flux due to changes in sea levels. By contrast, Lyell denied geological time's "directionality": he denied a pattern of progressive change. As with uniformitarianism, historians differ on how to read Lyell's exact position. Jim Secord, for instance, interprets Lyell in terms of epistemic caution: claims about progress in the fossil record required an "unphilosophical assumption which ignored the patchy nature of the fossil evidence" (Secord 2014, 147). By contrast, Martin Rudwick (along with many of Lyell's contemporaries) sees Lyell defending what Rudwick calls a "steady-state" model of geology. That is, Lyell took the Earth to be in a kind of dynamic equilibrium. "If processes observable at present are representative, in degree as well as in kind, of all those that have acted in the past, there cannot have been any overall directional trend in the history of the earth, which must therefore be in a 'steady-state' condition" (Rudwick 1975, 8). This steady state was maintained by multiple interacting, often mutually-compensating processes. Lyell applied this steady-state approach to the history of life: the fossil record did not represent a series of progressivist changes through some ladder of being. According to Rudwick, this view was not popular.

> To most geologists this seemed to be particularly implausible. It involved a complete rejection of the clear trend of palaeontological discovery, which seemed to be uncovering a broadly progressive history from the simpler or lower organisms, through the successive appearance of more complex or higher forms of life, to the geologically recent creation of man

> himself. It is well known that Lyell circumvented this evidence by attributing it to differential preservation, and that he asserted that even the higher forms of life (e.g., mammals) had existed at the remotest known periods. (Rudwick 1975, 557)

Although Lyell denied his view was literally cyclic, as Rudwick points out, he was quite understandably interpreted as such by his contemporaries. In the *Principles of Geology* and in various correspondences, Lyell proposed that when the right ecological and environmental conditions arise, so too will the same critters. As Lyell states, quoted by Rudwick: "Then might those *genera* of animals return, of which the memorials are preserved in the ancient rocks of our continents. The huge iguanodon might reappear in the woods, and the ichthyosaur in the sea, while the pterodactyl might flit again through umbrageous groves of tree-ferns" (Rudwick 1975, 558). The combination of a steady-state, homeostatic view of geology with the thought that appropriate species (whether literally the same or not) could arise given the appropriate environmental states, "led [Lyell] inexorably to a view that very similar species must have been produced repeatedly in the past and would continue to be produced indefinitely into the future" (Rudwick 1975, 558). Hence, the topic of De la Beche's caricature, whose aim was to "help laugh [Lyell's ideas] out of court" (Rudwick 1992, 45).

Did Lyell in fact adopt a cyclic view of geology and biology? Or, as Secord (2014) suggests, was it instead a theoretically open thought experiment designed to illustrate how little grip we really have on patterns in the deep past? Happily, for my historically instrumental purposes, we can avoid this question: We're interested not in a historical question about Lyell, but in the philosophical differences between Professor Ichthyosaurus and Prof. Ichthy.

So, (and at long last), we come to the third difference between De la Beche's Professor Ichthyosaurus and my Prof. Ichthy. They take opposing stances on a question regarding the nature of history. Professor Ichthyosaurus, representing Lyell (fairly or not), takes there to be a *past* but in a sense no real *history*: There is change, but no real direction or story to tell about that change. Geological forces are in a dynamic, steady-state equilibrium, and the relevant biological kinds simply arise as the relevant environments cycle through. By contrast, Prof. Ichthy emphasizes not simply the past but History (capitalized again!). The Earth is not cyclic but is, in some important sense, contingent: what cycles there are, themselves have a history. For Professor Ichthyosaurus, the deep past reveals the *continuity* of the past and present (and presumably future); its data can be used to underwrite regularities and laws, that is, the rules governing the steady-state system of the Earth (see Gould 1988 for a similar discussion). For Prof. Ichthy, the deep past reveals *discontinuity* of past and present (and presumably future); it shows the limitations of regularities and laws. As we'll see, focusing on lost worlds will help us identify and explore variations and middle-grounds between these two extremes. That is my task in chapter 8.

Despite Lyell's long historical shadow, the birth of geology and palaeontology relied on ideas and rhetoric much closer to Prof. Ichthy than Professor Ichthyosaur. To highlight loss' importance, let's turn to those ideas.

1.2 The Discovery of Loss

Martin Rudwick has beautifully chronicled the slow realization of the Earth's deep history: how natural philosophers "burst the limits of time" in discovering a long pre-human past (Rudwick 2005; 2014). Being pre-human, these pasts differed crucially from the contemporary world. As Ralph O'Connor has shown, 19th-century geologists adopted literary styles that emphasized these differences; rhetorically they "pulled off an imaginative coup—*a coup de théâtre*—by giving their public tantalizing glimpses of an Earth history far longer and stranger than the story of a literal six-day Creation which had held sway over much of that public at the turn of the century" (O'Connor 2008, 2). Thus, the differences between now and then, and the narratives these afforded, formed a central component of the rhetorical strategies these scientists adopted in describing their work and arguing for its importance. This is true regardless of whether they saw clear and clean breaks between the past and present, or whether they viewed the connection more uniformly.

For an early example, George-Louis Leclerc, Comte de Buffon, the 18th-century polymath, attempted to provide an overall picture of a pre-human past by postulating seven epochs of the Earth's history. Our species only appears in the seventh epoch, after a long series of environmental and geological cataclysms and shenanigans. As Rudwick puts it, "The whole of human history was confined to the most recent portion of a far longer temporal sequence. Buffon claimed that since this last metaphorical milestone, there had been few further changes of any significance in the physical world" (Rudwick 2005, 146). That is, the epochs before our arrival were in many ways more dramatic and changeable than our own. In a sense (and in stark contrast to Prof. Ichthy's post-mammalian perspective), the endgame of humanity was the end of (natural) history. While in the 19th century geologists distanced themselves from the speculative "theories of the Earth" produced by Buffon and others (Rudwick 2005), they nonetheless retained their rhetorical emphasis on "loss."

Chris Manias' *Age of Mammals* (2023) chronicles the role of mammalian palaeontology in the developing geosciences through the long 19th century. He provides multiple examples of appeals to differences between the past and present, appeals which aim to underwrite the value of knowing the past. For instance, Richard Owen's interpretation of the enormous ground sloth *Mylodon* relied on the idea that South America's tropical fecundity had been pumped up to massive scales in the Pleistocene, and this required enormous animals to tackle the verdant growth. For Owen, *Mylodon's* morphology was built to uproot enormous trees; it "was assigned the office of restraining the too luxuriant vegetation of a former world" (Manias 2023, 2). So, while in some ways South America's present was continuous with its past—peopled with jungles and sloths—its past was an exuberant, exaggerated version of its present. *Mylodon's* previously unseen traits were explained in terms of, and partially evidenced, the idea that South America's past differed in surprising ways from South America's present.

On Manias' account, the idea that the past was in many ways fundamentally different from the present is crucial for 19th-century apologies for palaeontology: "The idea that the world had formerly been inhabited by now-disappeared animals has been highlighted as one of the principal conceptual contributions of the study of fossils. . . . Palaeontology depended on former worlds having vanished, and new forms arising" (Manias 2023, 55).

For another example, Manias recounts Leidy's description of Wyoming as "a place of dead antiquity," perceived as "the wreck of another world which was once luxuriant with vegetation and teemed with animals (Manias 2023,153)." Here, Wyoming is understood as a kind of ruin, evocative of past glories.

Thus, 19th-century geologists invented a non-human history full of change, often but not always understood in progressivist ways: "Deep time was not incomprehensible, but could be understood as a series of worlds, each passing (by admittedly mysterious forces) into the next" (Manias 2023, 170).

In popular treatments, fossils were presented as what O'Connor calls "sublime relics of a legendary past" (O'Connor 2008, 33). Both visual (Rudwick 1992) and textual (O'Connor 2008) descriptions of geological phenomena were used to generate evocative visions of these legends: "Geology's main claim to poetry consisted in its ability to call up dramatic images of vanished worlds" (O'Connor 2008, 357). In the 19th-century texts O'Connor describes, various literary techniques are used to draw narrative from geological description. For instance, in textual versions of Victorian diptychs, poets would overlay then and now, as in this Tennyson stanza from *In Memoriam*:

> There rolls the deep where grew the tree.
>
> O earth, what changes hast thou seen!
>
> There where the long street roars, hath been
>
> The stillness of the central sea.
>
> (Tennyson, quoted in O'Connor 2008, 365)

Geology reveals a past of change—no steady state but a historical trajectory—wherein oceans cover where forests once stood and busy human streets bustle where oceans once lay. Aesthetically speaking, the gift of geology was the ability to see natural formations in terms of their history. Derek Turner describes this evocatively with the notion of a "stereotemporal experience": "Like a stereoscope that uses two images of the same item to create an impression of depth, stereotemporal experience merges our current perceptual experience of the landscape with our understanding of its history" (Turner 2019a, 7). Knowledge of the past allows us, too, to observe the changes the Earth has seen.

It is in debate concerning the reality of natural extinction that the historical importance of loss is most clearly felt, especially prior to the 19th century. As Rudwick puts it: "Most naturalists felt an almost gut revulsion against the idea that extinction might be an ordinary feature of the natural world. For it implied that some of the wonderfully diverse species, the fossil remains of which they took such delight in collecting and describing, might have been lost forever" (Rudwick 2005, 243).

The theological upshots of accepting extinction were potentially profound, and at least for some, loss through extinction could be denied. Perhaps, for instance, the thus far largely unexplored (in the colonial sense anyway . . .) New World could be a source of migrated critters and living fossils (we'll discuss living fossils in chapter 4).[4] Despite apparent evidence in the fossil record

4 Chris Manias points out to me that there is much disagreement between the historical actors on this point. Cuvier, for instance, doubted the new world housed fauna lost to the old.

of once-common and thriving lineages now unseen (ammonites, for instance), many natural philosophers still doubted that anything was truly "lost."

For those accepting the possibility of natural extinction, the possibility of patterns of extinction was sometimes extended into the future. Manias quotes Falconer, a 19th century colonial researcher of India's geology: "We appear to be gradually losing all the larger forms of creation. The elephant and giraffe of the present period will in all probability share the same fate as the Mastodon and Sivatherium of former eras, and be only recognized in the proofs exhibited by the researches of the geologists" (Manias 2023, 106). A differing past, one revealing new worlds and patterns stretching into the present (and into the future), makes knowledge of that past uniquely powerful—and poignant.

Beyond the 19th century, loss again came into prominence, with increasing focus on mass extinctions during the paleobiological revolution (e.g., Jablonski 1989; Raup 1992), coupled with the rising awareness of current plummeting biodiversity due to anthropogenic extinction (Sepkoski 2020). Mass extinctions mattered for many reasons. They represented a case where a signal from the fossil record revealed a surprising, apparently biological pattern with major consequences for understanding the history and nature of life at the macro-scale, a pattern amenable to statistical analysis, simulated study, and so on. It was thus important for establishing the paleobiological revolution's claims to scientific and theoretical legitimacy (Sepkoski and Ruse 2009). It also, of course, linked the history of life to currently occurring biological cataclysm, if perhaps tenuously (Bocchi et al. 2022). Further, and speaking more to the heart of this book, the occurrence of mass extinctions has been called upon to argue for major discontinuities in the history of life. This, it has been suggested, evidences that macroevolution is not just microevolution scaled up, but works according to its own rules (Jablonski 2005; McConwell and Currie 2017a; Turner and Havstad 2025) and further, that these rules might have changed over time.

At base, the eventual acceptance of extinction and pre-human time in the 19th century, and of mass extinctions towards the end of the 20th century, involved accepting that the past was in many ways not like the present. That is, accepting lost worlds. In a sense, then, my aim in this book is to return philosophical attention to loss. In the next section, we'll examine recent philosophical work on the sciences of the deep past. These have, by and large, focused on "erasure," that is, traces or their lack thereof. At least if I'm right, they've paid insufficient attention to loss.

2. Trace-Based Reasoning

> Prof. Ichthy's graduate student, now getting into the swing of things, deftly replaces one transparency with another. This new slide shows Ichthy's fossilized discovery from several angles. Ichthy gestures at the slide:
>
> "We see here this wonderful specimen, the only trace we have of Broadfoot. And here, Ichthy gestures at the skull, "a plaster cast of its skull. Our knowledge of the past is built upon objects like these. Sometimes I think of them as a bit like documents. That is, they are something to be read, interpreted, and decoded. But fossil specimens are not just code—they're not mere vessels for some abstract 'information.' Broadfoot's mineralized body is also a physical object. Its materiality matters crucially for our understanding of it. The fossil bears information, but that information is laid bare based not only on our understanding of how it formed, but also on how we have manipulated and structured data generated from it. As I have said, it is only in virtue of this wonderful, remarkable, wildly improbable trace, and our discovering it, revealing it, and caring for it, that we know of Broadfoot at all. And our study of it goes far beyond 'reading.' So, an understanding of Broadfoot as a trace, that is, as a fossil, is necessary. But, I think, that is not where the story of our science of Broadfoot can end."

The history of the Earth sciences has been understood as the history of a metaphor: *reading the historical record* (Sepkoski 2019; Currie 2023a). Fossils are not literally texts, but we discuss them as such so naturally that the metaphor is nearly, but not quite, dead, and that metaphor has been used extensively more or less since the Earth sciences began. Fossils make up records: records of traces that, if properly read, form the basis of our knowledge of the deep past. These broad brush-strokes can also be applied to the much shorter history of the philosophy of the historical sciences. Debate in that arena has turned on whether various historical records (consisting of fossils, archaeological remains, etc.) are characteristically information retaining or otherwise; the epistemic challenges historical scientists face emerge from incomplete, degraded, and messy records. We can understand this focus in terms of "traces," and I'll call the processes that generate epistemic challenges from a lack of traces "erasure" in the next chapter.

"Traces" are contemporary objects understood in terms of (1) the causal sequences that led to their formation, which, if properly understood, (2) act as evidence for past states of affairs. For a clichéd example, fossils are products of various fossilization processes, which, in virtue of our understanding of those processes, evidence past organisms. This is "trace-based reasoning": inferring from a contemporary object to the past via an understanding of the historical processes by which that object formed (Kosso 2001; Jeffares 2008; Currie 2018a; Currie and Killin 2019). Understood as such, traces and trace-based reasoning are everywhere in science. For instance, experiments produce data which we draw upon to test hypotheses. In performing such tests, we rely on metadata concerning the data's provenance—how and when it was produced, say. Insofar as all data-derived evidential claims rely on knowledge of the data's provenance, all data-derived evidential claims involve trace-based reasoning. In a sense, then, all science is historical (Currie 2019a; Roth 2019). But traces have a special place in philosophical accounts of historical science, and for good reason: the reliance on historical records—traces—is an obvious starting place for thinking about the status of such sciences. After all, Prof. Ichthy couldn't have imagined Broadfoot were it not for that remarkably preserved, fortuitous specimen. Bruce Routledge characterizes archaeology "not as a discipline of the past, nor of things, but as a discipline of traces" (Routledge 2023, 2). For Routledge, because traces mediate archaeological knowledge of the past, we're making a mistake to characterize that knowledge as either about contemporary objects or the past *tout court*. He thus places traces at the centre of archaeological science.

In the past two decades or so, the philosophy of the historical sciences has been shaped by two interrelated debates, both conducted in highly trace-focused ways. First, what is the method of historical science, and how does this relate to its epistemic credentials? Second, what epistemic attitude should we adopt toward the historical sciences?

Regarding the first question, Carol Cleland (2002, 2011, 2013) argues that sciences like palaeontology and geology proceed via *smoking-gun reasoning*. This involves noticing puzzling contemporary correlations among putative traces, generating hypotheses about the past which, if true, would explain those correlations, and then hunting for further traces—smoking guns—which would discriminate among those hypotheses. Cleland takes historical reasoning to be a form of common-cause explanation. Inasmuch as the events posited in the past explain multiple traces, they unify them via a common cause, and those common causes explain the correlations among traces.

Cleland underwrites smoking-gun reasoning with a claim about the relationship between the present and past. She adopts a metaphysical view whereby causes and effects stand in a one-to-many relationship, and effects themselves are causes of further effects. As such, causal information about an initial cause diversifies and disperses as time goes on (see Fox 2023 for scepticism). As a result, we should expect the strategy of hunting for multiple effects to uncover past causes—developing common-cause explanations that unify those effects—to be a successful strategy.

In Cleland's view, the contemporary plays a crucial role in historical knowledge. For her, in a sense, historical science is in the business of explaining contemporary traces. That is, Cleland gives us a particular account of trace-based reasoning as I characterized it earlier. There are various views that disagree in some philosophical detail or other, but are similarly trace based. Where Carol Cleland emphasizes common-cause explanations, others have argued that consilience—roughly, evidence of differing types and sources pointing to the same conclusion—does the work (Wylie 1999; Forber and Griffith 2011). Still others emphasize the coherence and consistency of historical explanations (Kosso 1993; Kleinhans et al. 2005; Turner 2009; Currie 2017; Currie and Sterelny 2017), while others stress the pluralism and flexibility of historical approaches (Wylie 1999; O'Malley 2016; Currie 2018a), especially in light of their capacity to use innovative technology (Tamborini 2020; Currie 2021a) and speculation (Currie 2018a; Bonnin 2019; Tamborini 2020; Swaim 2024). Regardless, these views all to some extent rely on a particular methodological picture: *historical science is in the business of trace-based reasoning*. So, the philosophical action regards understanding that form of reasoning's nature. Further, an epistemic debate follows from this methodological claim. If historical science is fundamentally about traces, then *the epistemic fate of historical science turns on the richness and accessibility of traces*. Elliot Sober and Mike Steel's discussion is particularly explicit about this.

> We are interested in how the natural processes connecting past to present constrain our ability to know about the past by looking at the traces found in the present. An optimistic view of these processes is that the past is potentially an open book; all we need do is understand the connecting processes correctly and look around for the right traces. If the relation of the past state of a system to its present state were deterministic and one-to-one, this optimistic view would be correct. If only we could know the present state with sufficient precision, and if only we could grasp the true mapping function that

> connects present and past, we would be home free. (Sober and Steel 2014, 558)

Sober and Steel adopt a formal approach, arguing that (assuming particular temporal branching structures) information loss over time is nearly inevitable, concluding that "the view of evolution as an 'information destroying process' is basically right" (Sober and Steel 2014, 571). If evolutionary processes destroy information, then this limits our knowledge of the past. We've thus switched from asking after methods of reconstruction to related questions concerning the epistemic fortune of historical science: How much will we be able to know about the past? Sober and Steel tackle the question by asking just how much information will be retained, which is to say, how many traces will there be.

So, on to the epistemic credentials of the historical sciences: should we be pessimistic or optimistic about their epistemic fate? Should we be realists or anti-realists about historical knowledge?

Derek Turner's arguments in favour of anti-realism about historical science rely on trace-based reasoning (Turner 2004, 2005, 2007). He points to the oft-noted impoverishment of the historical record and argues that our background theories about the processes that form that record should lead us to doubt whether that impoverishment will improve. So, we won't have many, or good, traces to work with. He further points to our incapacity to intervene in the past: where scientists-of-the-now can generate new evidence via experiments, scientists-of-the-then cannot. So, in comparison to experimental science, historical science has worse evidence and cannot generate its own. For Turner, this should lead us to expect underdetermination problems to be rife: although historical hypotheses may, in principle, be empirically delineated—we can imagine smoking guns—we also should expect smoking guns to be few and far between. That is, much of the time, our available evidence will not decide between our hypotheses.

So, rampant underdetermination should lead to pessimism regarding our knowledge of the deep past. Turner's pessimism can be compared with Carol Cleland's appeals to the many-to-one relationship between causes and effects (Currie 2018a, chap. 5). Recall that she argues that as causes breed multiple effects, which themselves act as causes for further effects, we should expect the world's current state to overdetermine its past state. If that were right, we'd have some grounds to expect many traces to be available to test historical hypotheses; over time, then, underdetermination would be resolved. By contrast, Turner points to the degradation of evidence and our helplessness to improve that situation. He thus expects underdetermination to remain.

We have what appears to be an impasse: the epistemic fortune of the historical sciences turns on whether much information from the past has been retained or destroyed. If the former, perhaps we have grounds for optimism; if the latter, we have grounds for pessimism. Or perhaps there are further contextual elements that should lead to pessimism at some times and optimism at others, maybe depending on the kind of epistemic good that interests us (Currie 2018a, chap. 10), the stage of research (Tucker 2011), or the granularity of claims (Malaterre 2024). In *Rock Bone and Ruin* (2018a), I pointed to the creativity and pluralism—"methodological omnivory"—of historical scientists toward defending a (maybe cautiously) optimistic position. A major part of the argument involved pointing to non-trace forms of evidence (simulations, analogies, and so on) that historical scientists could draw on.[5] Although I argued that traces don't exhaust what there is to say about historical knowledge, the discussion was nonetheless rooted in that trace-centric tradition.

Joyce Havstad (2019) has argued that philosophers should be agnostic regarding optimism or pessimism about historical science (although, as Turner (2016) has pointed out, agnosticism might not be an option scientists can afford). For Havstad, no one has given sufficient reason to decide whether, at the end of the day, historical processes are generally information-destroying or information-preserving. Part of her argument involves considering situations where the retention of traces seems highly implausible. For instance, she asks us to imagine: "An unseen bubble, landing on the surface of a lake, and dissipating without any sort of ever-recoverable trace. Or maybe a fleeting, ancient pathogen—one that wreaks havoc on a population of soft-bodied organisms whose fossils, if any remain, show no accessible evidence of the biological attack" (Havstad 2019, 3).

In the parlance of the next chapter, we'd say Havstad has described cases involving *complete erasure*. The possibility of such circumstances, she argues, should motivate agnosticism about historical science.

A further argument for a sort of agnosticism (or at least quietism!) regarding our epistemic attitude toward historical knowledge begins with a

5 Recently, Franziska Reinhard has provided an expanded notion of "traces" that includes more than simply direct causal descendants. This expanded notion is, I suspect, another way of pushing against comparatively limited conceptions of historical evidence (Reinhard forthcoming).

turn from considering traces as *evidence* to considering traces as *data*. We can understand "data" as physical objects that, under the right conditions, might be "enriched" to act as evidence. That is, data is potential evidence (Leonelli 2016, 2018; Boyd 2018; Watkins 2024). Along this vein, I have recently characterized the palaeosciences as often facing *epistemic scarcity*: circumstances where data are rare and fragile, hard to manage and non-renewable (Currie 2021a). Some of these features relate to the traces themselves but others relate to how historical scientists manage the traces as data: their storage, handling, the applicable techniques and technologies of analysis, and so on. Beyond traces, there are also worries about metadata. For instance, what data gets recorded about extraction during fieldwork?. Thus, information might be lost across "phenomena time" by, for instance, geological processes, but also "data time," by, for instance, scientists misplacing or destroying specimens (see Leonelli 2018; Watkins 2024a; Wylie 2024).

The upshot of this focus on data is that, in addition to worrying about information loss from the deep past, we also need to worry about the fate of data in the record-making and managing practices of scientists themselves. In my 2021a paper, I argued that, once we take data into account, debates about the underdetermination or otherwise of historical hypotheses become intractable. This is because we now need to not only track information loss and retention on nature's schedule, but also need to track this across the practices of scientists. Notice still that the epistemic situation—scarcity—is characterized in terms of *data*, which, in the palaeosciences, typically means traces and records. So, the data-oriented view thus far is still highly trace-focused.

Alison Wylie's analyses of archaeology often involve rich accounts of the challenges from traces and how archaeologists respond to them. Consider, for instance, her discussion of "shadow-data": "[A]rchaeology as a discipline is defined by the challenges of working with gaps and absences in its primary data. The shadows archaeologists wrestle with are not just a function of the notoriously fragmentary and incomplete nature of the surviving 'data imprints' of past lives but reflect, as well, the paucity and instability of the inferential resources they rely on to 'bring data to light . . . to realize its value' as evidence" (Wylie 2017, 204).

Archaeologists often face not only an incomplete record but an unstable base to make inferences from that record. That is, their understanding of *how* the archaeological record has formed is also fragile. To grasp the epistemic challenges of uncovering the past, then, is to understand the strategies historical scientists adopt to mitigate and overcome incomplete data sets and incomplete understanding of how that data formed. To an extent, I don't disagree with this. But, I think, to understand the value of traces, the value of past knowledge, and the epistemic situation faced by historical scientists, we must attend to what I'll call "loss" as well as incomplete records.

So, debate about the epistemic and methodological prowess of the historical sciences has turned on questions about traces. To the extent that the record is impoverished, so much the worse, but to the extent that the record is rich—or historical scientists have the tools and resources to mitigate a lack of richness—so much the better.

Another way into trace-based reasoning starts by considering how scientists construct proxies and measurements. Instead of asking how traces might serve as evidence about the past, we instead ask how Earth scientists use the palaeontological and geological records to carve up the past and measure time (Boudinot and Wilson 2020; Bokulich 2020, 2021a; Bocchi et al. 2022; Wilson and Boudinot 2022; Watkins 2024b). This involves considering how we conduct measurements on a trace that might act as a stand in for measuring some quantity in the past. Palaeoclimatologists measuring the oxygen content of an ice core to reconstruct past temperatures, palaeontologists sampling fossil diversity to infer past biodiversity, and archaeologists measuring carbon decay to estimate burial date, are all engaging in proxy measurement. This literature is less focused on general questions about the epistemic fate of historical science, but nonetheless is focused on traces and, thus, on what I'll call "erasure."

I'm not all that interested in directly engaging with trace-centric views, or the philosophical positions discussed above (see *Rock, Bone and Ruin* Currie 2018a for that!). I'm rather arguing for a broader reframing. Those philosophers rightly emphasize traces, but nonetheless miss the critical importance of the contemporary in thinking about history. For Cleland, Turner, and others, current observations matter insofar as they are treated like traces, that is, remnants of the past. But consider Professor Ichthy's discussion of their remarkable find. Their discovery of Broadfoot was remarkable because there is nothing like Broadfoot in the contemporary world: its unique features were unprecedented in science. This is not a claim about traces, but a claim about what exists. It doesn't concern erasure, but loss. Because Broadfoot-like critters are extinct in Ichthy's reality, they are extremely hard for them and their contemporaries to investigate and understand. This, I'll argue throughout, transforms how we should conceive of historical knowledge and the nature of the deep past itself.

So, accepting my moral doesn't at all mean that much of the rich work already done in the philosophy of palaeoscience must be thrown out; rather, the view is that it has been too narrow. There is still a lot of work to be done on erasure, but so too is there work to be done on loss.

Before moving on, it is worth emphasizing (as Ichthy does) that traces are not mere bearers of information, but physical objects that scientists and others extract, examine, manipulate, store, and organize. Historical science often involves intimate and embodied interactions between specimen and scientist—specimens are both data themselves and sources of data—and the creative and rich ways that historical scientists manage the various challenges such obstinate objects present are crucial for understanding how trace-based reasoning proceeds (Wylie 2021; Chapman and Wylie 2018; Currie 2021a). Although in a big way this book is saying "don't just think about traces," it perhaps ironically also points out that once we take loss into account, why traces are so valuable becomes much clearer.[6]

Let's finish with two caveats.

First, there is an ever-present danger for "state of the literature" sketches like I've just provided. Of necessity—and design—these will foreground some elements of a discourse and background others (Currie and Walsh 2019). This can have deleterious effects if given too much authority, as any particular framing will place some work and questions as central, while relegating others to the fringe. And there are many other ways one might frame the philosophy of the palaeosciences: one might focus on the dynamics of explanations (Dresow 2021; Meneganzin and Currie 2022), or reduction (Grantham 1999), or start not with historical science as the focus, but the Earth sciences instead (Bokulich and Oreskes 2017; Ohnesorge and Watkins 2024). Overall, the argumentative focus of philosophers frequently makes us untrustworthy historians, especially when it comes to our own history, so I hope the reader takes the above survey with sufficient critical salt.

Second, it would be a little surprising if philosophical analysis of historical knowledge didn't bump up against something like loss now and then. And indeed, throughout the book, I'll point to overlap with previous work. Discussion of uniformitarianism and its limits is one place to look. Take Max Dresow's recent discussion, for instance: "The present may be 'the key

6 I think I could be fairly read here as conceding to Alison Wylie's (2019) argument that the distinction I drew between "trace-based" and "non-trace-based" evidence draws the two apart too starkly: You don't get trace-based reasoning without non-trace-based input, and vice versa.

to the past,' but there is no guarantee that this key will unbolt every door. Sometimes the past is truly alien—a 'foreign country,' as L. P. Hartley put it in The Go-Between. In such cases, ordinary reasoning strategies are prone to break down, raising the question of what limits exist to our knowledge of geohistory" (Dresow 2023, 407).

Here, Dresow puts his finger on the central insight I'm pushing: Our knowledge isn't only limited because the past has left insufficient records, but also because the past is unfamiliar. The past sometimes leaves faint traces; the past is sometimes weird. Both matter for our knowledge of it.

3. Roadmap and Strategy

This book is centred around a set of episodes in the history of life: the Avalonian Ediacaran, when the first (that we know of) metazoans arose. In chapter 2, I'll introduce some of that science and draw on it to explain the conceptual machinery that forms the book's core. The aim of that machinery, briefly, is to expand our usual conception of the epistemic resources and challenges available to historical scientists. Where philosophical discussion has focused on *erasure*—whether past information has made it into the present—we should also care about *loss*—whether what existed in the past exists now.

Chapter 2, then, raises a question: How do we discover lost worlds? Through chapters 3, 4, and 5, I develop an answer to this question by articulating a set of strategies scientists adopt to accommodate and overcome the challenge of loss. Each chapter is named for one of the strategies. *Artificing* involves constructing something similar to the lost target. *Grafting* involves linking the lost target to something in the present via lines of causal ancestry. *Anchoring* involves examining aspects of the lost target across a set of differing analogues. Table 1.1 provides a quick, rough first-pass at these. In chapter 5, I'll introduce Table 2.1, which is much more careful.

Through chapters 3 to 5, a general solution to the challenge of loss emerges, centred on a common thread through those strategies. First, what I'll call *strategic perspectivalism*: the capacity of historical scientists to characterize their systems of interest in ingenious ways, enabling them to gain empirical and theoretical traction on those systems. Second, *iterativity*: how historical scientists navigate between these various strategic perspectives in surprisingly productive ways.

Each chapter will include examples from our Ediacaran mainstay, but will be accompanied by other palaeoscientific exemplars. I intend these cases

Table 1.1. Strategies, an Initial Sketch

Strategy	Basic Idea	Broadfoot Example
Artificing	Construct a model of the lost world.	A puppet with Broadfoot's gait.
Grafting	Examine the lost world's downstream ancestors.	Other extant or extinct mammals.
Anchoring	Examine multiple independent instances of the lost world's properties.	Evolutionary convergences with Broadfoot's traits (e.g., other electrosensing taxa)

to accumulate throughout the book, particularly through chapters 2 to 5. Each will be called back to, and sometimes expanded upon, as we move through the book's argument. I will focus on palaeobiological examples—apologies to those more interested in palaeoclimatology, archaeology, human history, or the other rich areas of the palaeosciences that should divert our attention. As the charlatan says, I leave for future work considering to what extent my arguments extend beyond palaeontology. While on that topic, allow me a quick methodological insert.

I am officially agnostic as to what extent my arguments rely on the particulars of the sciences I attend to, and to what extent I'm describing science generally. My approach to philosophy is to think through particular instances of scientific reasoning, and, as such, I don't think it is particularly useful for me to make general claims. Rather, I see myself as providing philosophical insight to the kinds of sciences I know well, insights that I hope might be of use to those who know other sciences.

My aim is to correctly characterize the *strategies* scientists use to uncover lost worlds. For me, scientific strategies are abstractly described approaches to tackling epistemic challenges. As such, although they partly depend on correctly describing and interpreting scientific work, they are fundamentally normative: The strategies are *good* strategies when they are appropriate to the context the scientist works within and their aims. As such, I'm not committed to the truth or plausibility of any particular first-order scientific claim about the past. In principle, all the science I interpret could turn out to be mistaken, and my arguments could still go through (although somewhat embarrassing—see Turner 2016 for discussion of these bad philosophical bets).

From chapter 6, I shift to a set of applications and upshots of the view developed in the first half of the book, mainly aiming to provide a picture of the value and nature of historical knowledge and scientific imagination. These include the following. First, a defence of the modal nature of knowledge generated by the deep past, a kind of minimal modal realism grounded in a commitment to artifactualism (chapter 6). Second, a discussion of the nature and limits of scientific imagination with particular focus on palaeoart (chapter 7). Third, an argument for the "peculiarity" of biological and geological history and how this underwrites a disunified, "dappled" picture of the present (chapter 8). Fourth, an exploration of whether the strategies employed to understand lost pasts might be fruitfully applied to unprecedented futures (chapter 9). And finally, fifth, a conclusion that wraps up the book overall and turns to the relationship between history and wonder (chapter 10).

Overall, I hope you walk away with a new appreciation for the creativity of historical scientists, as well as a new respect for and understanding of the nature and value of the knowledge they produce, and of course, the worlds they reveal.

2

The Avalonian Ediacaran: Lost, but Not Erased

"To see what is so challenging about the science of Broadfoot," Prof. Ichthy says, emphasizing "challenging" with a wave of their flipper, "we can compare it to the science of the ancestors of more-familiar fauna."

With another transparency switch, Prof. Ichthy's grad student reveals a set of reptilian beings well-known to the audience. As with the mammals before them, these are linked both to each other and to the deep past via lines of ancestral descent. Where the lines meet to form ancestral splits, there are speculative illustrations of the precursors of modern forms.

"If scientists want to make sense of the fossils of, say, Ichthyosaurs, well, we have modern examples swimming all around us." Prof. Ichthy indicates themselves and their audience.

"We can look at these modern examples to test and generate ideas; we can use our understanding of evolutionary history to infer from these modern forms to those ancient forms. We needn't speculate, as evolution has done our speculation for us. In these circumstances, there is a mutually beneficial, mutually informing relationship between our knowledge of contemporary ichthyosaurs and our knowledge of past ichthyosaurs. Under these conditions, because we know so much, a single incomplete specimen—a single tooth, say—is enough to reconstruct a millions-year-extinct ancestor with confidence. Sadly, this is not so for Broadfoot . . ."

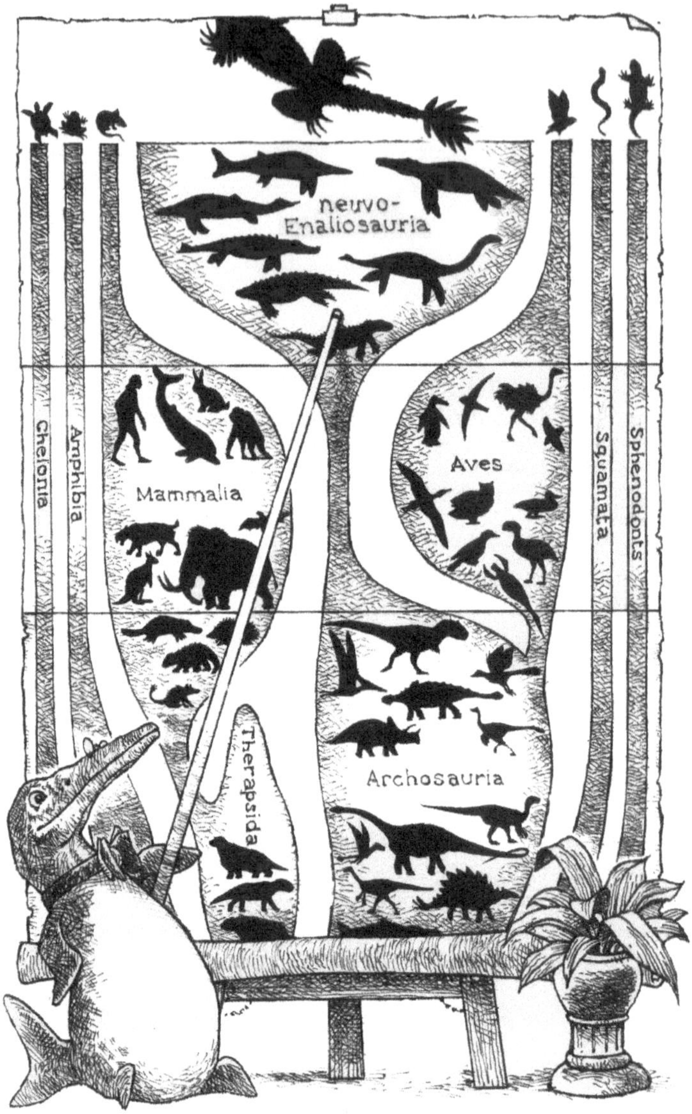

ILLUSTRATION 2.1 "Then Might Those Genera of Animals Return"

Ichthy pauses, glancing at their student, who quickly returns to the first transparency, again projecting the specimen and subsequent reconstruction, and Ichthy continues.

"Broadfoot diverges so radically from living mammals that *now* is, on the face of it, little use in understanding *then*. One of Broadfoot's teeth would tell us it is a mammal of sorts, and maybe tell us something of its size. Our understanding of fossilization processes, more generally, can get us some of the way. However, without the mooring of Broadfoot-like organisms around us, we are cast adrift . . ."

Ichthy stops for a moment, as if considering whether the "mooring" metaphor fits with the allusion to being "case adrift," before nodding to themselves and continuing. "In this instance, we are deeply lucky to have such a richly preserved specimen."

Ichthy pats the skull's case with a fond flipper.

"Broadfoot, we've discovered, has a plethora of unique features: from bill to venom to electrosense. When Broadfoot went extinct, so too disappeared those unique traits."

Prof. Ichthy repositions their spectacles further down their snout, glaring over them, out at the audience, leaning over the podium with a dramatic air.

"Broadfoot is lost. Lost, but not erased."

In the first chapter, I provided two motivating thoughts, one historical and one philosophical. The former highlighted the historical importance of establishing a non-human past that differed from the present, while the latter highlighted a philosophical focus on traces in analyses of palaeoscientific knowledge. In this chapter, I'll provide a conceptual framework that makes sense of what recent philosophy has missed, and of what 19th-century earth scientists were trying to establish. I'll develop a notion of "loss," that is, circumstances where some past entities (understood very broadly) no longer exist in the present. For example, from Prof. Ichthy's perspective, Broadfoot is lost to extinction. And I'll distinguish "loss" from "erasure," that is, processes by which traces degrade and disappear. For example, Broadfoot survived erasure by virtue of its fossil's preservation and discovery. In short, philosophers have missed the importance of loss, focusing instead on challenges from erasure, and early palaeoscientists, faced with traces of a surprising past, wanted to establish that loss had occurred.

Above, Prof. Ichthy compared the investigation of Broadfoot to past ichthyosaurs. Let's consider an analogous question: What is epistemically different about investigating *dinosaurs* and investigating *mammoths*? A natural answer considers these animals' traces. Non-avian dinosaurs are known, shall we say, mostly "geologically": their remains primarily take the form of funny-shaped rocks, mostly capturing previous dinosaur anatomy and morphology, as well as the occasional eggshell or trackway. The mammoth record, by contrast, includes much more than funny-shaped rocks. For mammoths, we have truly biological remains, stored and preserved in nature's arctic freezer. We have their hair, their tusks (sometimes even carved by the hands of our ancestors), and even their DNA. The comparatively short temporal distance between ourselves and mammoths affords a much richer record than that of dinosaurs. All of that is right, but it misses—I want to argue—something in the vicinity of half the story.

A major difference between studying dinosaurs and studying mammals is . . . (drum roll) *elephants*. We have currently existing examples of things extremely similar to (and closely related to) mammoths, and these epistemically and causally link the past to the present in rich ways. As Shoshani have put it:

> Unlike researchers who study dinosaurs and have to rely on distant living relatives, such as crocodiles and birds for indirect comparison, students of Proboscidea are fortunate to have living models with which to compare and contrast the extinct taxa directly. This advantage should not be underestimated, because knowledge of the living species helps one to interpret observations of extinct forms and vice versa. (Shoshani 1998, 487)

In the parlance I'll soon unpack, it isn't simply that dinosaurs are more *erased* than mammoths; dinosaurs are also more *lost* than mammoths. In this instance, this is largely due to the ancestrally retained similarities between extinct and extant pachyderms. Mammoth and elephant are both members of the clade *Elephantidae*, and, in this case, ancestral relations have preserved many similarities between them.[1] Although the clade *Dinosauria* is not lost insofar as birds persist, in this case, much less similarity has been retained.

---

1 Especially so for the Asian elephant: Molecular evidence strongly suggests that African species form an outgroup from mammoth and Asian elephants (e.g., Baleka et al. 2022).

So, many properties possessed by, say, *Diplodocus* are lost even though many others, for instance, their "pneumatized" hollow bones, have not.

I want to articulate the distinction between loss and erasure by discussing two cases from the Avalonian Ediacaran. So, providing relevant information about these is my next task. I'll then turn to the concepts themselves and set us up for the remainder of the book. This will involve applying the framework to the two motivating thoughts from the introduction, and finally, characterizing the epistemic challenge generated by loss.

1. The First Metazoans

It is commonly thought that metazoan life, that is, complex multicellular life, really got going in the Cambrian, a little under 540 million years ago (Briggs 2015). From that time, animals are largely as we know them today: critters with sensory organs, mouths for consuming, anuses for expelling, methods of locomotion, occupying roles including predators, prey, parasites, scavengers, herbivores, carnivores, browsers, grazers, and so on. But metazoan life didn't start in the Cambrian. Fossils record a pre-Cambrian history stretching backwards into the Ediacaran. The earliest examples (which we'll concern ourselves with here) are known from something like 600 million years ago, and that biota[2] has a striking set of differences from their Cambrian replacements (Narbonne 2005; Narbonne et al. 2012; Droser and Gehling 2015; Dresow 2025). These differences are most stark in the earliest deposits: the Avalonian Ediacaran (Liu, Kenchington, and Mitchell 2015). I'll provide some general information about the Avalonia before turning to two initial examples of loss.

1.1 Avalonia

The "Avalonia" is used to both pick out a past geographical location—now dispersed across England, Poland, and the northeast of the United States and Canada—as well as identify particular fossil assemblages hailing from the Ediacaran. The assemblage and timespan I'll refer to as "the Avalonia" are known from deposits in Charnwood Forest, Leicestershire, and in Canadian Newfoundland. A crucial spot is Canada's Mistaken Point Ecological Reserve

2 There is no happy term for these Ediacaran organisms: terms aligned with particular taxa ("animals", "metazoans") imply phylogenetic assumptions, "biota" implies something more unified than some are happy with. Perhaps the least controversial would be "progenitors of soft-bodied macrofossils" but that's such a mouthful that I'll just stick to terms like "biota" and hope the reader is sufficiently charitable.

FIGURE 2.1 Mistaken Point (Barrett & MacKay Photography, © Mistaken Point Ambassadors Inc., BYND)

FIGURE 2.2 Rangeomorph Fossil (Detail from Fig. 2, Liu et al. 2015). Creative Commons CC BY

UNESCO World Heritage Site, a series of dramatic cliffs around two hours' drive from St John's. There you'll find a bunch of fossil beds (on a guided tour, you can't just rock up on your own), which, despite the beautiful location, to the uninitiated might not exactly inspire fossil-fuelled awe (see Figure 2.1).

What are we looking at in Figure 2.1? Most familiar fossils are preserved "hard parts"—shells, bones, and teeth—which have either been preserved or "replaced": approximately, bones are held in place, retaining morphological information while their carbon matrices are, via water's slow work, replaced with other minerals (Watkins 2024a). By contrast, Mistaken Point's assemblages consist of the imprints of soft-bodied critters, which have been preserved through volcanic silt settling atop layers of mud. Something like 575 million years ago, this was a *benthic*, that is, seafloor, community, far below the reach of light. The denizens of this community left imprints in the sea floor's mud. These imprints were preserved by being coated in repeated layers of sediment rich in volcanic ash.

What kind of world do sites like Mistaken Point reveal? An odd one. The most common fossils are rangeomorphs, "bizarre frond-, spindle-, bush-, or comb-shaped" (Narbonne 2005, 426) creatures basically consisting of a central organic skeleton that supports a set of fractally growing "fronds" (see Figure 2.2). Some fossil forms appear to have been stuck to the seafloor, standing "upright", while others appear to lie flat.

The seafloor that rangeomorphs attach to is coated by a microbial mat, several centimetres thick (Kolesnikov et al. 2017). This relatively rigid structure both sealed the seafloor to erosion and gave the Ediacaran denizens something to stick to (Seilacher 1999). The ocean is significantly less oxygenated and less homogenously oxygenated than our own. Charlotte Kenchington has produced a haunting image of the fauna of Avalon, the rendition lit by an impossible light source (Liu, Kenchington, and Mitchell 2015, Figure 2.3). The image uses Emily Mitchell's maps (see below) to represent the organisms' size and spatial position based on data from the fossils in situ.

A surprising and, as we'll see, ecologically fundamental feature of Avalonian metazoans is their lack of mobility: the rangeomorphs are interpreted as firmly anchored to the seafloor. In later Avalonian deposits, there are features fairly interpreted as indicating a shuffling movement by some unknown taxa (Menon et al. 2013), and there is evidence of something similar at one contemporaneous site (Liu, McIlroy, and Brasier 2010). While there are always some mysterious geological features acting as centres of interpretative dispute, Liu et al. are firm on Avalonian sessility: "Rangeomorphs,

FIGURE 2.3 Charlotte Kenchington's Avalonian Scene, Depicting Upright and Flat Rangeomorphs. Note the Clever Use of a Positioned Light Source, Illuminating a Scene That Would Have Never Seen Light (Liu et al. 2015). Creative Commons CC BY

arboreomorph/frondomorphs, and the majority of the miscellaneous taxa abundant on these bedding planes are still interpreted as sessile and immobile" (Liu et al. 2015, 1370). The complex multicellular life of Avalonia was incapable of shifting from their spots, very different from the majority of metazoans today.

We'll be returning to the Avalonia, and the Ediacaran more generally, repeatedly throughout this book, so I'll save fuller discussion of what we know and don't know about these critters, and of various hypotheses regarding their feeding strategies, development, phylogenetic alignment, environment, and so forth, for when this is demanded. For now, and to motivate our conceptual engineering in the next section, I'll recount two stories, both of which highlight apparently unique features of the Avalonian world.

1.2 Pizza Disks

A long-standing puzzle from Avalonia concerns not rangeomorphs but another set of fossils: ivesheadiomorphs or "pizza disks" (Kenchington and Wilby 2014). Ivesheadiomorphs are mostly oval in shape (although they can be more distended), with very little internal detail other than raised lobes (Figure 2.4). They are distributed fairly randomly across Avalonian beds and are, as fossils, frustrating. Part of the frustration is that ivesheadiomophs are insufficiently alike: there is a distinct lack of regularity across their structure compared to more well-behaved fossil taxa. They are often asymmetrical, and their lack of internal structure makes reconstructing their morphology and phylogeny a baffling endeavour. As with many of Avalonia's denizens, they've been interpreted in a bewildering variety of ways. In 1979, Boynton and Ford interpreted one of the pizza disks, *Pseudovendia charnwoodensis,* as a kind of arthropod, grouping them with other animals sporting chiton exoskeletons, while other interpretations took them to be cnidarians. More recently, Laflamme et al. (2012) argued they were the remains of microbial colonies, while Wilby et al. (2011) suggested their irregularity was due to the differential effects of organisms collapsing in the bed above. In this discussion, I'll focus on Alex Liu and his collaborators' explanation, which seems to be generally accepted.

Liu et al. (2010) propose an explanation that ties ivesheadiomorph fossil forms to the putatively unique (indeed—perhaps *lost*) conditions of the late Ediacaran. They draw our attention to differences in decomposition between then and now.

Nowadays, an organism's death creates a scavenger smorgasbord. Were I to expire, the pattern of my decomposition would be influenced by the activities of various multicellular organisms: a retinue of bugs and creepy-crawlies would find me, seeking a food source (as potentially might my cat), various humans would likely move my decaying body about, perhaps much to their consternation (or at least complaining about the smell), and so on. But then-a-days, in the multicellular, sessile Avalonia, death and decay are different. Death in the Avalonia involves no such smorgasbord for metazoans, as they cannot get to you: Decay is left to single-celled organisms. As such, decomposition would take longer, potentially leading to differing patterns of decay. Add to this the possibility of the microbial mat acting to hold together partially decayed critters, and the less oxidized ocean, and we generate a decomposition pattern potentially unique to the Avalonian: "In such a scenario,

FIGURE 2.4 "Pizza Disk" Ivesheadiomorph (Fig. 1 from Liu et al. 2011). Reprinted with Permission from *Paleontology*

decay processes may have taken considerably longer than is observed in modern marine environments, implying that organic matter in the form of dead organisms could have been a common phenomenon at the Ediacaran sediment-water interface, and that impressions of these deceased organisms would have been preservable" (Liu et al. 2015, 1364).

The thought, then, is that ivesheadiomorphs are not true taxa after all, but instead a set of different lineages (presumably rangeomorphs and their allies) that have expired, decayed, and been preserved under Avalonian circumstances. Thus, "Ivesheadiamorphs and similar forms represent a morphologically disparate grouping of fossils, linked by their having experienced similar taphonomic processes rather than sharing true biological relationships" (Liu et al. 2011, 610). That is, it isn't that they are ancestrally linked, but instead that they decayed and were fossilized by a similar process. This matters not only for understanding organismic decay, but also the ecosystems themselves. As Liu put it in his doctoral thesis: "The late Ediacaran Period, at the time the Avalonian strata were deposited, is considered to be ecologically non-uniformitarian, in that there is currently no evidence for the pervasive bioturbation, macro-predation, or scavenging processes that rapidly break down and remove organic material in Phanerozoic ecosystems" (Liu 2011, 117).

If Liu is right, the processes of decay that produced the pizza disks are lost; that is, they only occurred due to specific, long-past Avalonian conditions. Let's turn now to a different case.

1.3 The Ecologies of Avalon

Here's another tale of Avalonia, also likely related to the lack of metazoan mobility. I mentioned that the Avalonian deposits, particularly those from Newfoundland, were formed in volcaniclastic events. That process of preservation is well-understood, and there is good reason to think they preserved the actual spatial relationships of living rangeomorphs. Treated carefully, then, these fossil beds act as spatial maps of Avalonian ecosystems. This is exciting because it enables the application of models also used to understand contemporary ecosystems, thus allowing us to quantitatively compare then with now.

Emily Mitchell and her team took careful scans of the fossil beds using laser-line probing technology, thus providing a digital representation of the fossils and their relative spatial relationships (Mitchell and Butterfield 2018). This enables the application of *Spatial Point Analysis*. Imagine a simple Euclidean two-dimensional space divided into space points. Now imagine randomly generating a scatter of occupied points across those dimensions; say, for each location, on each run, there will be a 50/50 chance of that location being occupied. Running this over and over and summarizing results will generate a kind of neutral model, a profile that represents the occupation patterns we'd expect if the location points were random. We could then compare this to the spatial distribution of a real ecosystem. Insofar as the observed pattern diverges from the null, we can infer that the ecosystem is structured in some non-random way, presumably by patterns of breeding, competition for resources, predation, and so forth.

Modern benthic systems (and other contemporary ecosystems) are not neutral in this sense: the kind of critter you are, and what other critters are about, make a difference to your spatial position. This is unsurprising, I take it, because ecological relationships—competition, mutualism, predation, food availability, etc.—all shape organism distribution across a modern ecosystem.

That is, spatial relationships are differentiated by *niche*. So, we'd expect the same for the Avalonian, right? I imagine you see the answer coming.

No.

Avalonian ecosystems appear to be neutrally structured, which is to say, the patterns Mitchell et al. generated converge on the null generated by spatial point analysis. Thus, Avalonian ecosystems appear to be dominated by neutral processes: What you are doesn't seem to influence much about where you are. It's worth noting how surprising this is from an ecological perspective. Ecologists examining ecosystems from the past 500 million years find niche-differentiated spatial arrangements, and no wonder, given the ubiquity, complexity, and importance of trophic relationships in metazoan life. Discovering a widespread, long-surviving assemblage of fauna that does not differentiate in this way suggests that a very different ecology is at play. Indeed, Mitchell and company consider a perhaps radical possibility emerging from this result:

> This study provides strong evidence that neutral processes structured Avalonian Assemblage palaeocommunities . . . These neutral-process-dominated community dynamics contrast with those observed in the modern marine realm, where neutral processes are typically rare across multiple spatial scales . . . This stark difference raises the question of whether Ediacaran early animal palaeocommunities had fundamentally different community dynamics to those of the present day. (Mitchell et al. 2019, 2034)

That is, if Avalonian ecosystems are uniquely neutrally dominated, might this mean that the usual rules of ecology don't apply? Presumably, this neutral ecological structure has something to do with the rangeomorph's sessility. We have remarkable records of the spatial relationships of Avalonian ecosystems, and we can infer their neutral domination. In light of this, Mitchell et al. ask whether neutrality is due to run-of-the-mill ecological dynamics acting in unusual circumstances or if Avalonian systems, with their wildly different organisms, environments, and so forth, simply play by different ecological rules (you'll have to wait until chapter 5 to find out Mitchell et al.'s answer . . .).

This dip into the science of the Avalonian Ediacaran yields two interesting cases of what I'll soon call "loss." First, metazoan immobility combined with the microbial mat and a less-oxygenated ocean generated never-seen-since patterns of decay and decomposition. Second, at least in terms of spatial composition, Avalonian ecosystems exhibit a near-unique neutral ecological structure. In the next section, I'll draw on these examples—and the more toy-like elephant case—to provide a conceptual scheme for thinking about these issues.

2. Loss and Erasure

> "When trying to make sense of Broadfoot, we are doubly challenged." Prof. Ichthy raises one flipper, "First, we've only one specimen, a paltry record." They raise another flipper, "Second, there's nothing like Broadfoot around for us to study." Ichthy lowers their limbs while a grin spreads across their maw, "Happily, this kind of scientific challenge is exactly what gets me excited . . ."

As we've seen, the difference between studying a mammoth and studying a dinosaur can be partly captured in terms of their differential preservation: for a mammoth, we have examples preserved in ice, samples of hair, and tusks; for dinosaurs, we have fossilized bodies, eggshell, and trackways. But attending to traces alone misses a crucial factor: elephants. A focus on differential preservation doesn't quite capture our two examples from the Ediacaran, either.[3] We have a fair number of ivesheadiomorph remains, and of great variety, but they are tricky to understand because they don't resemble anything we've seen before. Something similar could be said of the ecology of the Avalonian: We have extremely well-preserved spatial relationships, but an unfamiliar ecology. In these cases, the problem isn't a lack of traces. So, prima facie, attending to traces alone won't capture the challenges these scientists face, suggesting that philosophical focus on traces has been myopic. Further, as we saw in the first chapter, early geologists and palaeontologists emphasized their discoveries' importance by leaning on the idea that the past was crucially different from the present. It is time to provide some philosophical clothes for these ideas.

---

3 Recently, Max Dresow has suggested that a major difference in understanding the Cambrian and Ediacaran biotas is that the former has unusual features, but the latter has insufficiently preserved features. In my parlance, the Cambrian is lost while the Ediacaran is erased (and so might be lost, for all we know). This doesn't undermine my use of the Ediacaran examples; it is an excellent way into considering how loss and erasure interact (Dresow 2025).

These clothes will be largely stipulative. I'll begin by defining a "past world," and draw on that to further define "erasure" and "loss." These latter notions are intended to capture the epistemic challenges highlighted by bad preservation, on the one hand, and unfamiliarity, on the other. There'll be some philosophical finagling, but I hope the payoff throughout the rest of the book will justify it.

So, on to the notion of a "past world." First off, notice that "world talk" is evocative of a few different things. For many philosophers, it likely brings to mind David Lewis' analysis of modality. For others, it might raise a set of looser conceptions, "life worlds," or something similar. To an extent, my use of "worlds" trades on this ambiguity, but given that we'll be discussing possibility at length later in the book, I think it's worthwhile to connect my use of "worlds" to Lewis'. I do think that this machinery could be jettisoned, but I also think it brings some useful precision to the discussion.

Historical scientists are nominally interested in understanding some set of things in the past (although in later chapters I'll argue that they should have something much more modal in mind). Let's call that set of things a *Past World*.

> *Past Worlds* are subsets of the actual past; that is, they are *actual* but not *complete*. They consist of a set of temporally and spatially connected *denizens*: entities, processes, dynamics, relations, and properties, although which denizens the past world consists of is partly a matter of convention (that is, what scientists are interested in).

Let's tackle those italicized terms one by one.

What is it to say a world is *actual* but *incomplete*? We can understand a *world*, according to Lewisian analysis, in terms of spatial-temporal relations (Lewis 1986; Nolan 2015). If there are any temporal, spatial, or causal relationships between any two things, they are part of the same world. My left hand and the first rangeomorph are in the same world. However distant they may be, some spatial and temporal relationship exists between them. For instance, temporally speaking, my hand and the rangeomorph are something in the vicinity of 575 million years apart. If two objects lack such spatio-temporal relationships, then they occupy different worlds. For example, it might be possible that were true taxa after all (perhaps a kind of arthropod). A possible world semantics would let us understand this claim by saying that in at least that one world, there are arthropods.[4] There is no spatial or temporal relationship between true phylogenetic taxa and false taxa, so they occupy different worlds.

Worlds are *complete* in the sense that they provide an answer (that is, satisfy the truth conditions) for any meaningful proposition. Assuming that Liu et al.'s hypothesis is correct, presumably "are false taxa" is made true by the individual fates and decay of organisms in the Avalonian Ediacaran. If it is possible that ivesheadiomorphs might have been arthropods, then there will be a set of non-actual worlds containing arthropod ivesheadiomorphs. All worlds (according to Lewis, anyway) are possible worlds, and the actual is a possible world which is, well, here: *gestures around.*

Past worlds are subsets of the actual world. That is, they may contain some parts of the actual world, but do not contain all of it; hence, they are actual but not complete. They are, if you like, a bit of the world. So, when I say a past world is *actual*, I mean it is part of *our world*, the one we occupy. When I say it is not necessarily *complete*, I mean that it does not include every aspect of our world. It doesn't contain all it needs to ground all those truth conditions, only some of them. A past world is typically picked out or indicated by historical scientists via a set of *denizens* (or the traces of denizens).

I'm using the term "denizen" as a kind of catch-all, intended to include whatever kind of ontology you'd like. Your metaphysical inclinations may lead you to admit only simples and the void, or processes or structures, or entities with special causal powers, or whatever. However you cash it out, your metaphysical view will nonetheless need to (even if pragmatically) make sense of individual things such as particular rangeomorphs; processes such as the decay that produces ivesheadiomorph fossils; dynamics such as the neutral-driven ecology that structures Avalonian fossil-beds; relations such as those between individual rangeomorphs and the microbial mat they are embedded in. I hope I can remain neutral regarding matters concerning our views on fundamental ontology: at base, "denizens" are whatever entities, processes, dynamics, relations, and properties that constitute the past world.

There are potential downsides to the term "denizen," implying as it does a kind of "thing" or entity, as opposed to a process, say, or a set of dynamics. But I doubt any term will avoid bias in one direction or another, and I quite like "denizen." Again, I'm aiming to be as metaphysically relaxed as I can. If you

4 I'm not primarily interested in developing anything like a temporal analogue of modal semantics here (e.g., Rini and Creswell 2012).

believe that ultimately there are only relations, not relata, that is fine. For you, any entities and the like are merely derivative of those structures or processes, and vice versa for anyone committed to a more "substancy" metaphysics.[5]

When historical scientists tackle the Ediacaran, say, they more or less indicate a set of taxa, or an ecosystem, figuratively saying "I'm interested in *that* set of denizens." Perhaps more carefully, they indicate a set of specimens—traces—and thus indicate a set of denizens causally upstream of those specimens. So, what counts as being included in the world is at least partly indicative. This has two important consequences. First, there are *a lot* of past worlds: as many as there are denizens and combinations of denizens. Second, which world we are talking about is highly sensitive to description, depending on which world has been "indicated." The Avalonian Past World, for instance, is populated by rangeomorphs, the bacterial mat, processes of decay, and so on, but what exactly is included will depend in part on scientific practice.

Although some denizens are explicitly included in a past world, others might be included in less direct ways. Here, we can take our cue from the semantics of fiction. Some truths in a work of fiction are explicit, while others are implicit (Lewis 1978). Prof. Ichthy is an Ichthyosaur, has graduate students, is delivering a lecture, and so on. These are true in the fiction due to being explicitly stated in the text. But we might also infer from Ichthy being a professor that there is a university to which they belong, that marine reptiles can be PhD candidates, and so on. These truths are implied by both what is explicit in the fiction and our background expectations. Although past worlds are not fictions (or at least I'm not committed to them being so), there are also implicit denizens that, to some extent, fill the world out. For instance, although less complex multicellular organisms rarely figure in reconstructions of rangeomorph ecosystems, they likely haunted these worlds, and sometimes become explicit denizens when investigation demands it. The flexible boundaries of past worlds, as we'll see, are a crucial factor in how historical scientists investigate them. I'll call using this flexibility *strategic perspectivalism*, and it'll be introduced in the next chapter.

You might worry whether I'm talking about worlds as scientists represent them or worlds in themselves. In the past, some palaeontologists spoke of an Avalonian past that included a true taxon identified as "ivesheadiomorphs"; now they do not. It is helpful to distinguish between *putative worlds* (worlds as

---

5 And while I'm at it, if you've no metaphysical time for this "possible worlds" talk, don't despair: I'm only referring to the actual after all. Well, don't despair yet . . .

scientists represent them) and *true worlds* (the worlds themselves). Assuming we've got something right about the Avalonian, the true world doesn't contain an ivesheadiomorph taxon; that was simply a putative world. All worlds as scientists represent and discuss them, of course, are putative worlds, and they are true worlds when those representations are veridical or sufficiently so. Most of the time this distinction won't matter, but I'll wheel it out when necessary.

Okay, we have a notion of a *past world*: a set of denizens, indicated by scientists. Given some past world, what kinds of features make a difference to our epistemic access to it? One set of factors refer to our knowledge, technologies, and scientific skills and capacities. Another set, and what concerns us here, are at least partly features of the world itself. Recall the contrast between dinosaurs and mammoth. The latter has more and richer *traces*, downstream causal remnants. But the latter also has *elephants*: the denizens, insofar as they are members of Elephantidae, still exist. Dinosaurs, in the terms I'll now introduce, are both more "erased" and more "lost" than mammoths.

> Some past world is *erased*, relative to our current world, insofar as either (1) there is no downstream causal information of its denizens or (2) that information is not retrievable in the present.

Erasure is a partially epistemic notion. It may be that some past denizens literally leave no causal information (as Havstad suggested in chapter 1). Such denizens are permanently erased. If information persists, we need sufficient know-how and luck to be able to discover and make sense of it. As such, erasure is relative to our current knowledge states. Something might have been erased, but due to some new technological innovation or fossil discovery, become unerased. In this sense, erasure is not necessarily permanent. Dinosaur muscle does not leave direct causally downstream remnants in quite the same way as mammoth muscle sometimes does, although it is not completely erased, as we can infer it from features of preserved bone (compare with my definition of "trace" in Currie 2018a, chapter 3). Erasure, then, turns on whether causal information has been preserved and whether we can interpret it. Let's now turn to loss. A world is *lost* insofar as its denizens do not exist in the present.

Although mammoths and elephants are differing taxa, both are members of Elephantidae, and insofar as we are interested in understanding mammoths, understanding them as members of that still-existing group is an enormous boon. For dinosaurs, we have still-existing birds—the clade Dinosauria still exists; however, that is often not the aspect of the past world that interests scientists (that is, they don't pick out those denizens). Lineages like the enormous sauropods, the remarkably head-geared ceratopsids, or the larger predatory theropods have left no direct descendants, and birds have diverged from them dramatically. Of course, some aspects of mammoths—members of Elephantidae adapted to Arctic-like conditions, for instance—are lost. So, whether loss has occurred is in part dependent on how we characterize denizens. Let us, then, define loss in terms of denizens:

> Some denizen is *lost* if it doesn't currently exist or doesn't exist for all we know.

So, loss is a property held by a past world relative to now (in principle, it could be relative to other times as well), which is determined by taking the set of denizens constituting the past world and checking which are now present.[6] We'd call a world "lost" in case some (or enough) of its denizens are lost, or if sufficiently significant denizens are lost. Similar to how our characterization of denizens is sensitive to context—at minimum, what scientists indicate—whether a world is lost is also sensitive to context.

A fairly straightforward way to understand loss and its context-sensitivity can be drawn from the philosophical literature on similarity (e.g., Goodman 1972, 437–446).[7] Let's take a very broad conception of how we might group denizens: say, two denizens can be grouped so long as they are similar (I'll expand on kinds below). This proposal runs into an immediate problem. Without any limitations on similarity, every denizen is infinitely similar and dissimilar. For instance, a rangeomorph and my hand are both similar insofar as neither is the moon, neither existed on January 27, 1980, neither can speak Japanese, and so on. Similarity *tout court* is wildly profligate and unhelpful, to say the least. As such, similarity claims should be made in the context of

6 A way of understanding loss that continues the Lewisian theme would be to appeal to "alien properties" (Lewis 1986, 91–92). Whereas alien properties, for Lewis, are properties that do not exist in some world but do in another, we could understand loss as an alien property in a past world relative to ours.

7 Thanks to Tyler Brunet for pointing this out.

some set of properties or measurements. To say that mammoths are not lost because elephants exist is to say that elephants are similar to mammoths in specific ways. In particular, it is to say both that elephants are similar with respect to many pachyderm properties (they both have trunks, for instance) and, perhaps, that they are both Elephantidae. Thus, insofar as loss judgements are judgements of similarity, loss judgements are sensitive to context. That is, a denizen is only lost under some characterization.

Presumably, all past worlds are to some extent lost and to some extent erased. As I've said, denizens are profligate: They could include such insignificant aspects as, say, my spatial position in a room. That property is then lost when I move from that position. Going similarly fine-grained on erasure has similar results: presumably, some information from a few seconds ago is not retrievable.

Following from this, erasure and loss are both plausibly understood as *graded*, potentially along multiple dimensions. A past world will be erased insofar as information about that world is not contained in the present; a past world will be lost insofar as there are denizens of that past world that are not present now. On these definitions, holding fixed a set of denizens at least, one world might be more or less lost than another. For instance, mammoths are less erased than dinosaurs, as we have more, and more detailed, mammoth fossils. However, as we discover more fossils or discover more ways of interpreting them, dinosaurs might become less erased. So, erasure can be understood in graded terms. Loss is somewhat trickier on this front, as matters of existence are not typically considered to be vague: you either exist or you do not. However, mammoths, understood as members of Elephantidae, are not lost; but understood as the species *Mammuthus primigenius*, they are lost. This is sensitivity to description, not vagueness, but the speed at which scientists often switch between characterizations generates an effect very close to true vagueness. Further, under a more epistemic conception of loss, we might be uncertain whether some denizen is lost or not; so, we might have varying credence in loss. Further, considering loss as it applies across past worlds, differences in loss in those worlds' denizens will lead to gradation. There are senses, then, in which loss is graded as well.

On my view, in many circumstances, absences can become lost. That is, in principle, an absence could be a denizen, or at least a relevant property of a past world. Part of what makes the Avalonian Ediacaran mysterious isn't just what populates it, but what doesn't populate it. The lack of non-sessile complex metazoans marks an enormous difference between Avalonian and later times. So, one of the properties of the lost Avalonian past world is a *lack* of non-sessile complex metazoans. Including absences as properties, I think, is crucial for capturing elements of loss as it plays out in the epistemology of historical science. It also further increases the plenitude of lost worlds to infinity. However, I don't think this undermines the distinction: as with issues concerning similarity, loss being indexed to a particular set of denizens as partly determined by scientists' interests sufficiently constrains the notion in particular applications.

The upshot of this is that we shouldn't really call a world "lost" *tout court*. Rather, when I say a world is lost, I typically mean that a particular set of denizens is lost, relative to now, in some significant way. I don't think I mean anything fancy by "significant" here. Our own interests will do: The loss of the Avalonian Ediacaran is significant in part because it makes understanding that early example of metazoan life so difficult—and so enticing.

Because denizens can be characterized in so many different ways, and loss turns on the existence of denizens in past worlds and now, it will be helpful to say something about how we *kind* denizens. That is, how do we sort them into sets or groups? I can imagine a more restrictive view than mine that requires loss to appeal to natural kinds, say, or classifications as scientists use them. There is, indeed, a rich literature in the philosophy of biology and philosophy of science on classification (e.g., Richards 2016), and classification in the Earth sciences is often complex, with particular schemes playing multiple roles simultaneously (Witteveen 2024). Continuing with my (hopefully not too frustrating) non-committal approach to metaphysics thus far, I won't favour any particular story here about kinds or classification.[8] Nonetheless, there are three broad ways of characterizing denizens and linking them together, which are helpful for my purposes (I'm adapting these from Khalidi 2021; see also Currie 2019a). (In chapters 6 and 7, I'll draw on Catherine Elgin's notion of "exemplification" to explain artifactualist reasoning. She uses the language of "features" rather than "types," and I think that could work well here also, but the distinction between historical tokens,

8 My metaphysical leanings do tend toward fairly expansive, "promiscuous" ontologies (Dupre 1993), and it might be that some of that is being smuggled in here. I suspect that more restrictive ontologies could nonetheless accommodate much the same account of denizens and their properties as I have, it would simply result in some of these being considered mind-dependent or derivative of more fundamental ontological structures. I take it that all of these "denizens" would still be epistemically relevant, however.

types, and ahistorical types does the work I need it to.) So, let's list and discuss Khalidi's three notions.

> (1) Historical token (i.e., historical individuals: *that* mountain, *that* lineage, *that* organism)

Philosophers of biology often take many biological categories, particularly those linked with taxa, such as species, orders, etc., as *historical individuals* (Hull 1976). Historical individuals have identity conditions similar to the individuality of ordinary objects: they will have particular *persistence conditions*, often something like spatio-temporal continuity, and potentially particular *generation conditions* (see Ereshefsky 2014 for a particularly cogent discussion that emphasizes the "historicity" of individuals). For instance, the Dodo lineage began when it speciated from its pigeon-like ancestors and ended when the species went extinct sometime in the second half of the 17th century. My time as a postdoctoral researcher at the University of Calgary began with my signing a contract in 2014 and ended with my shifting to the UK in 2017. When we consider loss in the context of historical token individuals, we ask whether that denizen has survived into the present. Squanch, the individual mammoth, is lost due to her death. The mammoth species is lost due to extinction. The historical individual Elephantidae (which mammoth belongs to) survived into the present and thus is not lost. So far as we know, no rangeomorph taxa survived into the present—they left no ancestors.

Epistemically speaking, historical individuals often act as denizens as well as provide information about denizens; they can thus block both loss and erasure. Elephants, *qua Elephantidae,* and regarding their "pachydermy" properties, ensure that mammoths are not lost in those regards. In virtue of their ancestral relations, elephants also possess information about their common ancestor with the mammoth, and thus are partly traces of them. I don't see any particular problem with entities playing both roles.

A crucial split between ways of characterizing denizens is taking them as token individuals on the one hand and as types, sets, or categories on the other hand. I'll divide that latter division into two.

> (2) Historical type (membership requires the *same process type*, e.g., igneous rock)

Some classes are such in virtue of having undergone the same kind of generation process or, potentially, maintenance process (Millikan 1999). Igneous rock is a typical example: any rock formed of cooling volcanic lava is igneous, and will share various properties in virtue of this. That is all it takes to be igneous. A biological trait is an *adaptation* if it has been selected for in the past. So, we can group adaptations in terms of what they are an adaptation to. Wings are adaptations for flight for birds, bats, and pterosaurs. Consider particular adaptive radiations, such as that which spread our species out of Africa and across the globe, or that which took the Cambrian's metazoans into evolutionary prominence. We can consider both of these as individuals: the Cambrian radiation completed many millions of years ago, so it is now lost in that sense. But we can also understand them as historical types: an adaptive radiation is an event whereby some set of taxa, in virtue of a novel adaptation or set of adaptations, occupied new territory. Insofar as adaptive radiations still occur and are still occurring, that historical type is not lost. Forms of speciation are also historical types: cladogenic speciation, for instance, occurs when two populations split, that is, a branching event as one lineage becomes two. Insofar as this still sometimes happens, cladogenesis is not lost.

As a final example, consider Liu et al.'s explanation of iveshediomorphs. Where previously the similarities were understood in terms of historical individuality—the fossils represented ancestrally linked taxa—Liu et al. argue that they are in fact similar due to undergoing a similar process, namely, decaying under Avalonian conditions. Potentially, then, "ivesheadiomorph" is a historical type: to be one is to have decayed and been fossilized under those circumstances. And further, it seems to be a lost historical type.

> (3) Ahistorical type (membership turns on shared properties/ dynamics: e.g., possessing traits, ecosystem/ecotype)

Where historical types demand certain generation procedures for membership, there are various ahistorical types that rely simply on shared properties or characteristics. Typical examples of these include essentialist conceptions of objects: "gold" is anything with the relevant chemical composition, for instance. But consider as well the ecological dynamics that interest Mitchell et al. To have those dynamics would be to approximately instantiate a neutral-driven profile in an ecosystem. If their work is right, then there are no (or very few) systems like that around now. But further, recall their worry that Avalonian ecology "plays by different rules." Here, the worry might be not so much that there are no instances of that kind of dynamic around anymore, but that the models we use to understand contemporary ecological dynamics don't gain purchase in the Avalonian: ecology itself works differently.

So, historical scientists indicate putative past worlds and try to understand them, both via engagement with traces from those past worlds and by examining those historical individuals, historical kinds, and ahistorical kinds that have survived. The accessibility of a past world turns on the loss and erasure of that world's denizens relative to now. Denizens can be characterized as token historical individuals, and historical or ahistorical types at various levels of grain. Individual mammoths are lost, as is that taxon understood as a species, and they are lost when characterized as members of Elephantidae adapted to Arctic-like conditions. But understood as Elephantidae, or as cold-adapted mammals, or as far-roaming herd animals, they are not lost.

Before applying the framework in the next two sections—and throughout the remainder of the book—there's a central objection to tackle: Can we really distinguish between loss and erasure? If we consider loss and erasure's relationship, we find them intimately connected: Every case of erasure is a case of loss, and likely any real-world case of loss will include some erasure. For instance, a mammoth fossil being destroyed via subduction is a case of erasure, but it is also a case of loss, specifically, loss of that fossil. A mammoth's dying is a case of loss (of that individual), but it will also involve the erasure of various traces of living processes. The two are asymmetric in this regard. As traces can count as denizens, erasure can always be understood in terms of loss. However, in principle, a perfect recording machine of a mammoth would ensure against erasure despite loss. This is merely in principle, however: in the actual and near-actual circumstances that concern us throughout most of this book, loss will bring erasure. Do these features suggest that the distinction I've drawn collapses? If we expect these categories to be applied independently of scientists' interests and practices, then perhaps. However, appealing to the context-sensitivity of erasure and loss saves both the distinction and its function.

Once we fix a set of denizens, we can ask whether examples of those denizens have survived into the present, to what extent loss has occurred, and whether downstream information about those denizens in the past has been retained, that is, to what extent erasure has occurred. Elephants are, in a sense, both the same kind as mammoths (as members of the same phylogenetic family) and contain traces of mammoths insofar as they contain traces of their joint ancestry.[9] In Liu et al.'s explanation of ivesheadiomorphs as false taxa, that is, a phenomenon generated by the unique decay conditions of the Avalonian, one of the denizens included in the past world is the decay conditions themselves. The traces of the decay conditions have not been erased, but the decay conditions themselves have been lost. What makes this the case is partly what is included in the past world and what is not. To see this, let's examine Liu et al.'s argument in a little more detail.

Liu et al. rely on a careful interpretation of patterns of ivesheadiomorph morphology, linked to a particular picture of decay and preservation in the Avalonian. In short (see Liu et al. 2011 for details), they argue that ivesheadiomorph morphology reveals a pattern of more or less linear, effaced decay: from a rangeomorph-like beginning, you can arrange them into a pattern of increasingly decayed and distorted signals. From this pattern, a mechanism involving slow microbial decay coupled with preservation potentially occurring anywhere throughout the process is suggested. And this mechanism, they argue, would generate pizza-disk forms from once-living frondoid forms. Thus, it is concluded that pizza-disks are dead rangeomorphs and their allies (see Figure 2.5).

Liu et al. are not, in this instance, trying to understand a world occupied by ivesheadiomorph fossils, but one populated with the conditions leading to those fossils. That is, the past world includes rangeomorphs, a deoxygenated ocean, the microbial mat, and so forth. When scientists try to understand erasure processes themselves, they often do investigate worlds where things like fossils are taken as denizens—if we're trying to understand how subduction processes might have distorted or destroyed fossils, for instance—but we can identify which is which by attending to the interests of the scientists at hand: which denizens they indicate. Liu et al. aim to reconstruct rangeomorph decay processes. As such, the past world they indicate includes denizens relevant to understanding the decay of pizza disks.

Although interest-sensitivity saves the distinction from collapse, loss and erasure are still connected. For example, to the extent that loss has not occurred, then some erasure has not occurred. Currently living elephants contain information about past living elephants: they are (potentially, see footnote 9) traces of them. An upshot of all this is that if you really want to

9 There is an open question here regarding whether we take common-cause cases, such as the ancestral continuity between mammoth and elephants through their common ancestor, to count as traces. I don't think much turns on this for present circumstances.

FIGURE 2.5 Liu et al.'s (2011) Explanation of Ivesheadiomorph Decay, from Living Forms (Upper Left) to Highly Effaced and Pizza-Disk-Like (Bottom Right) (Fig. 13, Liu et al. 2011). Reprinted with Permission from *Paleontology*

think about all of this in terms of our accessibility to information, you could likely pull it off, but you'd need to recover a distinction along the lines of loss/erasure to capture the aspects of historical knowledge I'm interested in.

To summarize, if we take a past world in a sufficiently maximal way, loss and erasure nearly collapse; however, if we hold a set of denizens fixed, then we can identify loss and erasure in that context. Thus, the distinction is meaningful across the (many, many) contexts that are relevant to understanding historical knowledge.

So, we have the basic conceptual machinery on the table. Palaeontologists aim to reconstruct a past world and, in so doing, indicate a set of denizens. Their investigations are thus challenged by both erasure, that is, the loss of information causally downstream from those denizens, and by loss, that is, those denizens' cessation.

3. Tracecentrism and History Redux

In the last chapter, I claimed that the philosophy of the historical sciences has been too narrow, focused too much on traces. I also discussed how the early sciences of palaeontology and geology justified their importance by claiming that the past differed significantly from the present. The distinction between erasure and loss allows us to capture these ideas more carefully.

Tracecentric accounts of the epistemology of the historical sciences claim (or at least imply) that the central challenges of uncovering the past turn on the erasure of traces, and characterize the method of historical science as relying heavily on inferences from traces. There's no doubt that the past has had a causal effect on the present, and thus these effects provide an inroad to the past, which is a foundational insight that shouldn't be disregarded. However, to uncover the past we need, yes, the traces, but also an understanding of both how those traces form and the kinds of denizens that could leave those traces. Ichthy can tell us about Broadfoot not simply because they have the specimen, but because they understand fossilization processes, which they gained in part from study of similarly acting processes in the present, and because they live in a world with mammals, albeit mammals quite different from Broadfoot. If no such mammals existed, if they too were lost, then Ichthy's epistemic situation would be significantly more difficult.

Such loss-induced difficulties are pressing for unusual extinct taxa, such as some dinosaurs. Although we have a fair number of traces of, say, *Diplodocus* and other sauropods, without existing analogues, our capacity to make sense of them is diminished. Considered as historical individuals, sauropods are not completely lost because the Dinosauria clade they belong to survives in modern Aves. But birds depart so far from these massive, quadrupedal animals that they constrain our understanding insufficiently. Understood as an ahistorical type, say, as a large quadrupedal animal, we have some mammals to consider (elephants, for example), but it is often unclear how similar we should take elephants and sauropods to be. Or if sauropods lived in herds, we have plenty of animals in herds (an ahistorical ecological and behavioural kind), but it is difficult to tell how similar herd behaviour in sauropods is to herd behaviour in contemporary animals. The significant loss between sauropods and anything that exists now makes it hard to know which aspects of today's possible sauropod analogues can be projected back to sauropods.

In short, it seems as if successful investigation of past denizens requires (1) their traces and (2) those denizens not being completely lost. That is, something must still exist that can be linked to those denizens as historical individuals, historical kinds, or ahistorical kinds. Existing examples enable us to understand how traces form, and the kinds of entities that might form them. I'll further articulate this idea in the next section, but the point here is just that to the extent that an account of the method and challenges of historical science is focused on traces and decay only, it misses the critical importance of loss.

As Rudwick and other historians of science have shown, the discovery and articulation of a pre-human history revolutionized our understanding of the world and further provided a motivation for studying the deep past. Although other cosmologies had histories, they lacked an open, discoverable and surprising narrative about that history.[10] In a sense, these geologists discovered loss: the denizens of the past differ from the denizens of the present. In the last chapter, I contrasted Professor Ichthyosaur, a caricature of Lyell, with Prof. Ichthy. Professor Ichthyosaur sees the geological world according to a steady-state model. Although particular historical individuals might come and go, the general dynamics of geology, the environment, and biological responses to that environment are not lost but continue acting today. Although Lyell might not want to say that the marine reptiles that would reappear in Jurassic conditions would literally be ichthyosaurs, he nonetheless seems to be committed to them being ichthyosaurs in many of the ways that matter. By contrast, his detractors like De la Beche (and presumably Prof. Ichthy, if only they were to know of Lyell's views) see the world as being in significantly more flux, much more contingent and more path dependent. It thus potentially contains more loss. They deny that the processes and dynamics that apply today apply in the humanless worlds of the past.

Loss's sensitivity to description allows us to capture different positions regarding the relationship between the past and present. A steady-state perspective sees much loss, no doubt, but that loss is not epistemically significant: by examining today's entities and processes, we can recover all we need to understand the past. Individual losses can be reconstructed via an understanding of currently ongoing processes, and, indeed, the losses are likely insignificant due to the circularity of the Earth's processes. This thought underwrites Lyell's uniformitarianism. As our metaphysical picture shifts from a steady-state model to an increasingly contingent one, we see more significant

10 As Chis Manias points out to me, Christian cosmologies certainly include notions of loss!

loss, making the past more mysterious—uniformitarian reasoning becomes ineffective.

Recall Mitchell's question about the ecological dynamics of the Avalonian Ediacaran. It isn't simply that Avalonian ecosystems are lost; she also worries that the very ecological dynamics that they instantiated might be lost too: they raise the possibility that the Avalonian played by different ecological rules. If we want to understand the Avalonian, potentially we don't simply need to consider quite different denizens, but these denizens might break the general models we use to understand how ecosystems form and function.

Thus, the conception of loss I've characterized provides a challenge to philosophers interested in understanding historical science, and it has the resources to capture various positions on the nature of history. On one extreme, loss is never significant—the world moves in stable patterns, which we can uncover by watching those patterns unfurl. On the other extreme, significant loss is rampant, and history is wildly contingent and path-dependent. The challenge, I take it, is to understand how to navigate between these two extremes. I'll take up this challenge in chapter 8.

4. The Challenge of Loss

"Broadfoot is unique. And this uniqueness is baffling. Us scientists," Prof. Ichthy gestures magnanimously, including various graduate students, "deal in numbers: statistics, data, measurement. We take nature's chaos and find order with a ruler. And, of course, we can measure our specimen." Prof. Ichthy places their lecturing rod alongside the encased skull.

"My pointer here is about thirty centimetres long, and the skull is about one third of the length of the pointer. So, I can surmise a skull of approximately one third of that length, that is, in the vicinity of ten centimetres. But what are we to make of this measurement without anything to compare it to? Is ten centimetres normal for members of Broadfoot's species? Is this specimen an outlier or an infant? Should we treat it like a typical mammal in terms of head length, when we know it's not a typical mammal in so many other ways? And if we must ask these questions of something as simple as skull size, what are we to make of its apparent aquatic adaptations—the bill and webbed feet—to say nothing of its reptile-like features, or its plethora of other mysterious, unique properties?"

Ichthy pauses for effect, leaning their lecturing rod against the lectern.

I've stated that Broadfoot was venomous, that it hunted using electrosense, that it was the only egg-laying mammal we know of. But how could we possibly find that out? We interpret fossils against the backdrop of the wide variety of things we know, and that knowledge requires numbers. Again: statistics, data, and measurement. If some fossil feature is unique, it seems, we haven't any way of gaining the grounding required to make scientific sense of it.

Ichthy takes a step from the podium, lays a flipper on the encased skull, and says in a stage whisper that carries across the audience: "Might all my research merely amount to non-scientific speculation?"

Ichthy provides a conundrum: if scientific knowledge depends on examining multiple examples, how can we have a science of the unique? As we'll see, although loss doesn't necessarily generate uniqueness *per se*, it can generate similar problems.[11] The next three chapters could be read as responses to Ichthy's conundrum. To identify the challenge that arises from loss, I'm going to take a tangent into a related discussion: *N=1* problems as they play out in macroevolution and astrobiology.

An *N=1* problem, traditionally understood, holds when there is some entity or process we want to understand, but our sample set is one, that is, we have only one specimen. The term has currency in two intimately connected debates:[12] discussion of life's contingency at a macroevolutionary scale, and investigation of life beyond our planet's boundaries (Powell and Mariscal 2015; Powell 2020; McMahon 2021). I'll argue that a single specimen does not an N=1 problem make. Instead, N=1 problems are problems where we have insufficient specimens or data to establish the background theory required for legitimate interpretation. With N=1 covered, I'll argue that the challenge of loss is similar in character.

Questions about the robustness or fragility of evolutionary outcomes at a macro-scale have much energized palaeontologists, theoretical biologists, and philosophers of biology. For instance, as we'll explore downstream, the Avalonian world ended at some point, and the Ediacaran fauna were replaced by the funky, complex, and strange metazoans of the Cambrian. Was that

11 For discussion of epistemic problems related to uniqueness, see Tucker 1998; Buskell and Currie 2021.

12 As Rose Trappes points out, there are also N=1 cases in medicine. I'm not sure if these play out in the same way.

outcome inevitable or contingent? Did it have to happen, and, if so, what necessitated it? And, further, what were the factors that shaped that modal profile? Were rangeomorphs and their ilk out-competed by novel species? Were they fragile in the face of environmental collapse? Or did they simply get unlucky? Overall, we want to know what the choice-points in evolution's history were: where its path could have gone one way or another. Discussion of contingency, convergence. and what upshots it might have for the nature of the past has garnered a fair bit of attention from philosophers (for instance: Beatty 2006; McConwell and Currie 2017; Ereshefsky and Turner 2020), and we'll return to it in chapter 8.

Astrobiology aims to understand the diversity of life in the universe: if there is other life out there, how similar will it be to life on Earth? How might we detect extraterrestrial life? Can we take current life as representative of all other life or not? This matters for how we might go about detecting extraterrestrial life: how will we know it when we see it (Mariscal 2015; Cleland 2019)? Some connect macroevolution and astrobiology rather intimately: Rachel Powell (2020), for instance, argues that various convergences on Earth, and regularities drawn from them, can underwrite fairly robust predictions about what alien life will be like.

Both discussions of macroevolution and astrobiology run into a similar problem: on the face of it, we have only one specimen of life. Our N, then, is 1. Life on Earth has a common origin; so, its modal profile could have been shaped by quirks inherited from that common origin. In answering questions about contingency, we ideally want independent events that generate data to inform our hypotheses, but common origins mean the data is not independent. As Kim Sterelny has put it, even in instances where distantly-related lineages have evolved down similar pathways, "The envelope of evolutionary possibility for each lineage will depend in part on homologous developmental resources" (Sterelny 2005, 588). Similarly for astrobiology: without relevant, independent contrasts, it doesn't seem obvious that we can determine whether life on Earth is as it is because that is what life is like everywhere, or if there are particular quirks about life on Earth and Earth's particular conditions that determine its nature. Some argue at least, that common origins, and commonality in many of Earth's conditions across time, mean that N=1.

Although N=1 problems are often cashed out as due to a lack of specimens, that can't be the whole story, and this matters for understanding loss. There are many cases in palaeobiology and other historical sciences where N seems to equal 1, but these don't appear to have the same epistemic consequences as reported in macroevolution or astrobiology. Professor Ichthy's Broadfoot specimen is the only one available: it is the only direct trace of the animal's existence. But, as the Professor points out, we know enough about fossilization and mammals to be able to make solid claims about the past based on that single specimen. In real-life vertebrate palaeontology this is common: new species are identified through single teeth and other partial, singular specimens, fairly regularly (see Currie 2018a, chapter 1). This is made possible by the rich epistemic background palaeontologists draw on. Because we know so much about mammal teeth, a single tooth—N=1—is sometimes sufficient to underwrite well-evidenced claims about the past.

So, N=1 problems are problems not *simply* because we have insufficient specimens, but because we lack the epistemic background required to understand the relevant causal and modal structures that the "1" is embedded in. As Ben Jeffares has pointed out regarding our knowledge of fossils: "Our understanding of the palaeontological record is not built de novo from the observations of fossils, but constructed on the foundations of our knowledge of the contemporary natural world and the deployment of a general regularity" (Jeffares 2008, 474).

So, although for some fossil taxa we have only one specimen, because fossilization is so well understood, having a single specimen doesn't necessarily block knowledge.

This generalizes: traces are only traces, that is, evidence of past worlds, because we have some grip on regularities concerning how those traces form. We are only able to build on that knowledge because we have some grip on the kinds of dynamics and relationships that hold between these kinds of systems. To get that knowledge—at least it seems—we need to observe and study examples of those denizens. This is what makes N=1 problems tricky. With only a single example, we seem to lack the resources to develop the background knowledge required to make sense of counterfactuals concerning life's path or surprising extraterrestrial signals.

The apparent inability to build robust epistemic backgrounds is what makes loss an epistemic challenge. The denizens of the Avalonian Ediacaran, the strange fossilization processes due to the microbial mat and lack of metazoan mobility, the surprising neutrally-driven ecological relationships between rangeomorphs, as well as their uncertain phylogenetic relationships, uncertain metabolic and feeding strategies, uncertain developmental pathways, and so on, are lost. There is (for all we know) nothing like them around today, meaning that (putatively) there is nothing we can look to in building our background theories. As such, on the face of it, we lack *any* specimens with which to constrain our theorizing about them. This isn't a N=1 problem. In some sense, we're faced with an N=0 problem.

Of course, it is not literally N=0. We *do* have traces—the world of the Avalonian might be lost, but it isn't erased—so we have some partial access. Those interested in life's contingency and in astrobiology have extremely rich access to a sample of life—it is all around us—while those studying lost worlds often have extremely limited access.

To see the idea, let's briefly compare the astrobiologist and the palaeobiologist of the Ediacaran (we'll return to astrobiology in chapter 6). The former is often in the business of trying to detect biosignatures from exoplanets. The chemical composition of an exoplanet's atmosphere is retrievable from its wavelength, and some atmospheric compositions could signal the presence of life (heaps of oxygen, say). So, the light from exoplanets counts as traces, and the theory of biosignatures works analogously to a theory of fossilization. From where was that theory developed? Well, observations of our own Earth (also studies of early Earth and comparisons with early Mars, but let's not worry about that): to simplify, we know the kinds of biosignatures to look for—we have the basics of how to make sense of the relationship between atmosphere and biosphere—because we are intimately familiar with a particular atmosphere and biosphere. An N=1 problem is generated because we don't know how representative our planet is: it could be that Earth-like atmospheres can be generated by abiotic processes, or it could be that there are biota that have profoundly different biochemical signatures than ours. But still, that 1, the Earth, is a crucial lynchpin. But also note that N *isn't* in fact 1, at least regarding planetary atmosphere: there are many traces from those.

Now compare this to studying something lost, like the Ediacaran pizza disks.

The pizza disks are the result of conditions of decay and preservation unique to the Avalonian Ediacaran. There is no analogue of Earth's life for astrobiology here. Studying something lost from the deep past is a bit like trying to understand exoplanet atmospheres without being able to look at the Earth's atmosphere. There are—on the face of it—only the traces to go on. This is the challenge: how can we build the kind of robust background theory required to interpret traces when their sources are lost? Without sufficient denizens surviving into the present, such theories seem beyond our reach. Thus, we cannot make sense of lost denizens. With such limited information about those lost denizens themselves, analysis of their traces can do little to improve the situation.

As with Prof. Ichthy's rather staged admission of doubt, I'm surely laying it on a bit thick here. Clearly, loss doesn't seem to block scientific knowledge. We've already seen a few hints as to why. A lack of erasure can mitigate loss, and a lack of loss can mitigate erasure. Prof. Ichthy's impoverished, but still present, mammalian fauna underwrites their ability to make sense of the strange fossil specimen they discovered and hence infer many of Broadfoot's surprising properties. With a sufficiently rich set of traces, we can puzzle our way to understanding loss even without contemporary examples, as we saw with Mitchell's uncovering of the surprising ecological dynamics of Avalonian systems and Liu et al.'s arguments for ivesheadiomorphs' ichnotaxic status. But as loss becomes increasingly rampant, the challenge becomes pressing. What can we do then?

That question drives the next three chapters, whose aim is to abstractly capture three (non-mutually exclusive) strategies historical scientists adopt when faced with loss. One set of strategies involves emphasizing what in the target world has *not* been lost: they *graft*, that is, seek aspects that are historically continuous with the past world and seek to understand those. Another set of strategies is, in a sense, less direct. Scientists *artifice*, that is, construct a system with similar properties as the lost world. They resurrect it. And finally, they might *anchor* by navigating between a set of differing systems to construct a model of their target. These strategies are often used simultaneously, but characterizing them separately, I hope, helps us understand their underlying logic. As we continue, a characterization of an overall strategy arises: as we'll see, loss is tackled through a combination of *strategic perspectivalism* and *iterativity*.

Interlude: The Parable of the Lost City

Many years ago, there was a thriving city deep beneath the waves. It thrung with thoroughfares and parks, homes, temples, stores, and markets. In the city centre, the people built a great temple of carved, smooth stone blocks, arranged into an imposing, cube-like form, tapering towards the top. Throughout the temple were carved niche-lined tunnels, some opening into rooms, sleeping quarters, meeting chambers, and spaces for prayer. Outside, the walls were stained a deep purple and coated in seaweed of deep greens and blues curling in patterns reaching upwards towards the top. Atop the temple, the stone formed a flat square covered by a garden of fungi, anemones, and underwater flowers.

Each year, twenty children were selected by ballot from the city's orphanages. They were raised in the temple and given solemn tasks. They were required to learn the sacred mantras and celestial calendar, keep the history of the city carved in ornate script throughout the temple's corridors, crevices, and byways, and—of course—attend to the upkeep and proper care of the garden. When the children came of age, they were ushered into the city's bureaucracy in a grand ceremony. Those with the best memories for mantras and dates joined the priesthood; the best script-carvers became the city's architects and city planners; and the best gardeners became the rulers, judges, and executioners.

A city's order relies on relentless activity. It is, then, unsurprising when a city falls, and history's cruel processes of erasure and loss do their work. Without gardeners, sea life grew with abandon atop the temple. Ocean currents wiped the purple decoration from rock walls and eddied through the passageways, smoothing out the carved inscriptions. All about the temple, the city fell into ruin, partly from water, partly from fauna and flora, and partly from those who remained scavenging building materials.

ILLUSTRATION 2.2 The Lost Temple

And, so they say, for some further long span of years, the temple sat lonely. The city was abandoned. The topic of ghost stories, rumour and speculation. But eventually, people returned to the temple, neither to live nor worship, but motivated by something in between curiosity and greed. First came explorers, stumbling upon the city amongst the seaweed jungle. Second came raiders and treasure hunters, seeking to plunder whatever riches the temple hid. Third came surveyors, mapping out the jungle for "civilization," as they called it. Fourth came archaeologists, digging out segments of the old city, following the original tunnels through the temple, and sniffing out traces of the old gardens. Fifth came tourists, and with them vendors, tour guides, hawkers, and shysters. Thus, once again, the temple became culturally alive, no longer as a site of worship but as a site of surprise for explorers, adventure for raiders, cartography for surveyors, mystery for archaeologists, and holiday snaps and wonder for tourists.

And most who visit the old temple see it as lost—that is partly why they come—but see it as lost in different ways . . .

The romanticist, who cares about mystery, says, "This unique culture is forever lost—who knows what ancient wisdom disappeared along with it?"

The archaeologist, who cares about ancient culture, says, "Although the ritual purposes of this temple are lost, the site is full of clues, and we hope they are not erased."

The global historian, who cares about regularities across cultures, says, "This particular society may be long-gone, but its story represents a pattern repeated throughout time and continuing today; thus, it is not truly lost."

The architect, who cares about the design of buildings, says, "This culture is lost only partially—see here, the stately beauty of the temple's basic design still stands—and many more subtle aspects might be recovered."

The engineer, who cares about the how of structures, says, "The techniques of construction used by this ancient society are lost, but we hope to recover them."

The restorationist, who cares about returning the broken-down to original state says, "with such a rich set of traces of this temple, and our building know-how, we can return the temple to its past glory: what is lost might yet still be rebuilt!"

The heritage conservator, who cares about preservation, says, "much has been lost to time, but with care we might slow the decay of what remains."

The historical chemist, who cares about patterns of decay, says, "The ruins are not lost: they decay as most cities do."

The linguist, who cares about the lives of languages, says, "the script is in one sense lost and in another not: although we cannot decipher it, we see echoes of it in better-preserved ancient languages and, indeed, it seems to be part of the same linguistic lineage as our own."

The social anthropologist, who cares about the particular relations between people within cultures, says, "Whatever ways these folk lived and loved are lost, gone with their unique culture."

The evolutionary psychologist, who cares about evolved patterns across species, says, "Although some inconsequential specifics of these people's culture are lost, because we all share a biological heritage, much is not lost: they carried the same set of genetic packages as we do after all."

The politician, who cares about what the temple might symbolize, says, "These great builders are not lost, but are part of the grand story of our nation!"

And the descendants, who never truly left the city, living nearby for centuries, say, "This city and this culture is neither erased nor lost. Our stories and the bones of the temple keep it from erasure, and our lives and lifeways show that it is not lost."

3

Artificing

Shuffling on their lower flippers, Prof. Ichthy shifts to stand over the skull on the podium. "Broadfoot is so unusual, so unfamiliar, that many of its properties can't be straightforwardly inferred from its fossil. Because we've only one specimen,"—a gesture to the skull before them—"and because the specimen reveals an extinct animal so unlike today's animals"—a gesture to the ancestral tree projected behind them—"finding the answers to some questions seems difficult indeed. For instance, how did Broadfoot move? Modern mammals are largely arboreal, built for scurrying through trees, so are hardly a good guide for Broadfoot's apparently aquatic or semi-aquatic lifestyle. So, with limited remains, and no good analogues, what can we do?

"How about . . . " Ichthy pauses dramatically, allowing a little sparkle into their eye, "building a Broadfoot!?"

Prof. Ichthy gives an emphatic nod to a grad student, who steps rather gingerly onto the stage holding a small object, maybe two feet long. As they place it on the ground, it is revealed to be some kind of small, puppet-like, toy-like object, designed and painted to look a fair bit like Ichthy's reconstruction. Some kind of simulacrum of Broadfoot, then. The grad student looks expectantly at Ichthy, clutching a length of string attached to a small hook at the front of the model.

"Currently living mammals all walk with what we call an 'erect gait,' their legs pointing straight downwards." Ichthy performs a mime with their flippers, "Not so, it seems, for Broadfoot. This ancient lineage seems to walk with its limbs splayed outwards in what we call a 'sprawling gait,' quite similar to some of today's lizards." Ichthy's flippers shift from pushing directly in front of their body to a wider position, reminiscent of an ungainly hug. "Partly based on the aspects of Broadfoot anatomy we've puzzled together, partly informed by reptilian and mammalian analogues, we've built this little model. And . . ." (another dramatic pause) "lo and behold . . ."

Ichthy gestures grandly as their graduate student begins to pull the little model along by its string. Emitting a rather sad squeaking noise, it locomotes across the stage in an awkward (and, members of the audience agreed later, cute) gait. It first moves its limbs on the right side of its body and then the limbs on the left, creating a kind of rhythmic twist across its body. There is a small smattering of applause.

ILLUSTRATION 3.1 A Toy Broadfoot

"Excellent work." The professor nods to their student, who takes up the model Broadfoot and vacates the stage with some embarrassment.

In the last chapter, we discussed Liu et al.'s explanation of ivesheadiomorphs. They argued the fossils are not traditional taxa: instead of belonging to a particular clade, ivesheadiomorphs are the remains of a set of different rangeomorph (and other) lineages strangely preserved by the unique Avalonian conditions. Liu et al.'s argument relied on constructing a sequence of fossil forms, taking us from rangeomorphs to pizza disks in a smooth transformation, and describing a mechanism of decay specific to the Avalonian Ediacaran, which could explain that sequence.

You may have wondered, *how did they test that*? And you may also have thought—*oh, I bet I know how*. Perhaps they built an experiment that mimicked Avalonian conditions, introduced deceased organisms, and examined the subsequent decay patterns. As with Prof. Ichthy and their Broadfoot puppet, Liu et al. would have thus *artificed* a lost world, building a simulacrum and examining how that worked. In the face of loss, that is, a lack of Ns that you might use to construct a robust theory, you make some Ns for yourself.

If you thought something like this, you'd be exactly right.

In Alex Liu's doctoral thesis (Liu 2011), he describes a series of experiments. These involved filling a set of small plastic pots with seawater and a substrate of similar composition to Avalonian environments. The pots are populated with microorganisms collected from the seawater. These are placed in tanks blocked from both the sunlight and visible biotic intruders. Some pots are left for a few months, allowing the microbial community time to form something approaching the microbial mat of the Avalonian, while others are left for a few days. Afterward, body tissue from deceased metazoans is introduced: jellyfish, sea anemones, deep-water worms, and seaweed. These were left to decay, with timed photographs recording the process, before being coated in tuff (rock rich in volcanic ash) and left for a year. After this process, they were CT-scanned.

Liu highlights two important features of the results. First, the sea anemones did, indeed, form lobe structures analogous to those of ivesheadiomorphs (the worms and jellyfish decayed too quickly, and the seaweed decayed too slowly: sea anemones were just right). Second, decay occurred much more slowly in the pots that had been allowed a few months to settle; it seems that undisturbed microbial mats indeed affected decay rates.

Liu didn't necessarily need to artifice their proposed decay process to provide a fairly strong argument in favour of their interpretation; demonstrating a transition pattern from rangeomorphs to ivesheadiomorphs, coupled with a plausible mechanism, may have sufficed. Nonetheless, Liu's experiments go some way to both add plausibility to the ivesheadiomorph interpretation and provide a means to explore how that mechanism may have operated. Such experimental approaches are common throughout historical science (Jeffares 2008). My aim in this chapter is to characterize this strategy for tackling epistemic trouble caused by loss. I'll call the strategy "artificing."

"Artificing" involves conceiving of a denizen as a type (historical or ahistorical), and then constructing a token of that type (or, if you want, something analogous or sufficiently similar to that type). Liu conceived of ivesheadiomorphs as multicellular organisms that decay under Avalonian conditions (a historical type) and then set up those conditions to test various hypotheses concerning them. Thus, he artificed.

Artificing is easily linked with both philosophical work on modelling and experimentation. There is much familiar about artificing from the philosophy of science interested in modelling; and no wonder, artificing is, in a sense, a form of modelling.[1] Specifically, it is modelling with the aim of understanding lost worlds. The ubiquity of modelling and related scientific practices, such as idealization and abstraction, is well noted and well-studied by philosophers (e.g., Potochnik 2017; Frigg and Nguyen 2020) and certainly noted as important in the palaeosciences as well (Turner 2009; Bokulich and Oreskes 2017; Tamborini 2020; Currie 2024b). Despite this, I'm not interested in tackling any of the big topics surrounding scientific models in this chapter; rather, I'm interested in getting a grip on how strategies like artificing work to reveal lost worlds. So, artificing might be understood as a form of modelling (or experimentation!), but crucially, we should understand it as modelling or experimentation toward a particular purpose: tackling the challenge of loss. As such, rather than directly engaging with more general philosophical work on modelling and experimentation, I will pull on a different thread in the philosophy of science—*perspectivalism*—to capture crucial features of artificing, in particular, the importance of *characterizing denizens.*

Why the term "artificing"? I want to emphasize the active construction involved in the strategy: historical scientists respond to mysterious denizens by attempting to build them, or at least simulacra of them. As we'll see downstream, I also want to avoid overly representationalist approaches to the activities of scientists, and an emphasis on the construction of artifacts suits me—although one needn't agree with me on "artifactualism" (see chapters 6 and 7) to agree with the arguments of this chapter. In case any archaeologists are reading, note that by "artifact" I don't mean artifacts in the sense of traces in our species' material record, but rather artifacts that *scientists construct* to understand the past (the practices described in this chapter can be easily aligned with experimental archaeology; Currie 2022).

I'll first provide two contrasting case studies: one from experimental taphonomy (briefly leaving the Avalonian Ediacaran for the more familiar Cretaceous), and another using geometric modelling. I'll then provide an

1 I take this to be a classic example of the strategy of model-based-science as described by Godfrey-Smith (2006) and Weisberg (2007), although see Currie (2024b).

analysis of the strategy, giving a first pass at what I'll call "strategic perspectivalism." I'll close by discussing an extreme form of artificing: de-extinction.

1. A Dinosaur's Burrow

In his 2011 book *Convergent Forms Most Beautiful*, George McGhee notes that for any large, globally-dominant macrofauna, you'll find examples of lineages adapted to an underground life, that is, burrowers. Current mammal life boasts everything from moles to badgers to meerkats to wombats to aardvarks (and platypuses, of course!). "The number of independent convergences on burrowing morphologies is amazing", McGhee says, "particularly when it would seem that these morphologies are related to a rather restrictive and highly specialized way of life" (McGhee 2011, 26). Beyond current mammals, legless lizards, Carboniferous amphibians, various toads, and an army of insects and annelids are counted among the burrowing menagerie.

On this basis, we should expect burrowing non-avian dinosaurs. But for a long time, we didn't find any, and it's worth noting how surprising this is. Dinosaurs ruled the roost, as it were, for over a hundred million years, and every other successful lineage we know of has taken up a burrowing lifestyle. What gives? Here, we have a case of putative loss. Is there something about dinosaurs as a lineage, or their environment somehow, that meant that burrowers couldn't arise? In considering options, don't focus on dinosaur size. Although some were enormous, the average was around a middle-sized dog in stature (Farlow et al. 1995). Further, recent suggestions that *Lestodon* (one of the Pleistocene's enormous ground sloths that had Richard Owen speculating about a wildly verdant jungled South American past in chapter 1) might have been a burrower imply that size is no insurmountable obstacle to this kind of behaviour (Lopes et al. 2017). So, is there something special about dinosaurs or their environment that explains their apparently unburrowing nature?

This approach to tackling the question of dinosaur burrowing focuses on loss: What properties—which presumably no longer exist—explain the lack of burrowing dinosaurs? This isn't the only way we might explain the phenomenon, however.

A lack of burrowing dinosaurs could be more due to erasure than loss. First, we might think that there were dinosaur burrowers, but there was something about how they burrowed or the conditions of preservation during the Mesozoic that meant that they were not preserved. Here, we have a case of loss, but loss concerning the status of preservation. This mirrors what we saw concerning ivesheadiomorphs: traces are hard to recognize due to surprising patterns of signal decay.

Second—and much more directly—maybe there were dinosaur burrowers, but we just haven't found any yet. This is a simple case of erasure, and may well have been the case. In 2007, Varricchio et al. announced the discovery of a new dinosaur genus, one with a hitherto unseen lifestyle; they argued it lived in a burrow and, not only that, it also denned. That is, these animals raised their young in the protection of the earth. They named the genus *Oryctodromeus cubicularis* (the genus name meaning "digging runner" and the species "of the lair"), describing it as a fairly small biped. The find, from Montana's Blackleaf Formation (dating to around 95 million years ago), "Consists of the skeletal remains of an adult and two juvenile dinosaurs preserved within an infilled burrow" (Varricchio et al. 2007, 1361). Perhaps dinosaurs burrowed after all.

However, Varricchio et al.'s interpretation faces a major objection. As Woodruff and Varricchio (2011) summarize, other ancient burrows were identified "primarily on the presence of articulated or near-complete remains within the burrow." On the other hand, "the *Oryctodromeus* material differed in being (1) disarticulated and jumbled and (2) elevated in the fill above the burrow floor" (Woodruff and Varricchio 2011,141). A good, fossilized burrow, then, contains articulated skeletons in a sensible place—on the floor of the burrow where they presumably expired. The jumble of disarticulated bones positioned at the top of the burrow suggests that the animals didn't expire there, but were washed in, carried with the infill. Perhaps, then, the *O. cubicularis* specimens were not an unfortunate family caught expiring together in their burrow, but some bystanders who were carried in. Both tragic, but if the former is true, we have a much more significant discovery regarding the forms of life dinosaurs took up.

Let's get clear on how the machinery from the last chapter applies here. There are at least two possible cases of loss on the table. First, non-Avian dinosaurs might be the only successful, global vertebrates that didn't evolve burrowing; if so, then construed as such, we have a putative loss of that category. That is, the denizens *globally successful, non-burrowing taxa* are lost (recall chapter 2's brief discussion of absences counting as loss). Second, this particular find might represent an as-yet unseen pattern of burrow preservation; if so, we have putative loss, or at least rarity, of that pattern. On the former interpretation, loss attaches to an ahistorical denizen (non-burrowing globally successful taxa), on the latter, to the pattern of preservation. Woodruff and Varricchio hope for the latter: in a putative past world, we have a burrowing dinosaur that has left a pattern of burrow decay as-yet unseen.

So, what can we do? Given the topic of this chapter, I imagine you already have a good idea: artifice.

An extremely common mode of artificing asks after the conditions under which preservation occurs. In palaeontology, this often takes the form of "experimental taphonomy" (Briggs 1995). Various interpretations of specimens rely on not-yet-understood or unobserved forms of preservation. So, one approach is to bring about conditions analogous to those postulated and observe them, thus building a model of how those dynamics of information-preservation occur. In short, if you want to understand the processes by which traces form, build an experiment where something like those traces should arise. This is what Liu did in the chapter's opening.

Woodruff and Varricchio (2011) also take this strategy. They build a half-scale model of the burrow, fill it with rabbit bones in various positions, and use a PVC supply tube to fill the burrow. They do this over and again, varying the angle of the tube and the position of the bones. At base, the apparatus consisted of the tube, filled with a clay-sediment mixture (and, occasionally, rabbit bones), two wooden saw-horses, and the burrow model itself (see Figure 3.1).

The sedimentary composition and structure of the burrow was based on reconstructions of the Blackleaf Formation. Each run of the experiment included a single articulated rabbit skeleton, either "within the burrow chamber or just external to the entire burrow structure within the sediment supply tube or mixed with the sediments" (Woodruff and Varricchio 2011, 143). Variations tested included whether the sediment mixtures were poured through in a single go or in quick incremental pours, at various levels of "infill energy" (determined by the angle of the tube), various skeleton positions (starting in the tube or in the burrow) and at various skeleton "densities"—whether the bones were dry or wet. They ran thirteen trials, each time leaving the model for a few days before excavating it, much as they had excavated the fossilized burrow itself. This allowed them to more or less map the relationship between the initial (articulated) skeleton position and various states of infilling. Throughout their discussion, Woodruff and Varricchio are careful to note various points of discrepancy between the original specimens and their artifact, including differences between rabbit and dinosaur bones, variations in temporal and spatial scale, and so forth. In each case, they provide

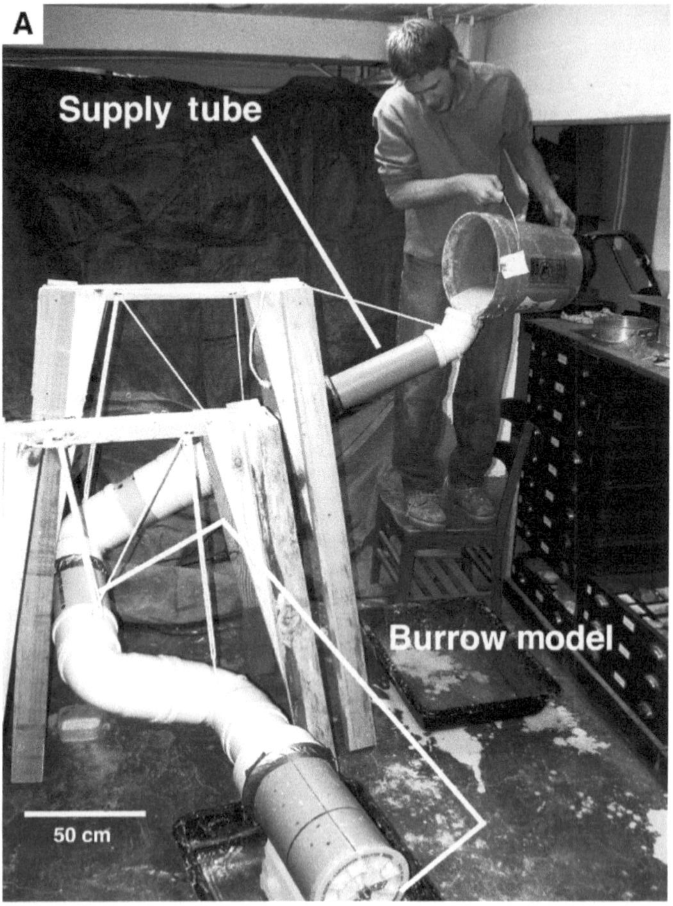

FIGURE 3.1
Burrow Model and Supply Tube (Detail from Fig. 3, Woodruff & Varricchio 2011). Reprinted with Permission from *GeoscienceWorld*

a set of reasons to think these discrepancies from the target shouldn't undermine their results.

The central result was that differences in final bone position "generally but not exclusively occurred in association with specific sedimentary deposits" (Woodruff and Varricchio 2011, 145). That is, the initial position of the bones didn't make a difference to the assemblage pattern, rather it was differences in infilling: "The angle of the supply tube and the mode of sediment delivery" (Woodruff and Varricchio 2011, 148). This result puts pressure on the interpretation that non-burrow-dwellers were swept into the burrow or otherwise deposited from outside. That claim, after all, relies on the idea that skeleton position provides important information about initial position as infilling occurs. This doesn't in itself prove that *O. cubicularis specimens* were burrowers, but it does remove an important objection: "the experiment failed to falsify Hypothesis 1: sediment infilling of a burrow chamber can produce a disarranged and elevated assemblage from bones sitting within the chamber. This lends support to the inference of Varricchio et al. (2007), that the Oryctodromeus assemblage resulted from burial of bones located previously within the burrow." (Woodruff and Varricchio 2011, 148).

The results do more than simply leave both options open. The trials where the bones began outside of the burrow generated results that differed importantly from the conditions of the find itself. These runs produced a differently mixed set of bones with, tellingly, many more breakages than seen in the specimens. Woodruff and Varricchio, then, take this result to show that the skeletons' position in the burrow is unlikely to be the result of their being deposited from the outside: "For a debris flow to cover a long distance and account for the *Oryctodromeus* assemblage would require what would seem like a very precise and fortuitous delivery of an associated bone assemblage into the burrow" (Woodruff and Varricchio 2011, 148).

Before drawing some lessons, a few caveats about the case. You might notice that the specimens being in the burrow at the time of infilling doesn't in itself show that *Oryctodromeus* was a burrower. This doesn't distinguish between "burrowing," that is, an organism constructing a burrow, and "denning," that is, the organism living in, and raising its young in, a burrow. Rabbits both burrow and den, while polecats den but do not burrow—they rather find existing burrows (typically rabbit burrows) and take them over. So-called "burrowing owls," which nest in squirrel burrows, are another non-burrowing denner. I suspect this is a case of the kind of productive ambiguity that I've argued is distinctive of some claims in palaeontology (Currie 2015): while current evidence is insufficient to distinguish between burrowing hypotheses and denning hypotheses, palaeontologists adopt a more coarse-grained hypothesis inclusive of both. One way of telling between denning only and burrowing is to point to morphological adaptations of burrowing, and indeed, Varricchio et al. 2007 do highlight some. Their find's morphology "Exhibits features of the snout, shoulder girdle and pelvis consistent with digging habits while retaining cursorial hindlimb proportions" (Varricchio et al. 2007, abstract only). I've abstracted this from the discussion for the purpose of highlighting artificing, but it is nonetheless a nice example of "recontextualization," which I'll discuss soon.

Building on Ben Jeffares' work (2008), I've previously described practices like experimental taphonomy as experiments that aim to establish dependencies between current traces and their upstream causal ancestors (Currie 2018a, chapter 3). Which is to say, these experiments seek to establish a dependency between a current variable (say, the arrangement of a set of fossilized bones) and a past variable (say, the speed of infill). Woodruff and Varricchio's results suggest that there is a dependency relationship between the speed of infill and the arrangement of bones, but that there is a very minor dependency, if any, between the initial position of bones and their resting place. Modelling these causal regularities enables historical scientists to make trace-based inferences from, in this case, a preserved burrow to a burrow-dwelling dinosaur.

Before returning to the Ediacaran, I'll draw out some important features of artificing from this case. These points delve a little deeper into the strategy experimental taphonomists adopt when trying to establish dependencies between now and the past.

First, notice how Woodruff and Varricchio conceive of their target. In these instances of artificing, the denizens are understood in terms of patterns and processes of preservation. *O. cubicularis* specimens were understood as something like "burrowing animals being buried in various forms of sediment within their burrows." More carefully, they were simply understood as "vertebrate bones being preserved in an infilling burrow." This allowed, for instance, for them to take the rabbit bones used in the experiments as the same relevant kind as the specimens. The basic bet is that there is a set of dependencies that are common across these elements, which the experiment captures and allows us to probe.

As I'll explain in the next few chapters, this practice of finding the right way of characterizing denizens in a past world is crucial for how challenges from loss or putative loss are overcome. Plausibly, Varricchio et al. (2007) face loss. There are no currently existing burrowing non-avian dinosaurs—they have been lost through extinction—and until now there have been no traces of dinosaur burrows; either they never existed, we've been unable to recognize them, or they've been erased. The putative burrow itself has idiosyncrasies that make its interpretation difficult: rather than nicely-behaved articulated skeletons familiar from previous fossil burrows, we have a mess. The response is to consider which properties might be isolated for the purpose of empirically testing the relationship between burrow preservation and bone position. As such, Woodruff and Varricchio identify sedimentary composition and vertebrate bones as properties of the past world that can be brought together in an artifact.

So, a crucial step in artificing is *characterization*: identify a set of properties in the past world that we can reasonably fabricate. As we'll see later in this chapter, processes of characterization—indexing our denizen as a kind under some description—are central features of strategic perspectivalism.

Second, although Woodruff and Varricchio consider properties of the past world in a kind of proxy system, they also *recontextualize* those results, in dialogue with the specimen itself and other aspects of theory, to generate an interpretation and justify aspects of their construction.[2] For example, their argument against the hypothesis that the bodies came from outside the burrow involved returning to features of the find itself, namely, their largely unbroken morphology and unsorted state, pointing out the differences between them and the results from experiments where bones began outside the burrow. If the find had been generated through that mechanism, they argue, it would likely have involved more breakages and been more carefully sorted, and thus would have been fortuitous indeed.

So, although various properties of past worlds are isolated to make it possible to generate tests, these tests are then brought back into conversation with particular, more fine-grained features of the specimens. I'll next note two further aspects of artificing.

Third, the construction and justification of the artifact are *bespoke*. Decisions made about how to design the experiment and what variations to perform are carefully guided by local, particular features of the specimen at hand. The specific kind of sediment involved, the decision to use rabbit bones, and the size and structure of the burrow (in particular, its "s-bend" shape) were based on the specimen and justified in terms of that. As opposed to the experimental paradigm being used "off-the-shelf," in this instance, it is constructed for a specific purpose.

Fourth, Woodruff and Varricchio's artifact is what I'll call *pluripotent*. That is, the results it can generate are used for multiple purposes, sometimes at the same time. It explicitly tests a particular interpretation of the specimen while also generating a set of dependency relationships—a kind of causal

2 A notion of "recontextualization" is important for Sabina Leonelli's account of data (2019). Data must be *decontextualized* from the context of its generation to be fruitfully used in, for instance, database practices (to "travel"), but then must be *recontextualized*, particularly using metadata, to be deployed as evidence. I take our uses to be related but different.

model—between skeletal remains and various kinds of burrow preservations and infills. As they put it, "This experiment specifically addresses the recent interpretation of a Cretaceous skeletal assemblage as a dinosaur fill . . . but we hope that results represent a first step in understanding the sedimentology and taphonomy within large vertebrate burrows" (Woodruff and Varricchio 2011,140). They take steps toward this by comparing their results to other fossilized burrows. For instance, a Permian synapsid (the ancient lineage that mammals eventually arose from) was found at the bottom of its burrow with a little disarticulation (Damiani et al., 2003). Woodruff and Varricchio point out that one of their trials, those with dense, water-saturated bones, generated a similar pattern: "This suggests that largely undisturbed skeletons can be preserved provided that bone density is sufficient to resist transport" (Woodruff and Varricchio 2011,149). Thus, their study might have consequences for understanding other burrows, past or present (or future).

So, although model construction is closely aligned with the particular specimen of interest (it is bespoke), its use and results are nonetheless extended across a set of other cases (it is pluripotent). Woodruff and Varricchio generate *both* an experimental test of a particular interpretation *and* a more general model of the dependencies between infill conditions and fossil position. Pluripotency, it should be noted, differs from omnipotency: scientists cannot build whatever they want, and their artifacts cannot do anything. In chapter 5, we'll touch on the possible dangers arising from the path-dependency of this kind of study. As we'll see, the artifact being at the same time bespoke but also extending beyond its target is a distinctive feature of strategic perspectivalism.

Artificing can be useful for understanding more than conditions of preservation—that is, circumstances where denizens include trace-forming processes—they can also be used to probe many other properties of past denizens. To see this, I'll examine some computational models of rangeomorph anatomy and growth. So, let's return to the Ediacaran to further explore artificing.

2. Modelling Rangeomorphs

In discussing experimental taphonomy, I identified a set of properties relevant to artificing strategies. In the face of loss or putative loss, Woodruff and Varricchio *characterized* properties of the past world in a way that allowed them to construct their experimental system. How the system was constructed and justified was *bespoke*, targeting specific features of the specimens. The results of their experimental work were *recontextualized* with features of the case itself, particularly specimen-relevant ones, thus generating further interpretations. Finally, the experimental system was *pluripotent*: testing hypotheses about a particular burrow, providing a system of dependencies between bone and infill properties, and allowing further recontextualized interpretations of further burrows.

Rangeomorphs are the most common multicellular denizens of the dark, benthic communities of the Avalonian Ediacaran. I've thus far focused on the apparently lost ecological and preservational status of those communities, a status partly generated by the lack of active movement. But that is only one aspect of this past world that seems significantly lost. Another is the morphology of the rangeomorphs themselves (e.g., Bamford and Narbonne 2009; Dunn et al. 2019; Taylor et al. 2023). The Ediacaran fauna contains many kinds of "frond" taxa with leaf-looking morphology. A "frond" is a structure most familiar from plants—their leaves—but is also found in cnidarians and algae. Fronds are formed of (to stick with botanical phraseology) a *root* (which attaches the organism to the ground), a *stem* (which either raises the frond up or increases its "reach" across the ground, while also acting as a hub for transporting nutrients), and, finally, *leaves* that sit along or atop the stem, often functioning in competition for resources (sunlight most familiarly, but also food particles in many non-plant cases).

So, when considered a member of the ahistorical type *organism with frond morphology*, rangeomorphs are certainly not lost, but members (presumably via convergent evolution) of a widespread brethren of all living things that are vaguely leafy. However, the details matter, and if we look closely, we find that rangeomorph morphology is considerably more mysterious. As Cuthill and Conway Morris have put it: "The branching morphology of Ediacaran rangeomorph fronds has no exact counterpart in other complex macroorganisms. As such, these fossils pose major questions as to growth patterns, functional morphology, modes of feeding and adaptive optimality" (Cuthill and Conway Morris 2014, 13122).

Rangeomorph morphology, like many fronds, follows a fractal pattern. A fractal is formed by repeating the same simple pattern over and over again, resulting in a pattern that is "self-similar," that is, it looks the same across various scales. In rangeomorphs, this consists of at least three levels of branching self-similarity (see Figure 3.2). Thanks to some rather remarkable

FIGURE 3.2
Rangeomorph Fossils Showing Three Levels of Branching Self-Similarity (Detail from Narbonne 2004, Fig. 2). Reprinted with Permission from the *American Association for the Advancement of Science*

three-dimensional fossils (Narbonne 2004), it seems clear that these develop more or less along two dimensions—they're flat.

This particular morphological arrangement is unique to the Ediacaran. As Laflamme, Xiao, and Kowalaweski (2009) put it, "Rangeomorphs are constructionally unique, adopting macroscopic morphologies that are uncommon or even unknown from modern or fossil metazoans" (Laflamme, Xiao, and Kowalaweski 2009, 14441). Given the litany of other uncertainties regarding rangeomorphs (no obvious feeding apparatus, unclear phylogenetic position, obscure developmental process, etc.), it is left quite unclear what to make of these strange critters. In other words, this morphology is lost (though not erased!) and, as such, understanding it requires overcoming some tricky epistemic challenges. We have, then, a classic challenge from loss. What to do? Given the topic of this chapter, you can guess: artifice. That is, characterize the past world in a way that affords the construction of a bespoke model and thus gains epistemic traction.

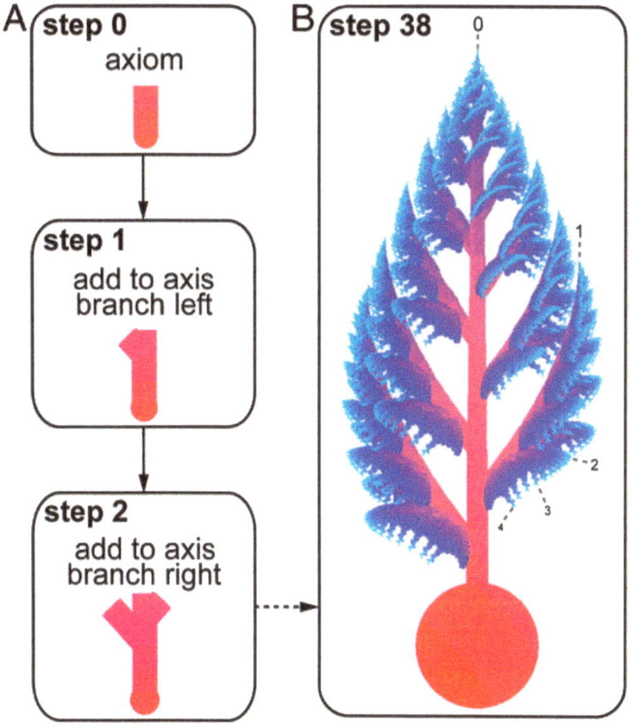

FIGURE 3.3
Explanation of Fractal Model (Left) and One Example of Final Result (Right) (Details from Cuthill and Conway Morris 2014, Figs. 1 and 2). Reprinted with Permission from the *Proceedings of the National Academy of Sciences*

Cuthill and Conway Morris (2014) present a fairly simple geometric model that constructs a rangeomorph-like fractal object by starting with a single "axiom" (the stem), adding a branch on the left, and then another on the right slightly further up, and then repeating this over a series of steps, where previous branches themselves become axes for further additions (see the left of Figure 3.3). In this way, they are able to generate shapes very similar to familiar fossil forms (compare the right image of Figure 3.3 with the fossil in Figure 3.2). They take the model to—at some level of approximation—fit the anatomy and indeed the development of rangeomorphs: "Reconstruction of overall morphologies that closely match the anatomy of the fossils . . . validates this model of branching and growth" (Cuthill and Conway Morris 2014, 13122).

So, we have our artifact: a geometric model of the three-branching fractal pattern distinctive of rangeomorph morphology. Cuthill and Conway Morris have characterized rangeomorphs as a historical kind: a lineage that grows via a three-branching fractal process. And they have thus constructed something like an instance of that kind. The model is bespoke, guided in its construction by particular rangeomorph features. As we'll see, it is also recontextualized in generating interpretations and, perhaps more importantly, is pluripotent in providing a basis for more general exploration of fractally-growing features, both in rangeomorphs and beyond.

What is the use of such a model? For Cuthill and Conway Morris, the model provides an inroad to the anatomy, phylogeny, development, functional morphology, and even feeding strategies of rangeomorphs. I'll now cover phylogeny, functional morphology, and feeding.

Let's start with phylogeny: at base, should we think that rangeomorphs are a distinct taxonomic group, separate from other Ediacaran denizens? That is, are all rangeomorphs cousins? Despite their similarly fractal morphology, it is not immediately obvious that this would be the case, as Laflamme and Narbonne have pointed out: "Many different morphological constructions are present [in Avalonian deposits], and this further substantiates the proposition that Ediacaran fronds represent an ecological adaptation rather than a taxonomic group" (Laflamme and Narbonne 2008, 175). The suggestion here is that fronding evolved convergently across various taxa, rather than being the domain of a particular ancestral grouping.

Cuthill and Conway Morris take themselves to have provided a "unified" model that captures the variety of rangeomorph anatomies via variation on a set of fairly simple rules. They take this to be evidence of rangeomorph ancestral relationships: "Shared possession of an approximately self-similar and fractal, axial, apical, and alternate body plan supports a rangeomorph clade" (Cuthill and Conway Morris 2014, 13123). The thought is that the diversity of rangeomorph anatomies in the Ediacaran can be explained by their following a particular developmental regime. However, this regime cannot be extended across the Ediacaran fauna, suggesting that "the Ediacaran biota was far from homogenous but instead included diverse phylogenetic lineages and body plans" (Cuthill and Conway Morris 2014, 13124). That is, we should think of Ediacaran multicellular organisms as complex, evolutionarily and ecologically divided, and—as we'll now see—well-adapted to the particular environments they found themselves within. So, in addition to providing information about rangeomorph-as-fractal-growers, the model also informs ancestral interpretations. Which is to say, it is pluripotent.

To explore how Cuthill and Conway Morris' model applies to functional morphology, we'll take a tangent into a fairly radical, but increasingly supported, model of rangeomorph feeding. Here we see an example of recontextualization, bringing the model back into conversation with local rangeomorph features, and a further feature of the pluripotent nature of artificing: being combined with other artifacts.

Laflamme and Narbonne (2008) interpret the various frond-like rangeomorph morphologies of the Avalonian as evolutionary convergences adapted to drawing nutrients from the water column. This involves seeing rangeomorphs as obligate osmotrophs. Osmotrophy involves feeding by absorbing carbon from the local environment. In single-celled critters, this basically involves interaction between the cellular wall and ambient carbon. Under those conditions, the size-to-surface-area ratio is critical. There is a trade-off between the energy demands of getting bigger and the relative amount of "wall" (compared to total mass) available to absorb that energy. Further, as there is no mechanism to distribute carbonic energy through the body, single-celled osmotrophy relies on passive diffusion, so the larger the body, the less efficiency on that front. Thus, there are size limits to obligate osmotrophy. We do see osmotrophy in some multicellular organisms—it is the basic metabolic driver of most fungi (and why they are decomposers par excellence)—but this is supplemented by other ways of gaining nutrients. Obligate osmotrophs as big as rangeomorphs would indeed be remarkable, and unlike anything we've seen.

In another case of artificing, Laflamme, Xiao, and Kowalaweski (2009) construct a theoretical model aimed at supporting an osmotrophic model of feeding. They focused on comparing the body-size-to-volume ratios of rangeomorphs and the largest ("megabacteria") obligate osmotroph bacteria (their use of megabacteria as comparisons is an instance of "'anchoring," which we'll meet in chapter 5). As should be becoming familiar, Laflamme et al. characterize a particular set of traits to compare past-world denizens with better-known contemporary analogues. At base, because rangeomorphs (and another Ediacaran denizen, the Erniettomorphs, but let's not worry about them) have that modular, fractal structure, it is fairly easy to model the trade-offs between overall size and surface area to size ratio: you represent the length, breadth, and overall size of the modules (in other words, the fronds). Although the results don't demonstrate bacteria-like surface-area-to-volume ratio efficiency (how could they?), they *do* underwrite Laflamme et al. exploring morphological adaptations in rangeomorphs that could facilitate this.

Drawing on older work by Gould (1966), Laflamme et al. (2009) point to two morphological strategies that reduce the costs involved in size versus surface-area-to-volume trade-offs (here characterizing the denizens as morphologies facing that kind of trade-off). First, *constructional flattening.* If you want to increase size, but minimize cost, then increase size in one dimension: worms get longer without getting thicker, and leaves are typically two-dimensional. And indeed, we see this in rangeomorphs: fossils that preserve the third dimension don't have all that much to preserve. The second strategy involves going fractal. Because rangeomorphs are constructed from modular, self-similar fronds of varying sizes, the use of tiny, complex fronds keeps a reasonable surface ratio despite increases in size. Laflamme et al.'s (2009) results suggest that "fractal branching in rangeomorphs results in significant increases in SA/V ratios that are comparable to some giant bacteria" (Laflamme et al. 2009, 14442). In other words, rangeomorph anatomy achieves impressively efficient surface-area-to-volume ratios despite its size. Third, fill out your increasing body size with stuff that doesn't have metabolic costs, such as non-organic sediment or dirt. Although fossil evidence for this is limited—but not ruled out!—if Laflamme et al.'s modelling results are correct, this suggests that to be efficient at those sizes, rangeomorphs likely did incorporate non-organic, non-metabolic material into their growth.

So, by characterizing some of the geometric features of rangeomorphs and comparing them to megabacteria, Laflamme et al. (2009) are able to explore the constraints individual rangeomorphs would face if they did turn out to be obligate osmotrophs. Where Laflamme et al. focus on the efficiency of particular rangeomorphotypes, Cuthill and Conway Morris' capacity to generate a set of varying morphotypes allows them to compare them to each other, as they are represented in Figure 3.3. If we follow Laflamme et al. in imagining rangeomorphs as osmotrophic, what ecological and evolutionary information might we gain by comparing their morphologies as generated by Cuthill and Conway Morris' model?

Rangeomorph body types can be organized along the usual spatial dimensions (width, depth, and height) to compare them and hunt for patterns, as in Figure 3.4 Cluster analysis of this information reveals three strategies: go for height, such as *Trepassia wardae*, find a trade-off between height and volume, such as *Bradgatia linfordensis*, and maximize width, as in rangeomorphs spread across the ocean floor, such as *Hapsidophyllas flexibilis.*

This information is then brought into discussion with yet another model (the details of which we'll spare ourselves!), which examines the effects of dense rangeomorph canopies on water flow. The model suggests, in a way somewhat reminiscent of plants competing for sunlight, that a dense rangeomorph canopy would stymie water flow, so that there could be selective

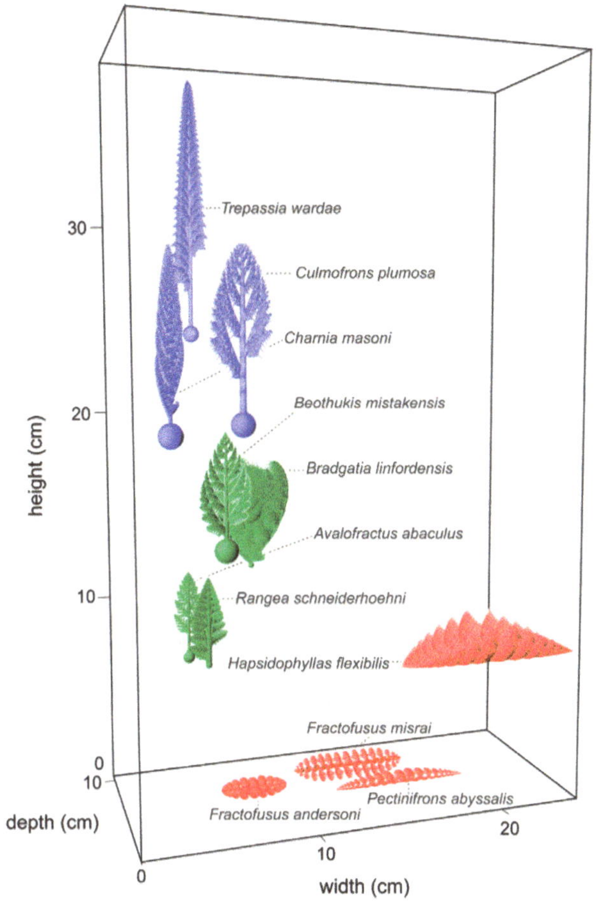

FIGURE 3.4 Comparative Rangeomorph Space-Filling (Fig. 3, Cuthill and Conway Morris 2014). Reprinted with Permission from the *Proceedings of the National Academy of Sciences*

pressure for increased height. Some rangeomorphs reached above the canopy where less-restricted water flow would increase the efficiency of osmotrophy. Cuthill and Conway Morris weave all of this into an evolutionary narrative:

> Among the in situ communities of the earliest Avalon Assemblage, rangeomorphs are by far the most diverse and abundant macroorganisms. Furthermore, establishment of the major rangeomorph space-filling strategies preceded the appearance of a much wider range of macroorganisms . . . This suggests that early diversification of rangeomorph branching patterns effected a radiation in the key characteristics of body-size and

> microhabitat . . . and was driven by ecological competition between rangeomorph taxa. (Cuthill and Conway Morris 2014, 13125)

In other words, early metazoan ecosystems were dominated and shaped by an adaptive radiation of rangeomorphs, a radiation largely due to that lineage exploiting the various space-filling strategies available to its body plan. In Cuthill and Conway Morris' and in Laflamme et al.'s (2009) models, we see a similar strategy as in the burrowing case: characterize a set of properties, build a bespoke model that instantiates those properties, generate results, recontextualize them with other information, and use these as a basis for further models. Let's think about this strategy yet more abstractly.

3. Strategic Perspectivalism

"The real secret to my success . . ." Prof. Ichthy glances at their students, "Well, our success—relates to how we *look* at Broadfoot. We don't just think of it as a fossil specimen, nor a mammal. Instead, we think about it from many, many different perspectives."

Ichthy begins to gesture a little erratically, their flippers tracing abstract geometries in the air.

"I like to imagine the animal as a kind of puzzle, a puzzle consisting of multiple complex dimensions. To solve the puzzle, you need to focus on just a few dimensions at a time, thus generating partial, small-scale solutions. But then, with those partial solutions, you get hints for how to solve aspects of the puzzle across other dimensions. Imagine a crossword puzzle, but instead of just a horizontal and vertical axis, puzzles work across three, four, many axes. And imagine that questions and solutions aren't simply linguistic, but are sometimes pictograms, sometimes sudoku, and at times purely geometric. Understanding such a puzzle would require looking at it in many different ways and considering how these ways of looking and the partial solutions they provide relate to each other. Just like that, each way we look at Broadfoot gives us crucial information, and that information bears on the other ways we look at Broadfoot." They bring their flippers around in a final circular motion.

"All this perspective-switching can be dizzying sometimes, but you get used to it."

How do historical scientists uncover lost worlds? How do they solve challenges from loss? One strategy is artificing, and we'll meet more in the following chapters. Each of these strategies belong to a larger category: the creative and opportunistic use of what I'll call *strategic perspectivalism*. That is, the adopting of a perspective that allows characterizing apparently lost denizens in ways that enable strategies like artificing to gain traction. In this section, I'll characterize this via a discussion of perspectivalism more generally.

We've examined two main cases of artificing. The first involves recreating something like the conditions of preservation—the processes of trace formation—the second involved recreating a past world itself. These two don't neatly divide: modelling the patterns created by collapsing burrows involves some understanding of the structure of those burrows and burrowing critters, and recreating past Avalonian conditions involves an intimate understanding of the conditions of its preservation. Regardless, these examples provide some clues as to how the artificing strategy works, and some features which, as we'll see, are common across the challenges of understanding lost worlds.

Across our cases, we saw a particular process. First, characterization: denizens are indexed via particular properties such as Avalonian-like conditions of preservation, burrow infill, and rangeomorph growth patterns. Second, these properties are instantiated or generated by constructing some scientific object: a scale model burrow or a geometric model, say. These models are bespoke both in terms of construction and validation. At base, this involves providing reason to think that the procedures and various pragmatic features associated with them will not bias the results as applied to the target past system, and will further make sense of the resultant artifact itself. Third, recontextualization occurs. Here, other features of the find and other aspects of context are brought into dialogue with the results of the study (Turner and Turner 2021). Roughly, an interpretation is generated and defended in various ways by bringing it into dialogue with other features of the past world. These bespoke models are pluripotent: providing information about other instances, as we saw with the burrowing case, and affording the construction of further objects, as we saw with the various models of rangeomorph growth, phylogeny, functional morphology and ecology.

I think we can understand the three-step process of artificing as an instance of a *perspectival strategy*. To understand what that is, we first need to get perspectivalism on the table.

Perspectivism, or perspectivalism, is at base a claim about the nature of knowledge. It claims that all knowledge is *from a perspective*, which is to say, there is no epistemic "God's-eye view" to be had. Instead, perspectivalists insist that knowledge is always had from some vantage point. How exactly to characterize "vantage point" is a vexed issue, as is the extent to which perspectives, and therefore knowledge, end up being private or subjective. Perspectivalism is a philosophical view with a grand history, which has fairly recently been a focus in the philosophy of science. These discussions are often in the business of threading needles between constructivist and realist conceptions of knowledge. Is it possible to reconcile the idea that all knowledge is from a perspective with the idea that the knowledge is of a single, concretely existing world? Philosophers of science developing perspectivalism often take their cue from visual imagery or other metaphors from the arts (see, in particular, Giere 2006; Elgin 2017; Massimi 2022). The views developed there bump up against other major traditions in epistemology and the philosophy of science, particularly standpoint theory and various forms of scientific pluralism.

As we'll see, my interest in perspectivalism is partly instrumental, as committing to some general epistemic view about the relationship between perspectival knowledge and reality is beyond my remit, as is committing to a particular story about what a "perspective" is. Nonetheless, recent discussion, particular Michaela Massimi's, is insightful regarding the nature and importance of what I've been calling "characterization." Crucially, "strategic perspectivalism" is *not* a philosophical view in the same sense as perspectivalism is: it isn't a claim about the nature of knowledge generally that a philosopher might adopt. Instead, it is a philosophical account of the strategies scientists adopt to overcome challenges from loss or putative loss.

Michaela Massimi identifies two kinds of perspectivalist representation (Massimi 2022). The distinction she draws is useful for us because it captures both perspectives in the sense of characterization (which denizens and properties we target) and in the sense of recontextualization and pluripotency (what can be done with our artifacts). Drawing from pictorial representation in art, she distinguishes between *perspectival*$_1$—"representation drawn from a specific vantage point [italics removed]" (Massimi 2022, 32)—and *perspectival*$_2$: a "representation drawn towards one (or more than one) vanishing point(s) [italics removed]" (Massimi 2022, 32). Often, images are intended to be viewed as if from the perspective of an onlooker. Consider Charlotte Kenchington's reconstruction of the Avalonian Ediacaran from chapter 2 (Figure 2.3). Their image provides a specific sense of where the viewer is located in space: some rangeomorphs are nearby, and others disappear into the distant gloom. Images having an implied onlooker who has a specific spatial position are an example of perspectival$_1$. For Massimi, such perspectival features are relevant when the "content of the representation is affected by the very vantage point from which the representation takes place" (Massimi 2022, 34).

Perspectival$_2$, by contrast, is less concerned with specific vantage points themselves, and more with how representations open "up a 'window on reality' that extends well beyond the boundaries of the representation itself and where the depth, angle, and scale of the representation leave enough room to make inferences about the space and what's in it" (Massimi 2022, 39). Kenchington's perspectival effect is achieved by representing the rangeomorphs as if they were illuminated: a specific light source illuminates the closest rangeomorphs, and those farther away get progressively dimmer, providing the illusion of an expansive, somewhat ominous space. The effect generated creates a *vanishing point*, much like carefully placed lines can give the impression of a landscape's horizon. Enough information, and information directed toward a vanishing point, underpins our ability to make inferences about things *not* explicitly represented in the image. Adopting a perspective sanctions inferences beyond the explicit content of the representation. Kenchington's image allows us to imagine further rangeomorphs, not represented, beyond our light source. This is perspectival$_2$.

Let's briefly run these concepts through our burrowing dinosaur case. We conceive of the specimen as vertebrate bones deposited in an infilled burrow. The experimental apparatus instantiates the vantage point analogously to how the arrangements of lines in a pictorial representation create a perspective. The aspects of the target that are explored turn on the bespoke, particular features of that experiment: its material composition (PVC pipe, dirt, rabbit bones) and what the scientists choose to do in their experiments (varying pipe position, initial bone position, and so on). That is perspectival$_1$—how the denizen is characterized. However, the experiment also affords inferences beyond what it explicitly instantiated. Through recontextualization, it helps generate an argument that the specimens were preserved inside the burrow. Through pluripotency, it acts as a model for burrow preservation more generally and links the particular specimen to other cases. This is perspectival$_2$.

Massimi takes there to be a lesson here for philosophers interested in scientific representation: "Philosophical discussions of perspectival representations have to start by acknowledging the role of the human agent not just as a spectator, but first and foremost as an architect in arranging lines of composition and creating in painstaking detail the perspectival effects of a 'window on reality'" (Massimi 2022, 32).

The crucial point to note at this stage is that the content of scientific representations is affected by decisions scientists make regarding what to include, how to contextualize, and what to omit. Scientific agency, in deciding how to characterize and with what apparatus, is a critical component of our story. In the language I've developed thus far, scientific representations require characterization. And in artificing, this characterization and the construction of an object instantiating those relevant properties allow inferences beyond the object itself. Cuthill and Conway Morris' geometric model, when recontextualized, enabled inferences about rangeomorph phylogeny, for instance.

In characterization, scientists represent their targets of inquiry *as* particular things. *Representation-as* has a fair bit of currency in current philosophy of science and related discussions (Elgin 2017; Frigg and Nguyen 2017). Kenchington represents the Avalonian Ediacaran *as* a dark, benthic, rangeomorph-haunted ecosystem lit by a highly unlikely light source. They do this by incorporating various techniques—such as the progressively dimmer rangeomorphs. In virtue of this (and their labelling it as such), we take it as an attempted representation *of* the Avalonian Ediacaran. Similarly, Alex Liu conceives of the Avalonian as a set of preservation conditions, Varricchio et al. characterize their specimen as bones deposited in an infilled burrow, and Cuthill and Conway Morris take rangeomorphs to grow via a geometric pattern.

Perspectival$_1$—the effect of a representation being from a particular place or observer—and perspectival$_2$, the generation of vanishing points enabling inferences beyond what is explicitly represented, are both achieved using what I'll call *perspectival tools*: the representational strategies, technologies, and techniques that enable a particular perspective. A perspectival tool is something that enables a scientist to instantiate, represent, or explore the denizens characterized as having such-and-such properties. Liu and Varricchio et al.'s experimental apparatus, and Cuthill and Conway Morris' geometric models, are all examples of perspectival tools, and we'll meet many more in the next two chapters. Suffice it to say, "perspectival tool" should be understood in terms of epistemic function: it enables characterization.

I've emphasized how, in artificing, historical scientists adopt very particular characterizations of target denizens. In brief, the denizens in question are understood at a level of grain that, all going well, enables them to be enfolded into a particular kind of historical individual. Varricchio et al. characterize their specimen as *vertebrate bones preserved in a burrow*, Cuthill and Conway Morris characterize their specimen as (a particular set of) *fractal morphologies*. This is perspective$_1$: they represent them from the perspective of objects with those kinds of properties. This perspective enables further connections and inferences beyond what is explicitly represented in their model; thus, it is perspectival$_2$. How they do this—how the effect is achieved—is via the application of perspectival tools. For instance, Cuthill and Conway Morris' fractal growth model is a perspectival tool that enables their taking that perspective, and the fossils they examine are also perspectival tools. In artificing, this is achieved by particular properties being *isolated* and *built*—represented in some way. For example, by building a scale model of a burrow and adding disarticulated rabbit bones, or constructing a geometric model with fractal growth patterns.

The perspective—the characterization—of the denizen, then, is achieved via strategies of isolation, emphasis, distortion, and so on. These strategies themselves play out in the construction and application of specific scientific objects and techniques.

Massimi emphasizes that different perspectives take place against a "horizon of alternatives," that is, in light of a set of different possible perspectives that could have been (and might still be) taken. As she puts it, "Many models for the same target system represent selected features of the target system *as y*, or *as j*, or *as k*, where *y*, *j*, and *k* are different properties belonging to a 'horizon of alternatives' and often enough incompatible or inconsistent with one another" (Massimi 2022, 40). For instance, Laflamme et al. represent rangeomorphs in terms of size and surface-to-area ratio given a fractal geometric pattern (thus enabling comparison with single-celled osmotrophs), while Cuthill and Conway Morris represent rangeomorphs through a basic set of rules from growing fractal geometric patterns, variations of which generate various rangeomorph forms (thus enabling comparison between rangeomorphs). These, and various other options, such as representing rangeomorphs as ecotypes, as Mitchell et al. do, constitute the horizon of alternatives (Massimi draws the idea from van Fraassen 2008).

What sets the horizon of alternatives? Massimi is interested in how perspectives enable particular forms of inference, inferences that are not necessarily tied to that particular perspective. While this is a crucial aspect of my explanation of how strategic perspectivalism is a good strategy in the face of loss or putative loss, she hasn't much to say about what the properties of these horizons are. I want to emphasize how the horizon of alternatives emerges from the available perspectival tools. Scientists cannot meaningfully or productively take up just any old characterization of a denizen. They adopt characterizations that they have the tools to underwrite, or plausibly might develop tools to do so. As such, the horizon of alternatives is constrained and enabled by the available perspectival tools. As Christophe Malaterre has said of the characterization (or "granularity") of traces: "The instruments and background theories that are available to research teams at certain points of time certainly influence the granularity at which evidence and statements are formulated" (Malaterre 2024, 23). I think this holds more generally: how scientists might describe or characterize their targets is more or less a product of the perspectival tools at hand.

What makes this perspectivalism *strategic*? First off, the phenomena I'm interested in have a lot in common with those that animate perspectivalists. Drawing on van Fraassen's (2008) discussion, Catherine Elgin (2017) identifies the properties of a representation that generate perspectives. For her, representations are perspectival when "They can show how occupants of that [logical or physical] space appear from a certain vantage point" (Elgin 2017, 155). Further, they are *indexical*, that is, "They represent how things look from *here* (for some value of 'here')" (Elgin 2017, 153). I'm not exactly certain what this means (see below), but I take the indexical nature of denizens, as well as the selection and construction of perspectival tools, to be very close if not identical to the phenomena Elgin discusses. Perspectival representations are at a level of grain, and their effectiveness often turns on getting the grain right (see, for example, Borg's recent discussion of "level-switching" in geochronology; Borg 2025). We saw Cuthill and Conway Morris' and Laflamme et al.'s models switching grains from individual-morphology to community-level as the questions they asked switched from individual optimization to group optimization in an environment. Finally, perspectival models often involve "occlusion": that is, capturing some element of a system often precludes including others. In characterizing a past world, selecting a bunch of denizens, and bringing a set of perspectival tools to bear on them, historical scientists inevitably occlude some features they might otherwise have included.

So, the phenomena I'm interested in are close enough to be legitimately labelled "perspectival." However, as mentioned above, my philosophical target diverges from traditional perspectivalists. This is partly due to being interested in different questions: perspectivalists are often in the business of trying to reconcile how the perspectival nature of representation can be married to views on the reality or otherwise of the world represented, and of arguing that knowledge (or representation) is necessarily perspectival. Depending on mood, I'm either asking a different but potentially complementary question, or I think that debates concerning realism and anti-realism have become an unproductive research direction in the philosophy of science (Currie 2018a, Postscript). Regardless, our job here is to understand the strategies historical scientists adopt to grapple with lost worlds, and an instrumentalist (or strategic!) conception of perspectivalism is the relevant one.

So, in part, putting "strategic" in front of "perspectival" functions to sidestep some substantive issues. For instance, are *all* scientific models perspectival? Massimi appears to think not, as she identifies a particular class of scientific models as being perspectival (thus implying that some are not). Meanwhile, van Fraassen argues that because of the perspectival nature of the data-generating processes which go into building scientific representations, all are perspectival. For another example of an issue hopefully sidestepped, try as I might, I'm never quite sure what perspectivalists mean when they refer to indexicality: that there are non-perspectival models where "One need not locate oneself in the space of representation to understand what and how it represents" (Elgin 2017, 156), whereas other models do involve somehow situating *yourself* in the representation. I suspect the right way to understand this question is whether characterization, that is, identifying a set of denizens, is sufficient on its own to generate a perspective. I suspect not, as other features of scientific interest—the kinds of questions they ask and the kinds of answers they expect—might also be included in a perspective.

So, given my interests, aims, and perhaps philosophical failings,[3] I'm hoping to not take a stance on some of these questions here.

I also suspect there are more bona fide, first-order philosophical disagreements between myself and perspectivalists, especially in terms of the nature and importance of representation. In chapter 6, I'll argue for *artifactualism* about perspectival tools, which doesn't take scientific tools as primarily representational, but instead as material objects that scientists build and interact with. This won't involve me denying the legitimacy of representations (see, in

3 There's an apocryphal story that I've been unable to find a source for that goes something like this. Bertrand Russell is giving a research seminar, and during question time an interlocutor has come back several times claiming they just don't *understand* what he is getting at (philosophers should well recognize this form of "misunderstanding as somehow an objection" seminar question), to which Russell eventually replied, "Your philosophical failings are not my problem!" I suspect on this count that perspectivalists perhaps ought respond to me with the same . . .

particular, chapter 7's discussion of palaeoart), but it does involve me folding representational tools into a broader category of "artifact."

So, my view is perspectival-adjacent and certainly consistent with perspectivalism (at least thus far), but what makes it *strategic*?

In approaching a new piece, an artist might ask themselves, "What techniques and tools do I have at my disposal? What effects can I achieve with these?" In other words, what kinds of affordances are there to generate different $perspectives_1$—what are the available horizons of alternatives—and which would be the most effective given my aims? Similarly, historical scientists, when artificing, consider what techniques and technologies are available to enable them to construct a perspective. And they pay particular attention to what other objects these perspectives enable connections between. Laflamme et al.'s perspective enables comparison with single-celled osmotrophs and the incorporation of the geometric/metabolic logic of body size to surface area ratio; Cuthill and Conway Morris' perspective enables comparisons between various rangeomorphs, the incorporation of fluid dynamical representations of waterflow, and models of ecological competition.

Note that I've not yet said anything directly about what makes an instance of artificing good, successful, or justified. But we've had a few hints in my discussions of recontextualization and pluripotency, and Massimi's notion of $perspectival_2$. How much grip do perspectival tools give us on the case at hand? What links do they provide to other cases, and what further inferences do they enable? As we've seen, historical scientists pay careful attention to how their artifacts and other perspectival tools might diverge from their lost target. Liu's pizza-disk experiments, for instance, made use of contemporary biota, which likely differ in many ways from those composing the Ediacaran mats. Their artificing relies on the idea that similar metabolic processes are held between them despite those differences. In the next chapter, I'll more explicitly discuss the kinds of overlap we might have between perspectives, and in chapter 5, I'll describe the iterative relationships between them. So, if you can, put a pin in your scepticism for now.

In artificing (and, as we'll see, in grafting and anchoring), historical scientists adopt or construct perspectival tools that enable fruitful characterizations of past denizens. It is in the ingenious use of strategic perspectivalism, I'll argue, that historical scientists are able to recover lost worlds.

4. Coda: De-Extinction

> "We know a lot about Broadfoot", Prof. Ichthy continues, a faraway look drifting across their face, "but there are still so many unanswered questions. What were their mating habits? Were they territorial? Were they fully or semi-aquatic? Did they raise their young? Were they solitary or did they live in groups? Without a time machine, there just doesn't seem to be any way of finding out. Unless . . ." Ichthy's eyes come into sharp focus as they slam a flipper emphatically on the lectern and lean forwards eagerly ". . . Unless we can *bring them back!*"

Artificing can be understood as the construction of a lost world. This might sound metaphysically dramatic: are you literally bringing a past world back into existence? Is the constructed simulacrum the same thing or kind of thing as the lost denizen? Recall De la Beche's poking fun at Lyell, illustrating the absurdity of thinking that, were Jurassic conditions to return, so too would Jurassic fauna. We might think this absurd causally (the chances of the same fauna arising just because we have similar environments are tiny!), but also metaphysically: how is the Jurassic fauna and the new fauna *the same,* exactly? Does artificing itself also involve these apparent causal and metaphysical absurdities? This question is a nice opportunity to discuss the notion of "de-extinction."

As we'll see, artificing needn't involve anything so dramatic as implied by the above questions. Because denizens are sensitive to description—in particular, to the kind they are characterized as falling under—and because artificing involves strategic perspectivalism, "bringing back" a lost world is just a way of saying that something of *that kind* instantiating *those properties* has been fabricated. This point, it seems to me, sheds some light on discussions of de-extinction, a set of practices sold as a way of "bringing back" the once-lost denizens of past worlds, specifically, extinct species. My aim in this subsection is to introduce and critique the notion of de-extinction in light of my discussion of artificing. In doing so, I'll articulate a few important features of the notion of loss, in particular, how it can foreground value (as we'll return to in chapter 10) by, in a sense, putting the resolution of metaphysical questions partly in our own hands.

I'll understand de-extinction as an attempt to, in some sense, bring extinct animals back from extinction. Sometimes the term is used in tandem with notions like rewilding, which is to return an ecosystem to something resembling a past state. Further, sometimes animals are considered extinct if they only live in captivity, and so attempts to return them to the wild are a kind of de-extinction. Here, I'll stick to "de-extinction" as a set of projects aiming to reconstruct (artifice!) a lineage known only from traces.

Calls for de-extinction are often made against highly moralized or romanticized backdrops: if we could bring back the victims of anthropogenic extinction, say, the passenger pigeon or thylacine, would this not be some kind of act of atonement for our ecological sins? Would re-populating the Arctic Circle with mammoths return us to some pre-human balance of nature (Seddon 2017; Odenbaugh 2023)? But on the other hand, don't such efforts risk overplaying our domination of nature or further undermining genuine nature with artificiality (Lindquist 2020; Katz 2022)? Such normative considerations need to be understood against the empirical reality of de-extinction.

Mechanically, there are at least two paths to de-extinction. First, back-breeding, which makes use of artificial selection to walk a lineage back along its genetic and phenotypic history (most typically associated with aurochs, the recently extinct wild European cattle; Stockstad 2015). Second, cloning, which involves artificially inseminating an animal with an artificially-grown zygote on the basis of recovered genetic material (perhaps most well-known for the mammoth; Shapiro 2015). These are extremely limited from both a technological and a practical perspective. Cloning, for instance, depends on the availability of ancient DNA. So, anything more than a few million years old is highly unlikely (despite the important influence such ideas have had over both public perceptions of palaeontology and dinosaur science generally; see Jones 2022), and this is to say nothing of the enormous difficulties of bringing unusual zygotes to term in either a different species or an artificial womb/pouch/egg. Back-breeding has a somewhat less science-fictional tone, but also involves tricky assumptions about the reversibility of genetic pathways, is extremely labour-intensive, and is extremely limited. As Thomas Reydon (2022) has put it:

> The scientific and technological reality is that de-extinction in this literal sense is fundamentally impossible. What is possible is "de-extinction" in a scientifically (and technologically) informed sense. But this merely involves the creation of organisms that are phenotypically similar in a limited number of traits to those of an extinct species, and/or that can perform the same ecological role as an extinct species. (Reydon 2022, 114)

Alongside scientific and technical worries, philosophers and biologists have also pointed out conceptual and metaphysical issues. What does it mean to *bring back* a species, especially if it is understood as a historical individual? Some might argue that no matter how similar a cloned mammoth is to a true mammoth, as it forms a clade with (Asian) elephants (its mother is an elephant, after all), it is at best an odd elephant, not a mammoth. Others question whether a mammoth can truly be a mammoth if it is not situated in a mammoth-like environment—is there a mammoth without the Great Steppe? Philosophers have had several field days applying various species concepts to de-extinction (see, for instance, Turner 2017; Siipi and Finkelman 2017; Finkelman 2018; Lean 2020). And the general lesson is, well, it depends on what you mean: what notion of species you have in mind or—to put it more generally—which characterization of the denizen interests you. If species are historical individuals, and that individuality requires spatio-temporal continuity, then you can't bring them back. But that's reliant on the characterization and, as we've seen, others are available.

The conceptual machinery from the last chapter gives us some traction on these kinds of questions. If you want to artifice a denizen, you need to ask yourself how you want to characterize it: what properties are you interested in, and what "kind of kind" (historical individual versus historical type, etc.) do you have in mind? Further, consider strategic perspectivalism. De-extinction efforts and speculation are underwritten by our technological capacities: cloning, back-breeding, and so on. Our characterization (that is, perspective_1) is in part generated by what we can successfully build and represent, that is, our perspectival tools. So, we can't just characterize the denizens in any which way we want in this context.

Certain kinds of de-extinction are plainly beyond our technological—and perhaps conceptual—reach. If we used DNA to clone a passenger pigeon via (say) a rock-dove intermediary, we do not thereby generate something of the same lineage as passenger pigeons. But, if successful, we have brought *something* back: there will likely be some properties that were once lost and now are not. It is up to us to decide if we think it counts as a passenger pigeon according to what is morally, or environmentally, or epistemically, or aesthetically significant. Potentially, there is much to learn from de-extinction projects about the reversibility of genetic pathways, the capacities of cloning technologies, as well as extinct animals' morphology, behaviour, and so on. Artificing is a very productive strategy, after all.

But one of the lessons of thinking about lost worlds is that, in many senses, you can't go back: some elements of the past can be rebuilt, but some cannot. And the question is, which aspects are we actually interested in, what gains might we get from pursuing that path, and what costs are there? Given the enormous expense and very real animal suffering (Browning 2018; Odenbaugh 2023) that goes into developing and deploying de-extinction technologies, it is worth our time to ask if we can't learn these things through less financially and ethically costly means, and indeed whether they are worth learning at all. Artificing a lost denizen requires choosing a perspective, taking a stance on its characterization. There is no characterization-independent answer, then, to whether de-extinction is successful or should be attempted.

How, then, should we answer the metaphysical questions underlying this discussion: does artificing literally involve resurrecting a lost world—when successful, do we have a lost world, found? Excitingly, sometimes yes! However, less excitingly, the answer is, well, it depends on how you characterize that world and its denizens.

There's a further epistemic corollary to note. If you attempt some kind of de-extinction, how will you know you've succeeded? Given that we don't know many of the properties of the extinct organism, how can you be sure that your artificed creature has the same properties as the past denizen? The possibility of uniformity failures is one of the major challenges of thinking about loss—not whether loss has occurred, but just that it might have. If we resurrected a mammoth or a passenger pigeon, how would we be sure it behaved as its extinct brethren did?

In artificing, you tackle loss—a lack of "Ns"—by constructing some Ns. But how can you know that they, in fact, are Ns? The answer to this question will emerge as we examine further strategies in the next two chapters.

"Or, maybe not . . ." Ichthy says, deflating with a guilty look at their grad students.

4

Grafting

Prof. Ichthy returns to the slide with the speculative history of mammal life.

"Of course, we shouldn't exaggerate the differences between Broadfoot and extant mammals. They're still *mammals*, after all, and thus joined together via ancestry. In recognizing Broadfoot as part of a set of historical trajectories—narratives if you want—we open powerful inroads to understanding Broadfoot, those narratives, and today's mammals." Ichthy uses their pointer to trace the black lines from mammals past to mammals present.

"These stories are carried across genetic and developmental pathways, from parents to offspring, across millions of years. Thus, currently living mammals that share in those stories all help us understand Broadfoot, just as Broadfoot helps us understand other mammals."

Ichthy turns from the projector. "But why restrict ourselves to mammals? We and Broadfoot are cousins, after all: vertebrates. As such, we also share a lineage, another story, tracing back into the deep past. And further, Broadfoot shares a deeper ancestry: that linking all life that we know. As we tell Broadfoot's story, then, we also tell the story of mammals, as well as our own story, as well as the story of life. Life is a complex, messy thing, and, as such, there are many different lines of ancestry, many different stories we might try to tell. This complexity is a boon for understanding Broadfoot."

Again, Ichthy begins counting with their flippers. "There are two lessons here, I think. The first lesson is that even though Broadfoot is a weird mammal, its very mammalhood means that we have vital connections to it through its cousins. The second is that our knowledge of the past isn't always restricted to little fragments or slivers of time: by understanding the connections between things in the past, we can weave stories which, or so I hope, are at least sometimes a little bit true."

In the last chapter, I introduced *strategic perspectivalism*. In response to the challenge of loss, the scientist adopts a perspective—characterizes their target—such that certain properties of that target may be isolated and brought into dialogue with something that can be examined.

We've seen this strategy play out in *artificing*, where those properties were instantiated and probed in a bespoke model or experiment. In this chapter, we'll see the strategy take a quite different form, which I'll call *grafting*. Under this strategy, scientists focus on historical individuals that have been retained into the present, or at least that we have better knowledge of. When you think of a mammoth as a member of the still-existing Elephantidae, or when you examine birds to learn about dinosaurs, you're grafting.

A "graft" involves linking previously disconnected tissues so they can grow together. In medical contexts, a skin graft attaches healthy skin to damaged (for instance, burned) skin to encourage repair. In horticulture, grafting involves linking the properties of two plants. For instance, combining the fruit qualities of one tree with the more resilient roots of another. Real grafting, then, involves actively linking previously unlinked properties—combining what was once disconnected into one individual. My use of the term "grafting" is more metaphorical than the last chapter's use of "artificing.' When I use it, I identify a set of practices whereby historical scientists, faced with a putatively lost world, posit existing properties that might be historically continuous with some of the lost world's denizens, thus "grafting" them together. A downside of the metaphor may be that it implies a more constructivist take on the practice than we might like. But given this chapter's focus on phylogeny, the horticultural reference is too good to pass up.[1]

We can understand grafting as a general strategy that involves understanding a target as part of a causally connected token trajectory. Often (but

1 Somewhat to my chagrin, Hans-Jörg Rheinberger has recently used the term "grafting" in his discussion of experimental practice in the life sciences (2023). His usage is different but, I think, usefully so. For Rheinberger, "grafting" involves incorporating some existing experimental apparatus or procedure into some other apparatus or procedure, thus extending its capacity into new territory, but in a somewhat modular way wherein the grafted objects maintain their identity. This is much like how, say, a variety of apple grafted onto another variety of apple tree does not, as it were, merge with the tree, but keeps its independent identity. This is "a delicate operation, often requiring a considerable degree of adjustment and refinement" (Rheinberger 2023, 66).

not always!), grafting is rooted in conceiving of your target as a historical individual, that is, some token entity or process extended through space and time. But it is crucial to see that past worlds are often—perhaps always—characterizable as many different kinds of historical individuals. Rangeomorphs aren't only construed in ecological, geometric, or morphological terms. They're also characterized as species, genera, orders: as taxa. That is, they are *lineages*.

As a first pass, we can define a lineage as a kind of biological individual, one that's constituted by further individuals connected through relationships of ancestry and descent. For instance, culture generates a plethora of lines of descent. Philosophers and other academics often delight in tracing their lineages: my PhD supervisor (Kim Sterelny) was supervised by Bill Bonney, making him my academic grandfather, I suppose; but I was also a postdoc with Marc Ereshefsky, so I might trace another line of ancestry from him. And this makes Celso Neto, whom Marc supervised as a graduate student, something like a step-sibling, I guess. So, cultural lines of descent might be drawn in multiple ways, although it is certainly not that anything goes. Neto (2019) has argued that biological lines of descent are similarly complex and plural. First, biological categories are often hierarchical, and at different levels we can identify different descendants and ancestors. Parent and offspring genes, homologous traits, individual organisms, microbiomes, as well as groups (consider termite mounds, prides, and so on) might, under the right conditions, be considered lineages, that is, individuals forming lines of descent. Second, lineages sometimes form within us. For instance, through development, our cells form lineages. Third, inheritance is not always vertical, but sometimes horizontal (most obviously in horizontal gene transfer). Horizontal "ancestry" means that biological history is not always best represented by tree-like networks, but sometimes (often?) requires reticulate networks.

For Neto, the upshot is that "There is, then, no single answer to the question, 'what is a lineage?'" (Neto 2019, 1109), and this pluralism makes room for a plethora of options regarding how we might graft.[2] When faced with a potentially lost world, we ask, well, what historical individuals or lineages might *not* be lost? Although there are no platypus descendants for Prof. Ichthy to make use of, there are plenty of cousins.

2 Alison McConwell (2019) has made a similar argument regarding the contingency of evolution (see also chapter 8) and the plurality of the biological individuals that might be under selection, and I think similar lessons can be drawn here.

Look out for two crucial lessons in this chapter. First—echoing Prof. Ichthy—if lost worlds can be intertwined into broader historical narratives, they can be made sense of in those terms; second, although perspectives on a particular target might differ, they also often overlap in various ways. This enables them to be compared, contrasted, and used to test one another. This second point underwrites "iterativity," which I'll introduce in the next chapter.

As with the last chapter, we'll start with an example from dinosaur biology, before returning to the Ediacaran. I'll then consider disagreements and overlapping perspectives, and finish up with a discussion of living fossils. The examples in this chapter will all draw on a particular subset of grafting strategies, namely, those involving phylogenetic reconstruction. It is important to note that not all grafting involves phylogeny: in the next chapter, we'll see grafting applied to reconstructing the relationship between complex metazoan evolution and the emergence of various global oceanic processes. It's also worth noting that trace-based reasoning is itself a form of grafting, as it involves linking a token trace or set of traces with their causal ancestors. I'll focus on phylogeny here because it is a particularly salient example in palaeobiology.

1. Do Dinosaurs Feel Pain?

"So, what was Broadfoot like?" Ichthy muses, "Some things we can be very confident of: we know they were vertebrates, like ourselves. So, they would have hearts and lungs, the usual battery of organs common to our lineage." Ichthy indicates relevant anatomical parts on themselves. "They would breathe air, metabolize similarly to us, seek out prey, and escape predators. There's a wide range of traits we can more or less safely assume Broadfoot would have had."

Ichthy again gestures to the mammalian history behind them. "Taking them as mammals, as I said, we can go further, but there are challenges. For instance, one might assume Broadfoot gave birth to live young, just like all living mammals do. However . . ."

Ichthy waves at their grad student, who quickly switches slides. The next slide shows some fossilized shards, immediately recognized by the initiated as the remains of eggshells. "You'd be wrong in thinking that."

Did dinosaurs feel pain? At first blush, this is a silly question. On the one hand, we might say: of course they do, insofar as any complex animal does (Browning and Birch 2022; Gibbons et al. 2022). On the other hand, we might consider this unknowable: feeling pain is a subjective, internal state that we have deeply limited access to. Put in these terms, it is hard enough to confirm pain in another human or our pets (due to not having access to those states), let alone long-extinct lineages (Harnad 2016).

These two interpretations of the question about dinosaur pain imply perspectives. The former implies something like the kind of phylogenetic inference we'll see below, the latter a set of philosophical views about the nature of mind and its accessibility. However, neither of these perspectives is very *strategic* from a scientific standpoint: in characterizing a denizen, the smart money is to adopt perspectives that don't make things too easy (dinosaurs feel pain because any complex animal does) nor too hard (to answer this, we need to solve the problem of other minds). The trick is to find a sweet spot where our interpretation allows us to use a set of tools that underwrite a perspective productively. In other words, neither of the above approaches are particularly *productive* for tackling loss or putative loss. I'll think harder about productivity and what makes a perspective a good one in future chapters. For now, let's examine another way of approaching the question of dinosaur pain.

Les Hearn and Amanda Williams (2019) have tackled pain in the Mesozoic. This is a useful case for us here, first, as a way of introducing grafting and, second, as a way of emphasizing how the evidential meaning of traces often requires grafting. Hearn and Williams' strategy is to consider how neontologists (biologists of the "now"; see Currie 2019b) think about pain in animals and consider how they might detect similar properties using the traces available to them. As we'll see, to some extent, the interpretative work they bring to those traces requires grafting, that is, thinking about dinosaur ancestry. To make this salient, I'll bury the lede by introducing their work without considering ancestry, and then I'll bring it in.

Biologists call the detection of painful stimuli *nociception*, and in non-human animals, it is inferred via two routes. First, the possession of particular internal structures, such as a central nervous system and the relevant receptors. We might ask: Does this critter have the organic machinery associated with pain detection? Does it have the infrastructure required for pain? Second, does it exhibit pain-avoidant behaviours, such as keeping away from pain-causing stimuli and sparing damaged parts? We might also ask: Does this critter behave in ways associated with being in pain? To apply this approach to dinosaurs, then, we need to know how to infer internal structures and how to infer behaviours.

Hearn and Williams focus primarily on pain-avoidant behaviours. They conduct a literature review, hunting for various dinosaur pathologies and other signals of potentially relevant behaviour. That is, they ask what damaged bones and other signs of bodily harm can be found in the fossil record. So, what would fractured dinosaur bones and the like have to do with inferring nociceptive capacities?

Hearn and Williams take their cue from the qualities associated with pain-avoidant behaviours: for instance, they don't simply look for fractures, but *healed* fractures. An injury healing requires the animal to avoid infection and further damage by altering normal behaviours to spare the damaged part. They also look for odd trackways suggesting pathologies, as well as evidence of social behaviour, particularly care for the young. What did they find? A fair bit: "67 [papers] dealt with healed or healing bone pathologies, mainly fractures, serious enough to have hampered normal functioning (all but eight in dinosaurs); eight described dinosaur tracks showing injury or altered gait; 82 covered dinosaur social behaviour potentially relevant to survival from injuries" (Hearn and Williams 2019, 2).

Hearn and Williams were specifically interested in cases where bones had healed from serious injuries: "Injuries (including subsequent infections) to limbs, shoulder and pelvic girdles, ribs, necks, backs, tails, and jaws, unless described as minor, would impair movement for a significant time, including for predation or escape, breathing and feeding" (Hearn and Williams 2019, 2). The idea is that such injuries indicate various forms of pain-related behaviour. If you just died from the injury, you wouldn't have much chance to spare the limb. Overall, they found that recovery from serious injury was fairly common, identified in 3% of dinosaur species (out of over 1100, so not a small number).

In addition to describing several particularly interesting specimens (for instance, Senter and Juengst 2016), Hearn and Williams pay special attention (rightly, I think) to Ishigaki and Lockley's (2010) discussion of a set of trackways in Morocco's High Atlas Mountains. These belong to bipedal therapods and appear to exhibit a distinctive "limping" gait. Potential limping is noticeable due to an irregularity in the distance between the tracks on one side and the other, a property that is typically regular in therapod trackways. Examining the tracks, Hearn and Williams identify two possibilities:

> This evidence suggests that the animal could not support the body weight fully with the digital part of the right foot, and thus it was obliged to use the heel part to support the body weight. A similar, alternative interpretation is to infer that the trackmaker was simply favouring one foot, leg or one side of the body while wading or "labouring" through rather soft mud that impeded its normal progression. (Hearn and Williams 2019, 106)

If indeed we have a limping dinosaur recorded in trackways (and, as they point out, this is contentious!), that seems reasonable evidence of sparing a damaged part. They also cite various pieces of evidence of potential protective behaviour among dinosaurs: herd behaviour, nesting and—as we saw in the last chapter—denning in a burrow.

You might start to wonder what any of this really has to do with dinosaurs feeling pain or exhibiting pain behaviour. Some examples, sure, can be linked to pain avoidance in fairly straightforward ways: dragging an injured leg seems to fit the criteria for sparing a damaged part. However, many seem circumstantial. What would denning have to do with pain-avoidant behaviour, for example? More generally, what does a bunch of data about dinosaur nesting sites, pathologies, and healing have to do with establishing pain-related behaviours?

As warned, I've buried the lede here. Hearn and Williams do not understand the relationship between pain-related behaviour and fossils independently of ancestry. First, from the get-go, they point to "evolved and highly conserved systems" (Hearn and Williams 2019, 1) deeply embedded in the history of life, that are related to nominally pain-avoidant behaviour, and they underscore physiological infrastructure related to pain. Second, their argument that various pain-avoidant behaviours in dinosaurs should be equated with feeling pain is based on thinking in terms of ancestry. They point out that "similar pain systems and behaviour among birds or responses to analgesia among reptiles, including opioid receptors in crocodiles, in terms indistinguishable from those used with mammals" (Hearn and Williams 2019, 1). The indistinguishability of pain behaviour and infrastructure between birds, crocodiles, and mammals is relevant because birds and crocodiles "bracket" dinosaurs—they are their closest living relatives, and the common ancestor of birds and crocodiles is also an ancestor of dinosaurs (together they form the group "archosaur"). "While systems and behaviours in extinct species cannot easily be ascertained, possible forms of these can be found in extant archosaurs" (Hearn and Williams 2019, 2). As dinosaurs are archosaurs as well, insofar as extant archosaurs are representative of that class, crocodiles and birds can be a guide to them.

What's going on here? The data that Hearn and Williams collate doesn't speak to nociception in isolation, but only does so when situated with information about dinosaur ancestry. They emphasize the phylogenetic continuity among dinosaurs, birds, and crocodiles as a way of inferring those properties. This is what I've called "comparative thinking": a distinctively biological way of doing business that involves situating a target in its evolutionary history (Baum and Offner 2008; Currie 2021b). This strategy informs thinking about dinosaurs as historical individuals: they are a token lineage that shares properties with currently existing token lineages in virtue of common ancestry; they are, from a certain perspective, parts of the same token lineage.

In virtue of these appeals to ancestry, Hearn and Williams *graft*. By carefully attending to inferred similarities between the past world and the current one—those aspects that are likely not lost—they are able to make headway on a putatively lost world. This is a rather non-serious case (Hearn and Williams frame their paper as a kind of "thought experiment"), but it acts as a comparatively simple example of grafting and allows us to somewhat expand our account of strategic perspectivalism from the last chapter.

Let's define grafting more carefully. I'll provide two definitions, one being a subset of the other. *General* grafting appeals to token causal links, while *specific* grafting appeals to historical individuals in particular. I introduce the latter because it is more specific to practices like phylogenetics.

> Grafting *generally* involves positing and exploring token causal links among denizens, other times, and the present.

We might understand the general strategy in terms of narratives. Consider the metazoan radiation that occurred during the Cambrian (the "Cambrian explosion"). Some scientists have speculated that the explosion is partly explained by earlier events in the Earth's history, namely, "snowball events" in which the globe partly or wholly froze over. The thought is that the earlier events constrained metazoan evolution, and that the arrival of balmier conditions triggered the radiation. In this instance, we're not considering the two past worlds—neither the snowball events nor the Cambrian radiation—in isolation. Rather, we've linked them in a particular causal trajectory (I discuss this in Currie 2014; see also Currie and Sterelny 2017). To do this, we *characterize* denizens as parts of a token causal trajectory, as playing a part in a story, for instance, snowballs acting as constraints on metazoan evolution.

In this chapter, I'll highlight a more specific version of grafting, which generates links by identifying various kinds of historical individuals:

> Grafting *specifically* involves characterizing a denizen as a historical individual such that it is continuous with some currently existing historical individual(s) or other past individuals we know about.

Where the general version of grafting concerned token trajectories or narratives, this more specific version concerns a subset of these trajectories: those concerning being part of a historical individual. We should understand "continuity" in terms of whatever persistence conditions we think relevant to that kind of individual, and this might well vary across cases.

Dinosaurs are weird. Many of their properties are lost. But they're not *that* weird: they share many properties with the birds and crocodiles we know and love. Hearn and Williams adopt a perspective, taking dinosaurs as *potentially pain-avoidant* animals. This enables inferred behaviours from fossils, such as a limping gait, to be relevant to their possession of nociception. But they also adopt another perspective, taking dinosaurs as archosaurs, recognizing continuity among themselves, birds, and crocodiles (just as we took mammoths as Elephantidae, and thus continuous with living elephants). These joint perspectives—seeing dinosaurs as *potentially pain-avoidant archosaurs*—make data and interpretations from fossils evidentially relevant to dinosaur pain avoidance.

Note two things. First, consider the relationship among inferred similarities, the continuity in the graft, and the evidential meaning of the data. The graft identifies a historical individual, say, archosaurs. That enables phylogenetic inferences, for instance: we know dinosaurs have opioid receptors because crocodiles do. And further, the graft makes sense of fossil data: fossils bearing healed pathology are evidence of pain avoidance, partly in light of the healed pathology signalling the same in living animals. It is comparisons with living relatives (and much of the rest of life) that make sense of specimens and what we can infer from them. There is, then, an interdependence among the perspective (how the denizen is characterized), how it is grafted (what it is taken to be conjoined with as a historical individual), and the relevance of data (fossils, in this instance).

Second, notice that the various perspectives—characterizations—intermingle and overlap. Hearn and Williams characterize dinosaurs as *pain-avoidant*, an ahistorical kind, and as an archosaur, a historical individual. But these are not independent characterizations: dinosaurs are *pain-avoidant archosaurs*. Under the right conditions, perspectives overlap, are combinable, and make empirical contact with one another: this is another element of their pluripotency.

Finally, grafting is closely related to a much more discussed feature of historical science: common-cause explanation (e.g., Cleland 2013). A common-cause explanation makes sense of some correlation by, well, positing a cause common to the correlates. No doubt explanations referring to common causes, and inferences following common causes, are important ways of characterizing some of the epistemic properties of grafting, particularly in its specific historical-individual guise. However, my focus here is less on providing an abstract characterization of the justification of such inferences. I'm interested in how the strategy underwrites tackling loss and putative loss.

2. Stems and Failed Experiments

"Understanding Broadfoot as a mammal challenges some common misconceptions about mammals. We biologists sometimes identify traits as 'primitive,' but this isn't supposed to be derogatory. It just means 'basal', that is, present in some common ancestor. However . . ."

Ichthy pauses, glaring over their spectacles with the air of a disapproving (not angry, just disappointed) parent.

"Some call mammals 'primitive' in a stronger sense, implying they are 'failed evolutionary experiments,' dismissed as nature's mistakes. Some say they were inevitably replaced by us sophisticated, clever, and dynamic reptiles. But discoveries like Broadfoot put these claims in their place. Mammals were diverse and well-adapted. They were, like us, sophisticated, clever, and dynamic creatures. Indeed, I see no reason, considering mammalian diversity, cognition, or form, to think them incapable of any of the same evolutionary achievements as us reptiles."

ILLUSTRATION 4.1 Prof. Platy Regards the Ichthyosaur

Prof. Ichthy allows themselves a whimsical smile. "I've sometimes imagined what a mammal with ichthyosaur-like intelligence might be like. Maybe in some alternative reality, a mammal, perhaps Broadfoot even, is engaged in the scientific study of a fossilized ichthyosaur! A rather unscientific idea, of course. But I don't think a sentient mammal is an impossible idea. Nothing about mammals says they couldn't or wouldn't evolve reptile-like intelligence. My thought is just this: categorizing mammals as a failed experiment is to undersell their diversity and evolutionary potential. That we are here and they are so diminished, it seems to me, is not due to something lacking on their part, but due to the whims of evolution and history.

"So, contextualizing Broadfoot with other fossil and living mammals helps us see how rich that lineage is—and thus push back against preconceptions of a 'primitive' evolutionary past."

I've introduced grafting as a strategy applicable to lost, or putatively lost, worlds. To further explore this strategy and the nature of loss, I want to take a little conceptual tangent into the notion of a "failed evolutionary experiment." As we'll see, grafting strategies have the capacity to transform both how we conceive of lost worlds and their epistemic value.

In the last chapter, Cuthill and Conway Morris emphasized the phylogenetic continuity of rangeomorphs while simultaneously asserting the richness and diversity of the Ediacaran fauna more generally. It's worth quoting them again: "The Ediacaran biota was far from homogenous but instead included diverse phylogenetic lineages and body plans" (Cuthill and Conway Morris 2014, 13124). What makes this diversity worth noting? Part of the story, I suspect, is analogous to Prof. Ichthy's pushback against mammals being characterized as "failed experiments." No, Ichthy insists, the mammals were not a dead end, were not a failed experiment. They were a thriving, rich lineage worthy of understanding on their own terms. Ichthy is here mimicking what has been said of Ediacaran fauna: these strange, often sessile communities were an evolutionary mistake. After all, the metazoans hadn't yet got their act together, and whatever these failed experiments amounted to, they were washed away by the sweep of the Cambrian explosion when metazoans really got going. Interestingly, the Cambrian and the Avalonian have been considered analogous insofar as both are "evolutionary failures": "the Avalon explosion represents an independent, failed experiment with an evolutionary pattern similar to that of the Cambrian explosion" (Shen et al. 2008, 84).

"Failed experiment" claims typically involve a phylogenetic element. As Dunn and Liu have put it, "Members of the Ediacaran macrobiota have been . . . deliberately set apart . . . by suggestions that they were either 'failed experiments' in the history of life, or members of long-extinct higher-order clades" (Dunn and Liu 2019, 512). What is it to be a "long-extinct higher-order clade"? Let's dive into some of the basics of cladism, which will allow us to better understand the notion of a "failed experiment."

One way of understanding "clades" is via *bracketing*, a notion we met in the last section. Take two extant lineages (say, birds and crocodiles), draw two lines backwards in time, along parents, grandparents, etc., to their common ancestor (around 240 million years ago; Green et al. 2014), and then move forwards again, including every lineage that also shares that common ancestor. Performing this procedure with birds and crocodiles produces the category "Archosauria": enfolding not only birds and crocodiles, but also many charismatic Mesozoic denizens, such as dinosaurs and pterosaurs (see Figure 4.1).

As we saw in our discussion of dinosaur pain, dinosaurs being bracketed by crocodilians and birds plays an important role in inferring various properties of theirs. Membership of that ancestral grouping—the historical individual Archosauria—allowed various features of dinosaur fossils (such as healed pathologies), and behaviours inferred from fossils (such as denning) to be made sense of, and made evidentially relevant to, pain-avoidant behaviour in dinosaurs. A little more poetically, this grouping also enables dinosaurs

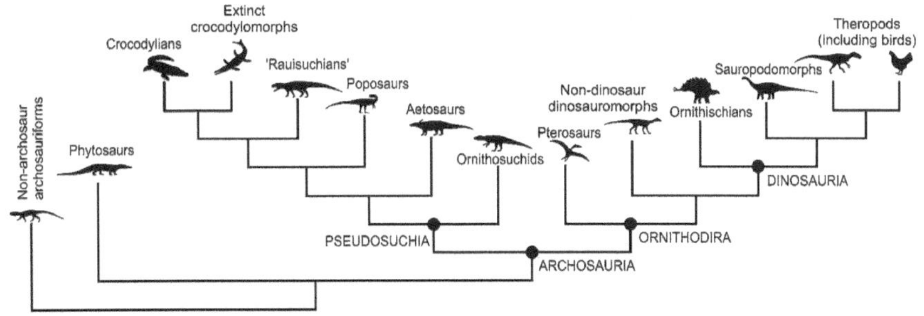

FIGURE 4.1 Bishop et al.'s (2020) Simplified Archosaur Cladogram (Fig. 1 in Bishop et al.). Creative Commons CC BY

to be part of life's story: in virtue of their placement in Archosauria, they are relevant to understanding and constraining the evolutionary and developmental narrative of the archosaurs.

With (very) basic cladism under our belts, let's return to Dunn and Liu's argument that the Ediacaran fauna have been done wrong by scientists. Consider the upshots of (1) taking the Ediacaran fauna to be a single group which (2) belongs to an extinct higher-order clade. We end up with something like Figure 4.2. The fauna of the Ediacaran cluster together as one clade, and, as a "higher-order" clade, they drifted from the main evolutionary narrative very early on. Here, the Ediacaran fauna are more or less excluded from the history of life: it isn't obvious that they can play a role in informing how we understand current life, and current life has very little to tell us about that fauna.

Figure 4.2 is, in a sense, a cladistic representation of a failed experiment: the Ediacaran fauna are taken as a single group that left no descendants. Further, they are not only lost but fatally disconnected from life more generally: *prima facie*, they have nothing to tell us of real interest about the history of life. Disvaluing extinct fauna has a storied history, from reinforcing a progressive view of history by taking dinosaurs as paradigmatically "extinct" monsters (Rieppel 2020), to reinforcing sapiens exceptionalism by representing Neanderthals as primitive lumbering brutes (Stiner 2017; Meneganzin and Killin 2024). As I'll argue throughout, but particularly in chapter 10, taking loss seriously often involves challenging not just how we think the past was, but the ways in which we value it.

FIGURE 4.2 Informal Cladogram Depicting the Ediacaran Fauna as a Dead End or Failed Experiment

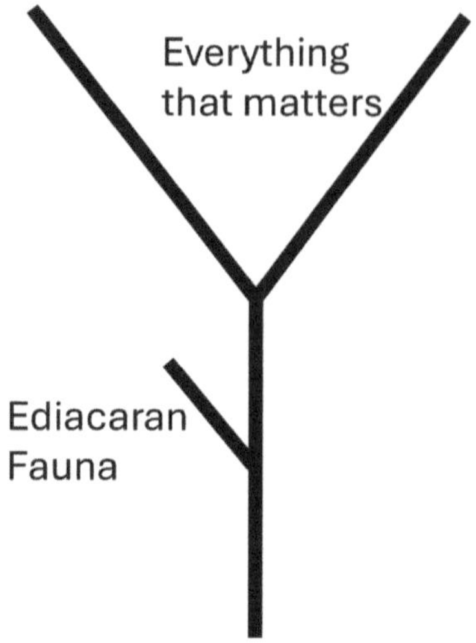

Dunn and Liu argue that representing the Ediacaran fauna in ways similar to Figure 4.2 is a mistake, and understanding why is illuminating of grafting. Here's how they state their position: "We advocate abandoning the failed experiment perspective and embracing phylogenetic thinking in order to make progress in determining the phylogenetic positions of these organisms and realizing their evolutionary significance" (Dunn and Liu 2019, 512). But what could this mean? Figure 4.2 *is* a cladogram (well, of sorts), so it does seem to constitute phylogenetic thinking of some kind. To see what Dunn and Liu have in mind requires a small—but surprisingly important—conceptual addition to how we're thinking about clades.

A clade consists of some common ancestor and all of its descendants. A common way of identifying clades, as illustrated earlier, involves identifying at least two extant lineages, their common ancestor, and all other descendants sharing that common ancestor. This conception of "clade" in fact picks out what palaeontologists would call a "crown group." Add to your conceptual repertoire another notion, that of a *stem group*. Recall that all life on Earth comes from a single origination event. This means that for any two organisms, there will be a common ancestor: for instance, there will be some common ancestor of ourselves and the Ediacaran fauna. A clade's *stem group* includes some ancestor and all of its descendants *minus the crown group* (Briggs and Fortey 2005; Bryse 2008). I find that definition difficult to

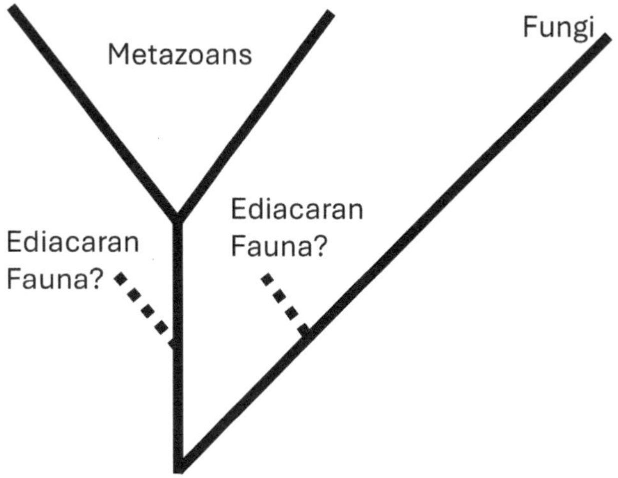

FIGURE 4.3
Cladogram Depicting Two Possible Stem Groups for the Ediacaran Fauna

wrap my head around. Here's one way of getting it. Let's add a third lineage to Figure 4.2, say, fungi. We thus imagine adding the common ancestor of fungi and metazoans, representing a larger clade that includes both fungi and metazoans (the Opisthokonts, in case you're interested, which will include non-metazoan organisms). With this in place, a further question arises: are the Ediacaran fauna on the line leading from the Opisthokont common ancestor to the fungi, or to the metazoans? We can represent that question with something like Figure 4.3.

Notice that how we answer this question makes a big difference to how Ediacaran fauna fit into the history of life: if the denizens of the Ediacaran are stem-group fungi, then they are part of that narrative, and can tell us something both about the base of the Opisthokonts and about how the fungi evolved. But if they are stem-group metazoans (which, as we'll see, palaeontologists now think many of them are), then they could be crucial for understanding the evolution of metazoans. After all, they are part of the story, dividing fungi and metazoa. So, simply by drifting somewhat further back in time, by adding the notion of a stem group to our phylogenetic repertoire, we transform how we characterize the Ediacaran fauna, what might be relevant for understanding them, and how they might help us understand broader-scale evolutionary patterns.

Further, as Dunn and Liu point out, considering the Ediacara as a possible stem group has occurred alongside palaeontologists taking a more fine-grained taxonomic perspective. These more recent studies often suggest, as we saw in Cuthill and Conway Morris' analysis last chapter, that the taxonomic status of these critters is neither as simple nor as monolithic as Figure 4.2 represents. So, they do not form an easily dismissed phylogenetic clump.

In the next section, we'll look at a case in more detail, but I want to wrap this section up with a few lessons moving forward.

First, the strategy I've described—of switching from thinking about a lost lineage as a failed experiment to thinking of it as a potential stem group—is an example of *grafting*. After all—and as we'll see in more detail in the next section—to determine where some lineage might sit in a stem requires emphasizing points of continuity between the putative stem and the relevant crown group. We might ask, what characters do the Ediacaran fauna and the metazoans have in common that sets them apart from fungi (and vice versa)? So, in light of an apparently mysterious lineage, one apparently unlike anything we've seen, we seek to find principled strategies to determine ways those lineages might not be lost, and thus be incorporable into an overall narrative.

Second, notice that we didn't make any directly empirical moves in shifting between Figure 4.2 and Figure 4.3: we added another node to the cladogram, and thus generated another common ancestor. You might say, we simply changed perspective. This is a general feature of phylogenetic thinking: although there are facts of the matter about which lineages are more related to which, whether two lineages are part of the same clade turns on which clade we're talking about. Considering the crown group including ourselves and crocodiles, we're in the same clade as the dinosaurs; considering the crown group including ourselves and cats, we're not in the same clade as dinosaurs. So, when using this kind of thinking to make sense of various lineages, we can—as it were—zoom in and out of various levels of description to get the best grip on the traits we're interested in.

You don't get much more strategically perspectival than that!

Third, and as a perhaps-too-late caveat, my description of phylogenetic, cladistic thinking has been purposefully casual: apologies to any phylogeny nerds out there. I've not said anything about the inferential strategies that go into constructing phylogenies (e.g., Sober 2002), nor about the complex analyses required to understand characters (let alone what counts as a character: DiFrisco, Love, and Wagner 2020; Meneganzin, Ramsey, and DiFrisco 2024). Further, I've only hinted at the rich ways that these underwrite comparative methods in biology (Currie 2021b). Max Dresow and I are developing an argument that the palaeontological innovation of the stem group transforms how fossils can be treated in phylogenetic contexts, and that involves a much more sophisticated treatment than I can manage here.

Regardless of all that, we can begin to see the grafting strategy as it plays out in evolutionary palaeontology: use phylogenetic thinking to emphasize the continuity between lineages with lost traits and better-known lineages, thus finding lines of ancestry between them, seeing them as historical individuals, and thus providing a perspective to start building understanding. Let's turn to a more specific case to see this in action.

3. *Dickinsonia*

Let's wind the clock forward, past the Avalonian, and later into the Ediacaran. As the clock's hands spin, we wave goodbye to the microbial mat around the same time as burrowing metazoans arise and the oceans start to become more reliably oxygenated. The sessile Eden of the rangeomorphs gives way as worm-like creatures evolve. As they eradicate the microbial mat, we must also bid farewell to pizza disks in the subsequent fossil record. We'll return to the fall of the Ediacaran in the next chapter, but for now, let's focus on perhaps the most well-known of the Ediacaran denizens: *Dickinsonia*.

The Ediacaran period is named after the old rocks of the Ediacara Hills, nestled in the Flinders Ranges of South Australia. It was there in the 1940s that *Dickinsonia* was identified by Reg Sprigg. Its fossils are oval, nearly perfectly bilateral imprints with rather beautiful lines radiating from a central axis (Figure 4.4). *Dickinsonia* varies in length from a few insignificant millimetres to a mighty near meter and a half. Its fossils have also been found in later Ediacaran beds in Russia, Ukraine, and China (Brasier and Antcliffe 2008; Ivantsov and Zakrevskaya 2022).

As is typical for Ediacaran fauna, *Dickinsonia* has long been a phylogenetic mystery (Sprigg originally thought it was a jellyfish, but over the years it has been described as everything from a fungus to a worm to something entirely unknown). Gold et al. (2015) describe the basic problem with trying to place *Dickinsonia* in the history of life: "Part of the difficulty in classifying *Dickinsonia* comes from its morphological simplicity—which would be consistent with its placement as a primitive organism—but also means that it lacks morphological synapomorphies to ally it with any living (crown) group" (Gold et al. 2015, 315).

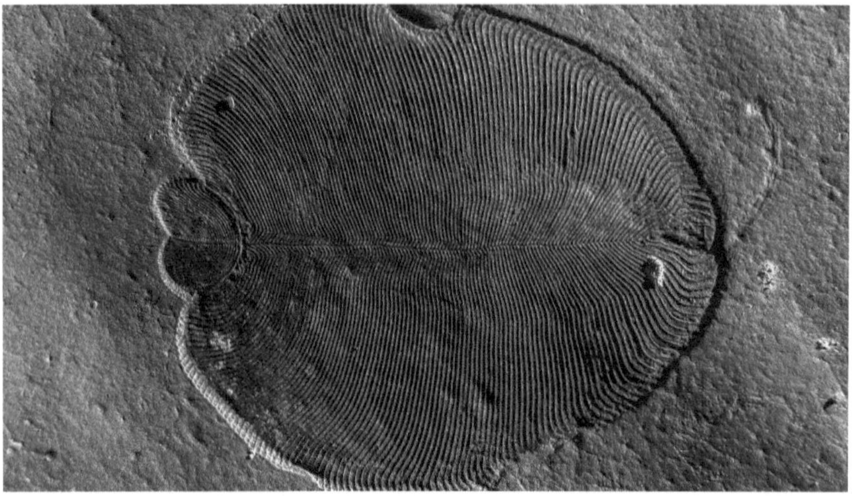

FIGURE 4.4 *Dickinsonia* Fossil (Detail from Fig. 1, Bobrovskiy et al. 2018). Creative Commons CC BY

In the last section, I informally introduced us to phylogenetic thinking, which, for the purposes of this book, is most interesting as an example of grafting. That is, it allows us to conceive of various critters as historical individuals—clades—and thus find continuity among them, continuity that might underwrite various inferences. But to apply phylogenetic analysis, we need *characters*—that is, traits possessed by the relevant lineages that can be used to empirically inform who is more related to whom. Dinosaurs and birds share various characteristics, such as three-toed feet and pneumatized bones, which is evidence of their shared ancestry. Now, look again at *Dickinsonia* in Figure 4.4. What morphological characters does *Dickinsonia* have? On what basis might we think it similar to or different from any living animal? As Gold et al. reference in the quote above, in modern phylogenetic analysis, we need to identify "synapomorphies." A synapomorphy is a character *uniquely* possessed by a common ancestor and its descendants—it is thus a trait that can distinguish clade membership from non-membership. For instance, the human chin has long been identified as a distinctive morphological trait among the primates; that is, we have it, but no other primate does. It is thus, potentially at least, a character that might be a synapomorphy, dividing our own taxon from that of our relatives (Meneganzin, Ramsey, and DiFrisco 2024).

What potential synapomorphies does *Dickinsonia* have? Well, to answer that question, we need to first ask another: What *characters* does it have? With so few features preserved, identifying meaningful morphological characters is extremely challenging. Gold et al.'s strategy is to find characters by appealing not to morphology, but to ontogeny: how *Dickinsonia* grows. That is, instead of looking for a morphological character, they seek out a developmental character (perhaps it should be called a "synapontogeny" . . .).[3]

The first step is to recognize that *Dickinsonia* grows via serially repeated structures. This enables us to characterize *Dickinsonia* as a critter that grows via serial repetition. This perspective underwrites a set of comparisons with various eukaryotes that also grow that way. As we'll see in the next chapter, this is a form of *anchoring*, so I'll leave a thorough description of that aspect until then. Suffice to say, Gold et al. argue that *Dickinsonia* grows in a very particular way: "A form of terminal addition, a developmental mechanism currently exclusive to bilaterian animals" (Gold et al. 2015, 316). If so, this suggests that *Dickinsonia* should be phylogenetically aligned with those animals, that is, "terminal addition"—a developmental character—is a synapomorphy among bilaterial metazoans.

Gold et al. generated a set of phylogenetic trees and ancestral state reconstructions, the former capturing patterns of relatedness and the latter identifying which traits belong to which common ancestors (let's not worry about the technical details here). In accordance with many other studies, the crown group of bilateral animals bottoms out sometime in the Cambrian; that is, the common ancestor of all currently living complex animals lived sometime after the Ediacaran and *Dickinsonia* had passed. Crucially, "All analyses agree that the probability of terminal addition existing in the last common ancestors of animals is less than 5%" (Gold et al. 2015, 319). So, the common ancestor of all animals didn't grow that way, which is to say, that mode of growth is not a "primitive" (or "basal") trait inherited by animals (and lost in some), but is rather "derived." However, their results do "suggest there is strong support for terminal addition being the ancestral condition of the bilaterian animals, and that there is little support for this developmental trait existing at deeper nodes of the eukaryote tree" (Gold et al. 2015, 320). In other words, basal bilaterian animals grew by terminal growth, and this emerged within that clade, not earlier.

---

3 Kirsten Walsh takes the blame for this neologism . . .

What does all this tell us? Growth via terminal addition in animals evolved in the bilateran animals, not before. *Dickinsonia* is at the very least allied with animals, and it grows via terminal addition. So, unless it evolved that character independently, it should be placed in the stem of bilateral animals. It is part of the story (even if going on something of an evolutionary tangent) of how we, bilaterally symmetric animals, got our start. Terminal-growth addition is a developmental synapomorphy in bilateral animals and *Dickinsonia*.

What is important about getting Dickinsonia's phylogenetic placement right? For Gold et al., "Our inability to place Ediacaran fossils onto the eukaryotic tree has impeded our attempts to constrain molecular clocks, assess the intensity of Earth's earliest extinction events, and interpret the significance of the Cambrian 'explosion' of animal life" (Gold et al. 2015, 315). Molecular clocks (roughly, figuring out the time between extant lineages' ancestral splits from one another using estimates of the pace of mutations in their respective genomes) need to be calibrated against timing derived from fossils. If we don't know where to place fossil taxa in phylogenetic trees, they can't be used for calibration. Further, if we don't know where Ediacaran fauna place in the history of life, we can't investigate the effects of major extinctions and other macroevolutionary events across various branches. For instance, we might wonder whether the Cambrian explosion represents the descendants of a few lucky survivors from the Ediacaran, or if there was a smoother transition between the two. Gold et al. take their view to support something like this latter view: "Large bilaterans were an important component of the Ediacaran biota, thus connecting this fauna with the following radiation of animal life in the Cambrian" (Gold et al. 2015, 322). Thus, this specific attempt at grafting also underwrites a more general instance of the strategy: folding *Dickinsonia* into the narrative of metazoan life.

I've simplified Gold et al.'s work for the purpose of highlighting grafting. In the face of a strange, unknown—putatively lost—lineage, Gold et al. hunt for particular traits possessed by the lineage that could be meaningfully brought into dialogue with taxa we are more familiar with. Morphological traits were unpromising, partly because of *Dickinsonia*'s apparently simple, lost body plan and because of the erasure of what useful morphology it may have had. But we do have information about how the creature grew (or at least Gold et al. argue that we have such information), and so we can hunt for ontogenetic synapomorphies instead of morphological ones. This echoes the strategy we saw for artificing in the last chapter: developmental features are isolated and investigated before being recontextualized, in this case using various phylogenetic methods.

So, when faced with a lost world, try to find aspects of that world that are not lost and weave them into a historical narrative. This brings denizens into productive dialogue with time periods we have better information about. That is the essence of grafting strategies.

With the ideas underlying grafting in hand, I want to turn to two related discussions. The first, in the next subsection, expands on strategic perspectivalism with an initial pass at disagreement. The second, and this chapter's conclusion, will turn to what might be a rather extreme example of grafting in biology (the companion to chapter 3's discussion of de-extinction): living fossils.

4. Conflicting Perspectives

I've been emphasizing the perspectival nature of palaeoscientific knowledge: what counts as a past world is at least partially up to scientists' interests; the distinction between loss and erasure turns in part on which denizens scientists indicate; how denizens are characterized turns on the perspectival tools at play. You might also accuse me of infusing strategic perspectivalism with a fair bit of epistemic optimism (Currie 2019c): I want to say that the clever use of perspectives and the tools that generate perspectives explains how progress can be made in light of loss and putative loss.

But, you might worry, there are two closely related epistemic problems with a perspectival account, even a merely strategic one. If the results generated by perspectives are tied to those perspectives, then (1) how can that knowledge travel beyond the contexts of those perspectival tools, and (2) how can apparent disagreements be resolved? This worry should be familiar to philosophers of science, echoing as it does the more extreme worries about incommensurability associated with Thomas Kuhn.[4] Less extremely, these worries often arise for pluralistic accounts of representation, as Massimi summarizes: "Whenever there are situations requiring a plurality of models for a given phenomenon, each model might accurately represent some aspects of the phenomenon at the cost of blurring, idealizing, or distorting other aspects. The outcome is that different models may deliver different,

4 Although incommensurability needn't be understood in such stringent ways (see Bocchi 2022, for instance).

sometimes incompatible, or even inconsistent images for the same phenomenon" (Massimi 2022, 53).

Notions of pluralism, perspectivalism, and what to do with inconsistent models are long in the tooth in the philosophy of science (Wimsatt 1987; Weisberg 2007a; Potochnik 2010; Massimi 2021). While I'll return to these themes, in this context I'll focus on how apparently conflicting perspectives might relate to uncovering lost worlds.

The possibility of inconsistency between models (or perspectives) and the related apparent difficulties of transporting knowledge from a perspective or resolving disputes and disagreements arising from differing perspectives share a common root. I'll call this root *perspectival isolation*. In some circumstances, particularly those favoured by more traditional perspectivalists and pluralists, perspectival isolation is not a problem, but a solution, due to the limitations of any particular representation or perspective. Here's Elgin's version of this strategy: "Different models make manifest different features of the target. Diverging models afford different perspectives on the same reality. And it is no surprise that different perspectives reveal different aspects of that reality. There is no optimal model for the same reason that there is no optimal perspective. Every perspective, in revealing some things, inevitably occludes others" (Elgin 2017, 180).

So, one—and, I think, the received—way of handling apparent disagreement is to follow this kind of divide-and-conquer strategy. Models do not truly disagree, but rather imperfectly describe different elements of the same reality. The model's representative capacities are thus, in a sense, isolated from one another. I'm not sure if this strategy will cut it in many of the circumstances I'm interested in. Scientists faced with lost worlds no doubt flexibly adopt various complex perspectives as they attempt to get a grip on lost denizens, but these perspectives are far too much in dialogue, and indeed sometimes in true disagreement, for this brand of ecumenicism to have wide applicability. So, toward the next chapter's discussion of iterativity, understanding how different perspectives might speak to one another and, yes, disagree with one another is crucial.

Insofar as perspectives are isolated, that is, self-contained and without overlaps with other perspectives, it is hard to see how disagreement might be resolved (or, for that matter, how perspectives might speak to one another in the first place). However, we've already seen in this chapter that perspectives are often combinable: dinosaurs are not only characterized as archosaurs and pain-avoidant critters in isolation, but as *pain-avoidant archosaurs*. In the next chapter, I'll argue that forms of epistemic iteration overcome perspectival isolation in practice. As a warm-up to that argument, let's look at a putative disagreement about the phylogeny of *Dickinsonia*.

As we've seen, Gold et al.'s analysis of *Dickinsonia* turned on its growing via terminal addition, which was inferred mostly via phylogenetic bracketing. Hoekzema et al. (2017) dispute this interpretation, arguing that the lineage grew in a two-step manner, first via insertion pre-terminally with some inflation, and then a second step dominated by inflation. That is, the early stage of *Dickinsonia's* growth mostly involves adding components, while the later stage largely focuses on increasing the size of those components. Crucially, their model denies that the lineage grew via terminal addition.

As opposed to phylogenetic bracketing, Hoekzema et al.'s evidence is based on examining a sequence of *Dickinsonia* fossils, interpreted as varying across ontogenetic stages. They use these to reconstruct and analyse the life history of *Dickinsonia,* thus testing different models of its development.

On the face of it, we have here the use of two different sets of perspectival tools. Where Gold et al. develop a novel developmental character and employ it via phylogenetic bracketing, Hoekzema et al. use an ontogenetic interpretation of fossils to generate an apparently conflicting story. If the arguments are made based on differing tools—and thus differing perspectives—can this avoid perspectival isolation? That is, is there a substantive empirical dispute between them? I want to insist yes, and to see why, we'll need to fill out Hoekzema et al.'s argument in more detail.

As we saw, Gold et al. argued for a bilaterian affinity for *Dickinsonia,* and Hoekzema et al. are explicitly in the business of disagreeing with their interpretation: "Recent studies into growth in *Dickinsonia costata*, and arguments for a bilaterian affinity based on ancestral state recognition, rely on assumptions regarding growth in this taxon that we here argue are incorrect" (Hoekzema et al. 2017, 2). How does this argument go? Everyone agrees that *Dickinsonia* grows via the successive insertion of units around a central axis, where each unit is, in effect, one of the lines crossing the centre of the distinctive morphology visible in Figure 4.4. In addition to the units, we can also identify two "ends" of *Dickinsonia*: the "deltoidal" region (or "D"), where the lines curve outward to form what is often interpreted as a head, and the "anti-deltoidal region" (or "AD"), which is typically interpreted as the end where insertion occurs (see Figure 4.5).

Hoekzema et al. measure three things across their specimens: (1) unit length, (2) unit count—the total number of units—and (3) unit "number."

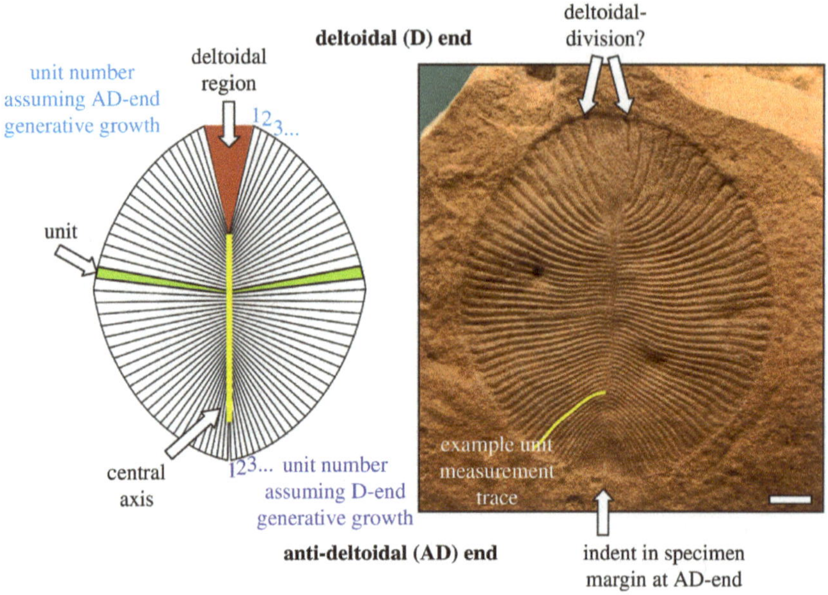

FIGURE 4.5 Measuring *Dickinsonia* Growth (Detail from Fig. 1, Hoekzema et al. 2017). Reprinted with Permission from *Royal Society B*

This last measurement needs some unpacking. They describe it thusly: unit number is the "progressive number of appearance of each unit, considering the possibility of generative zones at either the D-end or AD-end of the organism" (Hoekzema et al. 2017, 3). "Unit number" is the order in which that unit would have been added if the addition were at the anti-deltoidal end or the deltoidal region. The former, taking the growth to occur anti-deltoidally, is the assumption behind terminal-growth, and is supported by the fact that the smallest units are located there. But Hoekzema et al.'s measurements allow them to compare the relationship between unit length and count against two versions of unit number, one starting from the D-end and another from the AD-end, which can be understood through an overall ontogenetic model. At base, they can ask, which insertion point makes the most likely developmental sequence?

Hoekzema et al. compare the developmental sequence taking the (usually assumed) AD-end insertion route, that is, terminal addition, and the D-end insertion. The major difference between the two is the regularity of unit length and unit counts against overall size. We would expect the youngest unit, that is, the first one inserted, to be a similar size across specimens, but on AD-end insertion, we see nothing of the sort. By contrast, if insertion occurs at the deltoid, "New units neatly and consistently exhibit increasingly greater lengths as they are added" (Hoekzema et al. 2017, 6). On this basis, then, Hoekzema et al. argue that there is good reason to take *Dickinsonia* to grow not by terminal addition, but by deltoidal addition.

Hoekzema et al. abstract from their empirical measurements to generate a simplified growth model that they claim can generate a variety of *Dickinsonia* morphologies, a result that they further argue has phylogenetic consequences: "Although different *Dickinsonia* species have disparate morphologies, they can be rationalized by a common morphometric model, substantiating their coherence as a natural group" (Hoekzema et al. 2017, 6). We saw this kind of inference earlier in Cuthill and Conway Morris' unified developmental model for rangeomorphs in chapter 3. Producing a simplified, unified developmental model is taken as evidence for phylogenetic continuity. This is an example of how grafting and artificing can come together.

So, Hoekzema et al. and Gold et al., in a sense, have differing characterizations—perspectives—on *Dickinsonia*. Each incorporates different perspectival tools. Where Gold et al. incorporate (1) adult specimen morphology with (2) a presumed growth structure and (3) ancestral-state reconstruction, Hoekzema et al. use (1) measurements of *Dickinsonia* across ontogeny, which (2) allow a comparison of two growth models and (3) the generation of a theoretical geometric model. It might be tempting to think that these two perspectives—both made available in the horizon of alternatives thanks to those tools—are simply ones that scientists can select; and so, the disagreement is not truly empirical. After all, isn't it simply which perspective we choose to adopt that determines our endpoint? Massimi characterizes this as "perspectival disagreement": "Perspectival disagreement envisages the possibility that two or more epistemic communities from different scientific perspectives might agree about either the reliable experimental procedures or the justificatory principles, while also disagreeing on which scientific knowledge claims these in turn support" (Massimi 2021, 561–562).

I take it that perspectival disagreement occurs when scientists adopt the same characterization of a denizen—use the same perspectival tools—but disagree about, say, what a satisfying explanation looks like, or what the right kind of question to ask is. As mentioned in the last chapter, much discussion of perspectivalism targets this kind of problem. And I do think, if you're interested in resolving that question, it does present a philosophical problem. However, I don't see much—or any—perspectival disagreement in the cases I've been examining.

For instance, compare the differing characterizations of historical individuals we've examined in phylogenetics. Here, although there is a fact of the matter regarding the cladistic relationships between taxa once they are characterized, there is nothing at all stopping us from combining these. They are different perspectives insofar as they involve different characterizations of the lineage, but in this instance, a single representation can capture these differences, namely, a phylogenetic tree.[5] By contrast, Gold et al. and Hoekzema et al.'s differences diverge from this: they legitimately disagree. Their two perspectives are not simply talking past one another. The presumed growth structure in Gold et al. *is* more or less one of the growth models that Hoekzema et al. compare in their study. An initial point to realize is that although the perspectives differ, they also overlap.

If the two perspectives involve differing characterizations of the denizens, and the results they generate clash, we might ask whether the clash is due to more than differences in those characterizations. That is, can the disagreement be considered truly empirical? Yes, it can. Hoekzema et al. have provided arguments that undermine Gold et al.'s interpretations, ones that they or others could develop responses to. For instance, one might argue that Hoekzema et al.'s approach more directly tests the developmental sequence of *Dickinsonia*, while Gold et al. depend on more circumstantial tests, and we might then take it as stronger evidence. Further, as I'll explain in the next chapter, that disagreement itself can act as a launch pad for further studies.

Seeing why differences in perspectives needn't undermine productive empirical disagreement provides a vital clue to our story about how challenges from loss are tackled: we learn new things from adopting strategic perspectives, but that new knowledge is not limited to those perspectives and, as such, the knowledge is not perspective-bound, and the perspectives are not immune to empirical constraint or testing.

Recall Massimi's perspectival$_2$: perspectives enable inferences that reach beyond what is explicitly represented in that perspective. I take it that the

5 In instances of *discordance*, that is, when apparently nested lineages, such as species and parts of their genomes, have differing phylogenetic histories (Haber 2019), a single phylogenetic representation may be unable to easily capture the state of play. I don't think this is an instance of perspectival disagreement, however, just a case of limited representational capacity.

features of recontextualization and pluripotency that I articulated in the last chapter, and we see present in the case at hand, make further sense of how this might work. It isn't only that a perspective affords inferences; it is also that perspectives can be brought into dialogue: constraining, testing, and integrating with one another.

It is worth returning to the thought that under some conditions, perspectives may be combined. Again, Hearn and Williams didn't only conceive of dinosaurs as potentially pain-avoidant, or as archosaurs, in a sequential or isolated fashion: rather, they conceived of them as pain-avoidant archosaurs. Such mixed perspectives are common and crucial for understanding how lost worlds might be reconstructed. In visual cases (if we're to believe early 20th-century psychologists), a gestalt switch occurs as we swap between differing perspectives—call to mind the philosopher's favourite chimera, the duck-rabbit. But here we have a disanalogy with scientific perspectives. Our (or at least my!) visual perceptual tools disallow seeing the duck-rabbit as a duck and a rabbit simultaneously. But phylogeny allows us to represent both ancestral information about historical individuals—being an archosaur—as well as ahistorical kinds, such as having pain-avoidant behaviour. Thus, some perspectival tools not only enable perspectives to be put in empirical tension due to overlaps, but they can also sometimes act to integrate perspectives.

To summarize in somewhat of a rush: some facts are generated/graspable/confirmable from a perspective, but are not perspective-bound, as they may, for instance, be taken into (or even themselves form) new perspectives, as we often see in the "characterize-examine-recontextualize" process. This means that by shifting between perspectives, we can cobble together information that isn't available otherwise. In other words, even though the fossils are very distant from the past world, and the past world is very different from now (is lost), under the right conditions, we can nonetheless recover a lot of knowledge about it.

5. Coda: Living Fossils

The grad student switches transparencies again, this time revealing an organism familiar to the audience. The animal looks, at first glance at least, like a rather unremarkable lizard. The slide's appearance initiates a small cheer from a group of long, sleek, eel-like pleurosaurs in attendance. Prof. Ichthy gives them a small wave before continuing.

"I've been explaining that although mammals and Broadfoot share many properties, nothing too much like Broadfoot has survived. It's a unique extinct lineage. Which is to say, there's no analogy for what we marine reptiles have in our familiar old friend Peaks-on-Back."

Ichthy again nods at the pleurosaurs before turning to consider the image behind them.

"Peaks-on-back is a strange, rather wonderful animal. While clearly related to pleurosaurs, it nonetheless retains a number of highly primitive, conserved traits. Many of its traits are otherwise only known from fossils. Indeed, back in the day, the discovery of Peaks-on-Back certainly caused quite a scientific stir!" Ichthy chuckles to themself.

"If you'll permit me some rather unscientific speculation, were there scientists in the Age of Mammals, they may indeed have wondered after Broadfoot as we do after Peaks-on-Back. The mammals we know and love, those still surviving today, are what we call "placental mammals." But Broadfoot seems to be something quite different: only one of two genera we have so far found which represent the "monotreme" group. I can imagine a mammalian scientist, upon discovering a living Broadfoot, thinking that a stranger from an earlier time had visited them. Imagine, something known only from fossils just wandering out of the bushes. A lost world appearing before you. A fossil, but alive."

Throughout this chapter, a picture has begun to emerge concerning how grafting—that is, emphasizing historical continuities between past and current worlds—can make progress in cases of loss. In the face of a putative lost world—that is to say, a denizen with certain unique, unknown, alien properties—we seek out continuities in token historical individuals. This involves characterizing our target in ways that enable grafting: dinosaurs are bracketed by birds and crocodiles, rangeomorphs are stem lineages—but of fungi or animals?—*Dickinsonia* shares a growth strategy with animals. These practices of characterization and recontextualization enable points of comparison that further underwrite situating the denizen in some way: within the evolutionary narrative implied by a cladogram, within models of developmental processes, and so on. These situating practices extend the horizon of alternatives, that is, the perspectives available to scientists. But the components of these horizons are not independent: they partially overlap in ways that enable empirical dialogue between them. Thus, by flexibly switching between cases, the lost property can slowly emerge. In the next chapter, we'll concretize this further by developing a notion of "iteration," but before getting there, I want to briefly turn to "living fossils," as they might be thought of as an extreme example of grafting.

Although Prof. Ichthy could not possibly know, his speculations are right: the (European) discovery of the platypus at the turn of the 19th century caused various kinds of stirs. Darwin infamously took the platypus to be a living fossil (Gruber 1991; Moyal and Marks 2019):

> In fresh water we find some of the most anomalous forms now known in the world, as the Ornithorhynchus and Lepidosiren, which, like fossils, connect to a certain extent orders now widely separated in the natural scale. These anomalous forms may almost be called living fossils; they have endured to the present day, from having inhabited a confined area, and from having thus been exposed to less severe competition. (Darwin 1964, 107)

Darwin pointed to the platypus (*Ornithorhynchus*) and lung-fishes (*Lepidosiren*) as examples of forms that are anomalous insofar as they are very different from other living animals—widely "separated in the natural scale"—such that they most closely resemble taxa only known from fossils. Darwin's commitment to a linear, incremental, and progressive conception of evolution led him to explain these "living fossils" by appeal to "refugia": protected places where a lack of competition has enabled their less-adapted forms to survive. Further, Darwin's commitment to there being transitional forms made these "living fossils" potentially important for bolstering that aspect of his theory (Turner 2019b has a nice explanation of the importance of living fossils for Darwin, as well as the term's origin and pre-Darwinian use; see also Werth and Shear 2014).

What is a "living fossil"? It is more or less accepted that the uses of the concept across biology are unified at best by family resemblance. That is, the term picks out a set of non-equivalent notions—being rare, being morphologically similar to fossils, possessing evolutionary stasis, being at biodiversity risk due to smaller geographical range or populations, possessing basal characters, etc.—of which lineages called "living fossils" possess some but not all. There is disagreement as to the usefulness of the term, how the concept should be understood and deployed, and what the concept is for. Two quotes from Werth and Shear, I think, capture the problem and the promise. First, the problem: "We must not expect that all aspects of coelacanth or horseshoe crab biology have existed unchanged for hundreds of millions of years. Alas, this dashes Darwin's hopes that living fossil taxa might provide unsullied knowledge about the world as it existed long ago" (Werth and Shear 2014, 3).

That is, living fossils cannot be understood as fortuitous time machines. Darwin is making a mistake in thinking living fossils are "transitional forms." This is because they themselves are products of millions of years of evolution in their own right. A platypus, for instance, is not an ancient mammal, but a lineage that has followed its own evolutionary path throughout the Cenozoic. As Olivier Rieppel has recently put it, "As living fossils are characterized by the retention of some strikingly 'primitive' or ancient features, these are nevertheless mixed with specializations of their own" (Rieppel 2023, 6).[6] Nonetheless, returning to Werth and Shear, the promise: "No doubt living fossils offer, as Darwin anticipated, an unprecedented opportunity to study—with fascinating, attention-grabbing narratives—major questions concerning extinction, competition, and rates of evolutionary change, both morphological and genetic" (Werth and Shear 2014, 4).

So, as Lidgard and Love argue (2018), various perspectives associated with living fossils—apparent morphological stasis (or at least slow change), survival in refugia, and so on—are all questions deserving evolutionary focus. Love and Lidgard suggest that the purpose of living fossil concepts is to underwrite research programs targeting these questions. But how might living fossils be relevant for understanding lost worlds? Although they don't provide "unsullied" knowledge, can they help at all? Let's examine a case.

Tuatara (*Sphenodon* on formal occasions) is a lineage of lizard-like organisms that can be found on various islands on offshore Aotearoa (New Zealand). To my endless delight as an undergraduate student, they could also be found in a glassed enclosure near a walkover at Victoria University of Wellington.[7] They are named after their distinctive spines ("tuatara" approximately translates from Māori into "peaks on the back"). The lizardiness of tuatara is superficial. First, the basic lizard form appears to be basal for amniotes: the earliest amniote fossil basically looks like a lizard, so it is

6 Jack Ashby argues against calling Australia's menagerie of mammals living fossils, connecting this practice with colonial attitudes and with conservation issues (Ashby 2021, 2022).

7 I see that in 2017, they were moved into the University's science building. I hope they are happy there.

no surprise that some amniotes have lizard-like morphology. Second, there are major differences between tuatara and true lizards. Tuatara prefer cold weather, are nocturnal, lack external ears, possess a third eye (which is visible when they're young, but covered over once adults), and have extraordinarily extended lives compared to their lizardy distant cousins, averaging sixty years but sometimes hitting a century (Crook 1975). They have further unique traits: "*Sphenodon* is unique among modern tetrapods in having thoracic ribs split into jointed tubular proximal and flaring distal parts. Additionally, it is displaying gastral ribs, a feature shared in living tetrapods with crocodylians only" (Klein and Scheyer 2017, 3).

Tuatara are the last survivors of a long-lived and diverse taxonomic group, the Lepidosauria. Lepidosauria is the sister clade to the Squamates, an extremely populous group that includes lizards and snakes (over 9,000 species). In the deep past, the Lepidosauria included many tuatara-like critters as well as larger land-based carnivores and *Pleurosaurus*, 1.5-meter-long aquatic predators hailing from the Late Jurassic (Dupret 2004); hence their inclusion in Prof. Ichthy's audience.

On the face of it, the tuatara represents an incredible story of evolutionary survival: a phylogenetic, taxonomic, and morphological outlier that somehow outlasted all of its relatives. If it were to disappear, then many denizens would become lost. As such, an account of lost worlds might give us some way of telling us under what conditions we should value living fossils, and what their epistemic use might be.

Recently, philosophers have presented defences of the notion of a living fossil. As we saw above, Love and Lidgard present a defence appealing to pursuit-worthiness (Love and Lidgard, 2018). In broad agreement with Werth and Sheer, they argue that "living fossils" concepts structure various investigations: into rates of molecular evolution, morphological stasis, and so on. So, the term is useful insofar as it underwrites and partially links those investigations.

Derek Turner (2019b) has provided a somewhat more realist conception and defence, based on what he calls the *phylogenetic living fossil* concept. This consists of three criteria, which he argues "really do sometimes co-occur in ways that are theoretically and practically interesting": (1) morphological stability, (2) rarity, and (3) high phylogenetic biodiversity value (Turner 2019b, 23). He is particularly interested in emphasizing the conservation value of living fossils; hence, the emphasis on biodiversity value (Turner and Han 2023). Turner defends the tuatara as a paradigm phylogenetic living fossil:

"The group is species poor, and exhibits prehistorically deep morphological stability. The group is also evolutionarily distinct." (Turner 2019b, 23).

I'm not interested here in negotiating between Love and Lidgard and Turner (Lidgard and Love 2021). Nor am I really interested in defending the idea of living fossils against the various objections scientists have pushed, although I agree with Watkins (2021) that declaring a putative living fossil not so based on it lacking some particular feature, such as having a fast rate of molecular evolution, is to miss the point.

Instead, I want to ask whether phylogenetic living fossils like tuatara might provide some special insights into our knowledge of the past, something which Turner briefly mentions: "Horseshoe crabs and other taxa that exhibit morphological stability over the deep past might also have epistemic value to scientists interested in reconstructing the deep past" (Turner 2019b, 23). Turner gives the example of horseshoe crabs' unusual blood, grounding a basis for inferring similar blood in their ancestors. In other words, classic grafting: inferring a past property based on shared ancestry. Aja Watkins (2021) focuses more specifically on the kinds of inferences living fossils can ground, particularly about the relationship between phenotypic similarity and evolutionary stasis, arguing in particular that it isn't molecular rates of evolution that living fossils primarily evidence, but phenotypic and developmental change and stability.

I think there is more to explore concerning the epistemic value of living fossils. To see this, let's turn our attention to one of the tuatara's extinct relatives.

Palaeopleurosaurus could be understood as a transitional form between the Late Jurassic marine *Pleurosaurus* and its terrestrial ancestors from the Triassic. Like *Pleurosaurus*, it is a long, thin—somewhat snake-like—and potentially marine reptile. There is dispute as to whether they were obligate aquatic animals or partially lived on land. Klein and Scheyer (2017) examine bone histology, that is, bone microstructure, to figure out the critter's habitat. The underlying idea is that in most secondarily aquatic vertebrates, particular properties of bones are adapted to aquatic conditions, for instance, "Increase in bone compactness and/or deposition of secondary lamellar tissue within bone cavities during intense remodelling" (Klein and Scheyer 2017, 4). As such, by examining the histology of various secondarily-aquatic vertebrates, properties associated with aquatic adaptation can be identified. So, they characterize their target as a secondarily aquatic vertebrate, thus enabling comparison with, for instance, cetaceans (Gutarra and Rahman 2022). Because vertebrates that have returned to the ocean did so from similar morphological and histological starting points, it is thought likely that they might evolve parallel traits in light of those changes.

In addition to characterizing their target as a secondarily aquatic vertebrate, Klein and Scheyer's analysis is rooted in a further comparison: between the bone histology of *Palaeopleurosaurus* and that of its one remaining relative, the terrestrial tuatara. It is worth quoting and unpacking their conclusion:

> Aquatic Palaeopleurosaurus shows osteosclerosis (i.e. bone mass increase) in its gastralia, some osteosclerosis in its rib, and none in its femur. However, bone compactness in the (proximal portion of the) rib of the terrestrial Sphenodon is only slightly lower. There is no difference in femoral microanatomy between Palaeopleurosaurus and Sphenodon (Castanet et al. 1988), so that the former resembles the terrestrial condition of the latter. Thus, the microstructure of the studied bones of Palaeopleurosaurus does not reflect a clear aquatic lifestyle. (Klein and Scheyer 2017, 7)

In the above quote, Klein and Scheyer note similarities between the terrestrial tuatara and putatively aquatic *Palaeopleurosaurus*. Although in the latter there is some increase in bone mass and compactness, it is not so much more than seen in tuatara. And the femurs are basically the same. The idea is that if *Palaeopleurosaurus* was an obligate aquatic animal, its bones would be far more suited to that habitat, that is, they would diverge from tuatara in those regards. They go on: "Finally, we hypothesise that the life history of the aquatic *Palaeopleurosaurus* differs from that of terrestrial *Sphenodon* due to an irregular spacing of growth marks, the presence of non-annual rest lines and a possibly reduced lifespan in the extinct *Palaeopleurosaurus*, although this hypothesis needs further testing with more individuals" (Klein and Scheyer 2017, 7).

You can read an organism's growth, speed, and age by looking at a cross-section of bone, similarly to how you might read tree rings. It looks like *Palaeopleurosaurus* did not share tuatara's long lifespan. Klein and Scheyer further note that a proposed function for tuatara's strange ribs, that "the ventral portion of the thoracic rib with its partially overlapping, spade-shaped anterior and posterior expansions was supposed to form, together with the gastralia, a protective 'armor' for the viscera" (Klein and Scheyer 2017, 6), is undermined by the comparison with *Palaeopleurosaurus*.

So, although Klein and Scheyer deny that we should think of tuatara as living fossils (they call the term "erroneous"), their being a phylogenetic living fossil in Turner's sense is crucial for their analysis. Their study drew on three perspectives. First, taking *Palaeopleurosaurus* as a secondarily aquatic vertebrate, which enabled some general ideas about what aquatically adapted bones should be like. Second, careful examinations of bone histology in *Palaeopleurosaurus* fossils. Third, taking their target as Rhynchocephalia, that is, a member of the same ancient clade that tuatara hail from, enabled them to use tuatara as a relevant point of comparison. Tuatara's being terrestrial is here a boon, as it provides a point of contrast for exploring just how aquatically adapted *Palaeopleurosaurus* is. Note how this goes beyond Turner's suggestion of projecting tuatara-properties into the past. Instead, the tuatara acts as a crucial comparator for studying *Palaeopleurosaurus*: the two are grafted into the same phylogenetic story and thus become mutually relevant for understanding one another.

Also, note the coupled nature of the relationship between fossil analysis and comparative work with tuatara: the two are mutually informing. And indeed, we even learn something more about the tuatara themselves. The investigation demonstrates how palaeontologists bring together now and then in mutually-informing ways. It further demonstrates how flexibly adopting multiple perspectives and navigating between them can allow progress on long-dead, largely lost animals.

So, where does this leave the notion of "living fossil"? Scientists are right that we shouldn't literally treat such lineages as still-living fossil denizens. "Living fossils" retain some properties, yes, but evolution ensures they differ in many ways from their ancestors. But the way they differ from their forebearers can provide crucial information for uncovering those forebearers. Things usually identified as living fossils are certainly not special or unique in this regard: sharks are in no way rare enough to count as living fossils, but nonetheless, their morphological stability certainly underwrites much work in understanding extinct sharks.

Consideration of living fossils like tuatara demonstrates at least four things about the science of lost worlds. First, and I hope you're getting bored of me making points like this, whether or not some lineage is a living fossil depends on how you characterize it: what traits are you interested in? Second, living fossils bring the importance of loss into stark relief. If tuatara didn't exist, we wouldn't know some of the things we know about *Palaeopleurosaurus*. Third, it extends Turner, Watkins, and Love and Lidgard's points about the epistemic value of living fossils: they stand as examples of survival against loss with the potential to tell us a lot about the past. And fourth, their epistemic value lies partly in their capacity to act as comparators—members of the same lineage—as targets in the past.

5

Anchoring

Prof. Ichthy's cadence amps up toward their lecture's climax as the reconstruction of Broadfoot is projected behind them once more.

"Although Broadfoot is unique—lost—it also contains, and is, multitudes. It is a fossil specimen, and thus can be understood as we understand fossils and the processes that form fossils. It is a mammal, and thus may be compared and contrasted with its mammalian allies, both its rather unimpressive extant relatives and its fellow travellers from the Age of Mammals. But its mammalian alignment doesn't exhaust its phylogenetic properties. It is also a tetrapod, thus partaking in ways of life common across mammals and ourselves. It is also a vertebrate, a bilateral metazoan—an animal—a member of the great ancestrally-spun web of life. But Broadfoot isn't only its ancestry. It is also a predator, likely sometimes prey, it is a secondarily aquatic vertebrate, a venomous organism, an egg-layer, a user of electrolocation. It is a creature built of traits, cells, genes; biochemical, physiological, and immunological processes. And yet more abstractly, it is an anatomical, metabolic—even geometric—object. It is a set of ontogenetic, ecological, and evolutionary processes. Each perspective we take on Broadfoot brings lessons about it. They give us, as it were, a 'place to stand', and each of these platforms gives us information, information which then informs us about other platforms, other 'places to stand'. If I might switch metaphors . . ."

Ichthy pauses, as if expecting some objection from the audience,

". . . these different perspectives on Broadfoot provide different 'anchors' that, employed with diffidence and care, ensure safe passage on our journey to understanding this ancient, fascinating creature. Many anchors together provide safe harbour for our knowledge. *That*, I think, is the secret to our success."

The chapter has two central goals. First, to introduce our third strategy—"anchoring"—and second, to show how iterative investigation across artificing, grafting, and anchoring can reveal past worlds despite lost denizens.

Chapters 3 and 4 introduced their namesake strategies with sidesteps into non-Ediacaran examples. To introduce anchoring, by contrast, I'll return to an Ediacaran case: Gold et al.'s arguments concerning *Dickinsonia's* growth. They proposed a synapomorphy aligning that lineage with bilateral animals. I noted that a crucial step in their argument involved conceiving of *Dickinsonia* as a eukaryote that grows via *serial repetition*. Many eukaryotes grow this way: "serial-growing eukaryote" is a diverse historical kind. By conceiving of *Dickinsonia* as such, we can compare and differentiate it across various serially repeating developmental structures in nature. This is the central idea behind *anchoring*: by conceiving of the denizen as a *historical* or an *a historical* kind, we bring it into conversation with a set of other instances and examples—which I'll call *anchors*—allowing us to situate the denizen within the class, thus potentially establishing some of its properties, as well as potentially informing us about the class itself. The set of anchors may just include a single analogue, but, as we'll see, the use of multiple anchors is particularly powerful due to providing more opportunities for iteration. As a warm-up, let's briefly look at how Gold et al. (2015) anchored their analysis of *Dickinsonia*.

Serially repeated bodies are built from repeated, more or less identical segments. Perhaps the most familiar way of conceptualizing serial repetition is to start with segmentation, as seen in annelid worms, centipedes and millipedes, and so on. Each segment (or "annulus") of an annelid worm, for instance, is composed of the same parts of the various systems (e.g., digestive, excretory, metabolic, etc.) required for the worm to function (only the head and rear are not built of annuli). Thus, to increase length during development, the worm needs only to increase the number of segments. Beyond segmented invertebrates, plants are another familiar serial repeater. They sequentially add repeated parts—leaves—around a stem. In plants, we have two modes of growth: the stem does not grow serially (it gets longer) while the leaves are serial, added around the stem as modular components. Examples of serial repetition of various types are found in every living group that has evolved complex multicellular bodies except for fungi. Gold et al. represent some of these in Figure 5.1.

Gold et al.'s strategy, then, begins by characterizing *Dickinsonia* as a serially-growing organism. This enables them to compare *Dickinsonia* with

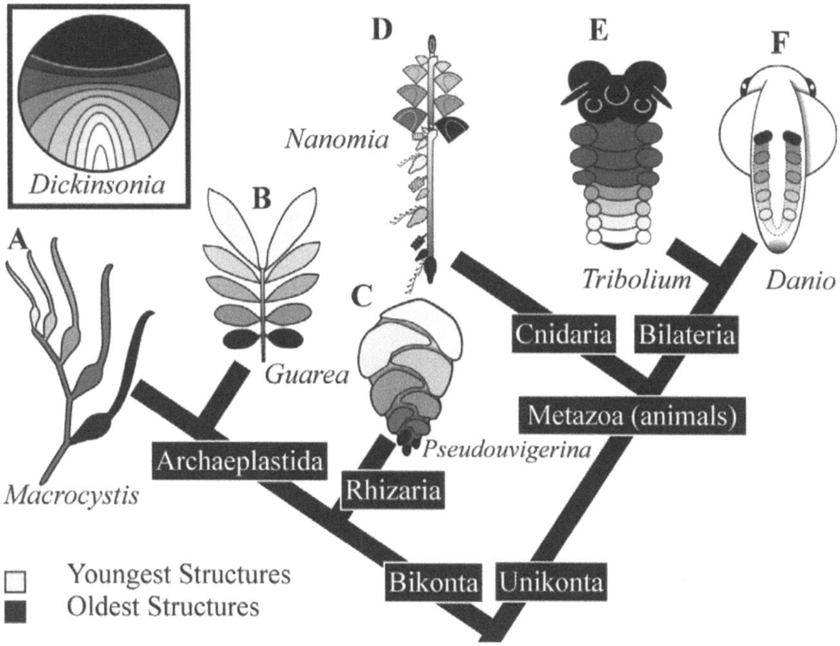

FIGURE 5.1 Phylogenetic Distribution of Some Serially Growing Lineages (Fig. 2, Gold et al. 2015). Reprinted with Permission from the *Royal Society B*

various other serial growers, thus navigating among other instances of that developmental phenomenon. These other instances are the *anchors*. Thus, various taxa that share in the historical kind "serially-growing organisms"—trilobites, seaweed, plants, and worms—each act as points of empirical contact for understanding that kind of development generally and for understanding *Dickinsonia* in particular.

For instance, Gold et al. point out differences between the bilaterian mode of growth and growth patterns in plants: "In bilaterians, the development of serially repeated structures often occurs in an anterior to posterior gradient, concurrent with posterior elongation of the primary axis" (Gold et al. 2015, 317). Because *Dickinsonia's* serial growth more closely resembles bilaterians, this speaks in favour of that alignment. They further point out affinities between *Dickinsonia* and what is found in complex metazoans. This is partially an example of grafting: Gold et al. link serial repeaters in clades to an argument about *Dickinsonia*'s phylogenetic placement. But grafting alone doesn't exhaust the strategy. Gold et al. are also *anchoring*, that is, characterizing their subject such that we can navigate between multiple systems (past and present) to build an integrated, comparative set of models concerning the relevant lost properties. Note that Figure 5.1 is another example of a mixed perspective, representing both phylogenetic information and historical kind membership ("serial growers"). These different examples of serial growers give Gold et al. (to crib from Prof. Ichthy) a series of "places to stand" when understanding and contextualizing *Dickinsonia's* growth and phylogeny.

So, in this chapter, I'll provide some examples and analyses of anchoring and introduce the role of iteration. I'll then turn to another Ediacaran mystery to tie together the last three chapters into an account of how lost worlds are found: specifically, through a combination of strategic perspectivalism and iteration. I'll close with a brief discussion of the relationship between loss and scientific laws.

As with common causes in chapter 4, there is a clear relationship between anchoring and analogue reasoning, which philosophers have discussed fairly extensively (e.g., Bartha 2022). This discussion has mostly involved identifying various possible schematic justifications of analogy as an argument form (although see Hesse's classic discussion of homology and analogy in biology; Hesse 1966). As I'm here interested in how anchoring works together with other strategies (namely, artificing and grafting), these schematic discussions are less useful. Paul Bartha has recently challenged practice-oriented analyses of analogous reasoning. As he puts it, "To advance the methodological debate, practice-based approaches must either make connections to these general models or explain why the lack of any such connection is unproblematic" (Bartha 2022). On my view, schematic views of analogy should happily fit the examples of anchoring I'll discuss, however (1) their schematic nature makes them inappropriate for bringing out how anchoring works in concert with other strategies, and (2) I'm not interested in the justification of token analogous arguments, but in the overall justification of the strategy in which anchoring plays a part. As such, I suspect I meet Bartha's challenge.

1. Ediacaran Ecology and Its Anchors

Recall Mitchell et al.'s discussion of the neutral-dominated spatial properties of Avalonian ecosystems. A neutral pattern was surprising because modern ecosystems, benthic or otherwise, are generally niche-dominated—that is, what kind of organism you are, or how you make a living, makes a difference to your spatial location. In light of this, Mitchell et al. raised a worrying idea. In the Avalonian world, did ecology just not work the same way as later times? As they put it: "This stark difference raises the question of whether Ediacaran early animal palaeocommunities had fundamentally different community dynamics to those of the present day" (Mitchell et al. 2019, 2034). The thought is as follows: in the oxygen-poor, sessile landscape of the early Ediacaran, the usual models that ecologists use just don't gain purchase. In the Avalonian, the laws of ecology are broken.

How should we approach this hypothesis? By anchoring. To do this, Mitchell et al. consider possible anchors and ways of characterizing Avalonian ecologies to make them relevant to one another.

Thus far, there are only a few single spatial studies of fossilized palaeocommunities. Jackson and Blois (2015) review examinations of Quaternary communities (a period spanning the last 2.58 million years to the present). Their analysis focuses on patterns in forests at the scale of thousands of years. Jackson and Blois emphasize the unstable nature of these communities, illustrating "how an ecological community is one small, ephemeral point in a roiling, dynamic unfolding of environmental change, distribution dynamics, and spatially aggregated ecological processes" (Jackson and Blois 2015, 4917). Their results emphasize the role of environmental change in driving niche differentiation. As in modern cases, these (comparatively recent) environments behaved very differently from the Avalonian.

Mitchell et al.'s strategy is to consider features of the Avalonian Ediacaran that might be responsible for a neutrally-driven ecosystem. This involves characterizing the ecosystem in terms of features that potentially could be associated with neutrality. These features can then be sought in other ecosystems. By navigating between these systems, they can test those ideas similarly to how Gold et al. did with serial repetition. Let's see this in action.

Mitchell et al. identify four features. First, populations in Avalonian ecosystems are small. Second, they are subject to regular disturbances, meaning that assemblies typically don't survive longer than three generations. Third, dispersal appears to occur over short ranges. Fourth, the resource constraints of the Avalonian ecosystem are minimal, at least compared to most modern ecosystems. Anchoring often involves building a model through which comparisons and contrasts between empirical cases are woven together. As such, Mitchell et al. use modern and Quaternary ecosystems that have some of those properties as a set of anchors that inform an artificed model.

Mitchell et al. draw on various resources to suggest that the features identified above should lead to neutrally dominated spatial ecologies. For instance, Fisher and Mehta (2014) modelled neutral and niche differentiation as a kind of phase transition, a framework that Mitchell et al. employ as artificing. They conceived of ecological systems as shifting from neutral to niche-dominated patterning in a fashion analogous to ice transitioning to liquid water. Using various simulations, they identify properties that will lead a community between these modes of ecological assemblage. As Mitchell et al. summarize: "Recent simulations show that community dynamics in small populations living in fluctuating environments are dominated by neutral processes, implying a lack of small-spatial scale environmental control on ecological dynamics in such systems" (Mitchell et al. 2019, 2035).

The suggestion is that small populations in insufficiently stable environments are unable to differentiate by niche because the environment keeps shifting, and the population isn't big enough for competition between organisms to have much effect. A lack of dispersal also limits competition because distant dispersal increases effective community size and thus increases opportunities for competition. Similarly, a lack of resource constraint limits competition: with plenty to go around, there is not much to compete over. Crucially, the ecological models are built in conversation with various contemporary ecosystems that exhibit some of these properties.

In effect, Mitchell et al.'s explanation is quite simple: an unstable environment means niches aren't around long enough for communities to adapt to them, and competition between organisms can't get going because various factors—e.g., community size and brief assemblage survival—dampen that.

So, do Avalonian ecosystems play by different rules? In a certain sense, yes, but not in a way that threatens the generality of basic ecological dynamics: "While the dominance of neutral processes within these paleocommunities differs substantively from the majority of the modern marine realm, the underlying dynamics are entirely consistent with models of assembly which include both niche and neutral processes, and are similar to those of modern communities subject to the same conditions" (Mitchell et al. 2019, 2035).

So, the Avalonian ecosystems are weird and, in some ways, unique, but not weird and unique in ways that belie ecological understanding: they, in fact, behave as we'd expect given the particular Ediacaran conditions preserved at Mistaken Point. In other words, they are lost, but we understand how and why.

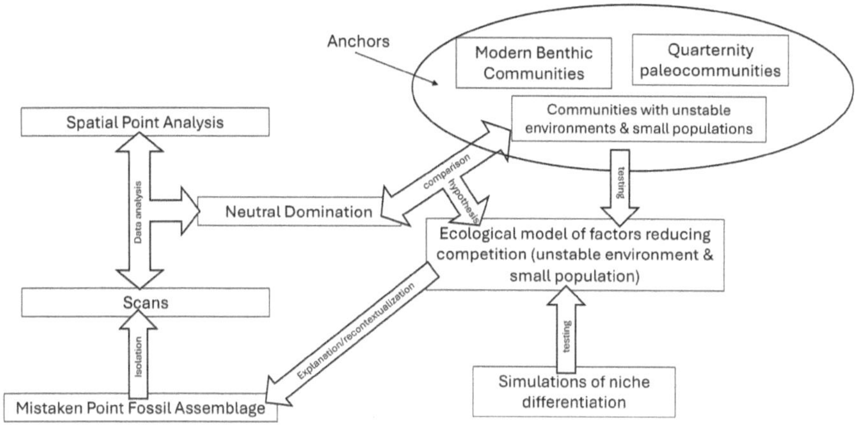

FIGURE 5.2 Approximation of Mitchell et al.'s Reasoning Process, Showing the Anchors in the Top-Right Corner

Let's summarize Mitchell et al.'s investigation to highlight anchoring. Mitchell et al. start with a series of fossils at Mistaken Point. They use scanning technologies (laser-line probing) and subsequent data analysis to generate a spatial and taxonomic map. Various taphonomic theories and studies of local context are then brought to bear to argue that the fossil assemblage can be read as something like a "snapshot" of the in situ community. This enables the application of spatial point analysis, which generates the surprising neutral result. We could understand the steps up to this stage as involving the characterization and decontextualizing of the Mistaken Point assemblage, considering only taxic and spatial distribution. This is carried out via various perspectival tools: laser-line probing, systems of data analysis, and so on.

With the neutral result in hand, Mitchell et al. reach for a set of anchors. They contrast Avalonian palaeocommunities with modern and Quaternary communities to identify factors that could explain Avalonian differences. These factors are investigated in various ways: for instance, observationally, as well as with various modelling techniques. Such techniques, together with further details inferred from Mistaken Point (for instance, confirming the short life of communities), enable an explanation of neutral niche differentiation. Figure 5.2 is a kind of messy and approximate sketch of the reasoning process.

Mitchell et al.'s strategy is a classic example of anchoring. In the face of a putatively lost world, they abstract a set of properties and look across many examples to compare and contrast to generate a set of explanatory factors that are then re-applied to the (recontextualized) case at hand.

At base, anchoring involves conceiving of your target as a token of a kind such that it can be brought into empirical and theoretical dialogue with other tokens of that kind. Mitchell et al. use perspectival tools, such as the assemblage itself, laser-line scanning, and spatial point analysis, to isolate the spatial ecological properties of the Mistaken Point assemblage. This isolation enables comparison with other ecologies in those terms. Let's characterize anchoring abstractly:

> *Anchoring* involves characterizing a denizen in terms of some set of historical/ahistorical kinds, identifying other instances of those kinds (anchors), and using these anchors to learn about the relevant kind.

Anchoring can be distinguished from grafting, that is, conceiving of your target as part of a historical individual or line of causal continuity. The difference is that in grafting, we "narrativize" denizens by linking them together as a token causal sequence, while in anchoring, we do not assume causal continuity; anchors are understood as tokens of types, not individuals. For instance, Kolesnikov et al. (2017) examine structures in modern biofilms to suggest new interpretations of fossils of Avalonian microbial mats. This doesn't require taking Avalonian mats and modern biofilms as part of the same lineage, As they put it: "It is not that a specific biological community has survived since the Ediacaran—it is that the biological response of microbial communities that manifested itself quite commonly in certain terminal Ediacaran and early Cambrian environments can still be found (seemingly in much more restricted settings) today" (Kolesnikov et al. 2017, 1).

Kolesnikov et al.'s investigation is anchored both by the fossils and by observations of modern phenomena that display relevantly similar properties. Thus, they take them as historical kinds (microbes responding to particular conditions), not historical individuals.

We can also approximately distinguish anchoring from artificing. Insofar as Kolesnikov et al. don't literally construct their analogues, they anchor rather than artifice. Similarly, Mitchell et al.'s identification of the relationship between, say, small populations and neutral domination doesn't rely on

Table 5.1. Strategies, Definitions

Strategy	Definition	Exemplar Perspectival Tool
Artificing	Denizen conceived of as an historical/ahistorical type, a token of that type is constructed and examined.	Geometric models of ontogeny.
Grafting	Denizen conceived of as part of an historical individual or part of a token historical trajectory.	Cladistic Phylogeny.
Anchoring	Denizen conceived of as an historical/ahistorical type, a set of different instances of that type are examined.	Identification of properties associated with spatial neutrality.

instances being causally continuous or being constructed. They examine different naturally-occurring instances rather than building them. As anchoring involves using multiple instances of a kind to inform us about that kind, it often goes hand in hand with artificing, where we might further explore our understanding of the kind by building a model of it. For example, when Woodruff and Varricchio (2011) artificed a model dinosaur burrow, they were partly influenced by examinations of mammal burrows, and indeed used burrowing mammal bones in the experiment—that is, various anchors. Table 5.1 summarizes the strategies, their definitions, and a paradigmatic perspectival tool for each.

Anchoring makes sense of and unifies several more specific proposals philosophers have made concerning historical reconstruction, so I'll point to several of these before turning to iteration.

In *Rock Bone and Ruin*, I describe a form of anchoring, which I call "exquisite corpse" modelling: in a nutshell, a single, difficult target is imperfectly modelled with multiple analogues. Perhaps if we squint just right, we could view Mitchell's investigation in these terms. Avalonian communities are understood as being a combination of various aspects of modern ecosystems: those that are small, that are frequently disrupted, and so on. Joe Wilson (2023) builds on this idea to argue that palaeoclimatology can provide partial analogues of modern and future climate systems (see also Watkins 2024c).

Meghan Page's demonstration that sometimes, as she puts it, "the past is the key to the present" (Page 2021; Dresow 2023) also involves anchoring. She describes scenarios where some regularity has been confirmed under present conditions, but we are nonetheless interested in the regularity's fragility. That is, is the regularity likely to continue to hold under different conditions? What is the range of the regularity's applicability? As she points out, by testing the regularity against conditions in historical records, we are able to further understand the circumstances under which it fails and under which it holds. By the past-present interactionist view I'm pushing here, we should expect the circumstances described by her account to be common. Mitchell et al. demonstrate that our ecological models of spatial distribution (perhaps a little enriched) can extend into the Avalonian world. They thus teach us something about the robustness of ecological models and the relevant regularities, much in the way that Page describes.

Finally, Aja Watkins examines how various scaling strategies—basically transforming data from imperfect analogues and data generated from the target—are required to gain evidential relevance in many historical investigations (Watkins 2023). She is particularly interested in how this occurs in palaeoclimatological contexts, where differences in scale between events occurring millions of years ago and those currently occurring (or soon occurring) threaten to undermine using *then* to understand *now*. In the terminology of this book, the ways data from analogues and targets might be scaled is a specific kind of characterization that enables comparison, particularly in anchoring.

What makes these accounts examples of anchoring is that all involve characterizing a denizen such that it falls under a kind that includes a bunch of other targets. This differs from grafting insofar as grafting involves unifying the target via a token causal trajectory—making them part of the same narrative, if you want—while anchoring involves taking targets as more or less independent instances of the same kind. It differs from artificing insofar as natural analogues are not constructed. The difference between artificing and anchoring is messy, depending on how much "construction" and modelling is required to connect the analogue and target. But I think the distinction is useful because the practices they highlight—building an instance of the relevant properties as opposed to navigating between various naturally occurring exemplars—differ sufficiently.

These strategies of characterization—strategic perspectivalism—enable iterative, mutually-informing investigations across perspectives. Under the right conditions, this can reveal past worlds despite loss. It is time to see how.

2. Epistemic Iteration

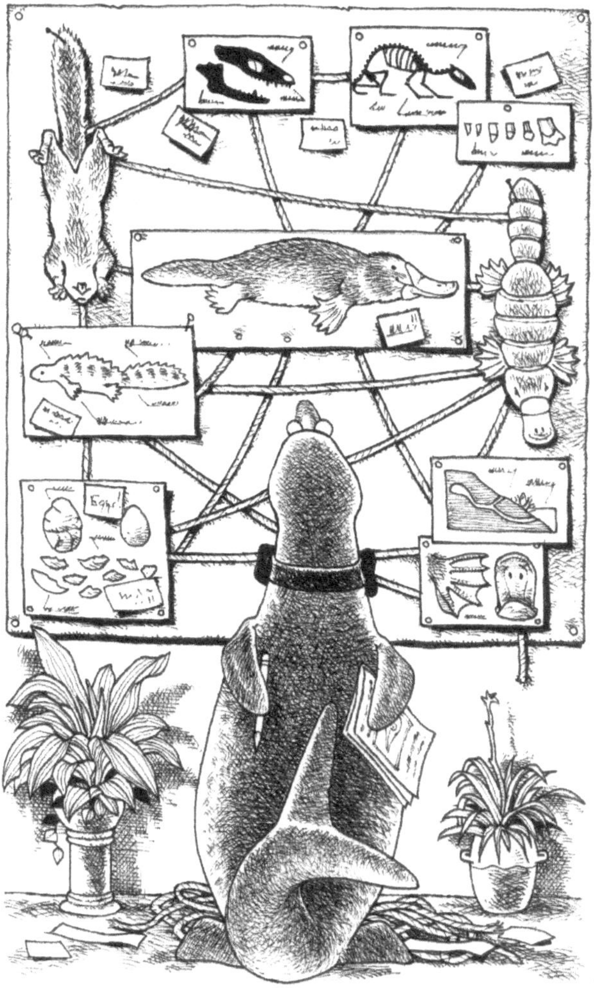

ILLUSTRATION 5.1 So Many Perspectives, So Many Anchors

> "With multiple platforms, or multiple anchors," Prof. Ichthy pauses as if struggling under the weight of juggling so many metaphors, "we can jump between perspectives, gaining a little insight here, discarding a once-promising idea there, all the while generating a richer picture of our target. Starting with a rather simple, likely wrong-headed conception of Broadfoot, we navigate between various ways of testing and probing that conception, and these are refined, made richer, and made more plausible. This jumping between perspectives, I say, is a self-correcting process, taking us from bad ideas to good ideas. Thus, iteration between perspectives is how we've discovered so much about Broadfoot despite their uniqueness, despite their loss."

What is advantageous about having multiple anchors? For Mitchell et al., what is critical about accessing a set of contemporary benthic ecosystems, the Quaternary system, the fossils themselves, and various simulations? One immediate answer—and certainly not a false one—appeals to induction. Multiple anchors mean more evidence, and thus a richer data set to draw on in support of claims. But I don't think a simple increase in enumerative inductive power captures the epistemology of anchoring specifically, or strategic perspectivalism in general. For this, we need to understand iteration.

Throughout the last three chapters, we have repeatedly seen a basic structure to applications of strategic perspectivalism, first laid out in chapter 3. First, decontextualize: characterize a set of properties using perspectival tools. As Helen Longino has discussed in the context of the study of behaviour, decontextualization is often required in scientific inquiry, since isolating properties renders them measurable and—crucially for our purposes here—comparable across contexts (Longino 2021).

Second, use these isolated properties to apply various investigative strategies. We might artifice, building something analogous to those properties; graft, linking them to a historical individual; or anchor, navigating between different targets that share those properties or contrast with them.

Third, recontextualize: return to the past world and attempt to make sense of the results.

This strategy is fruitfully understood via what philosophers have called *epistemic iteration*. Drawing inspiration from practices in mathematics, where iteration involves an approximate proof acting as a scaffold or springboard for a less approximate proof, Hasok Chang (2004) argues that similar processes occur in the development of measurements, in his case, temperature. He sees epistemic iteration as "a process in which we throw very imperfect ingredients together and manufacture something just a bit less imperfect" (Chang 2004, 226). As Kevin Elliot points out, such ideas are fairly common across the history and philosophy of science (Elliot 2012; see also work on scaffolding: Chapman and Wylie 2018; Currie 2018a; Walsh 2019; and the papers collected in Caporael, Griesemer, and Wimsatt 2014), and Daniel Swaim has suggested that the iterativity of speculative narratives has important epistemic functions in the historical sciences (Swaim 2024).

I want to emphasize how iterative practices interact with strategic perspectivalism to generate knowledge of lost worlds. I don't think what I have to say here clashes with these previous discussions, but I do think that the highly localized and particular ways in which iteration plays out in the examples that interest me differ from what occurs for more standardized products, such as measurement concepts and practices, that have energized folks like Chang. This picks up a major theme of my earlier work: that historical science is successful in virtue of *methodological omnivory* (Currie 2015, 2018a). As opposed to following a particular method (e.g., Cleland 2002; Kleinhans et al. 2005), historical scientists throw whatever methods they can against the wall, hoping to see what will stick. Methodological omnivory isn't simply the idea of using multiple threads of evidence, but actively and opportunistically generating often highly imperfect evidence in highly tailored, local ways. I think the combination of strategic perspectivalism and iteration, as I've described it, partly underwrites why methodological omnivory works.

Having multiple anchors is good not only because they provide more data for stronger inductions. Multiple anchors are useful because they provide multiple points for progressive iteration. Consider again Mitchell et al.'s work. The reconstruction and analysis of Ediacaran ecology relies on strategic perspectivalism—characterizing the fossils as recording *in situ* benthic spatial ecology—which enables the application of various anchors (the fossils themselves, simulations, other ecosystems, and so on, as approximated in Figure 5.2). But the relationship between Mitchell et al.'s anchors is iterative in the sense that knowledge generated from one investigation can motivate and be transported into another context, and thus be transformed and transported back to the original context. From spatial point analysis, we learn that the ecosystem is neutrally dominated. Based on a comparison with other ecosystems, we learn that this is unusual. This raises a question: what explains the exceptionalism of Avalonian ecosystems? This then leads Mitchell et al. to navigate between the Avalonian and current ecosystems, thus identifying various properties that appear to correlate with neutral dominance, such

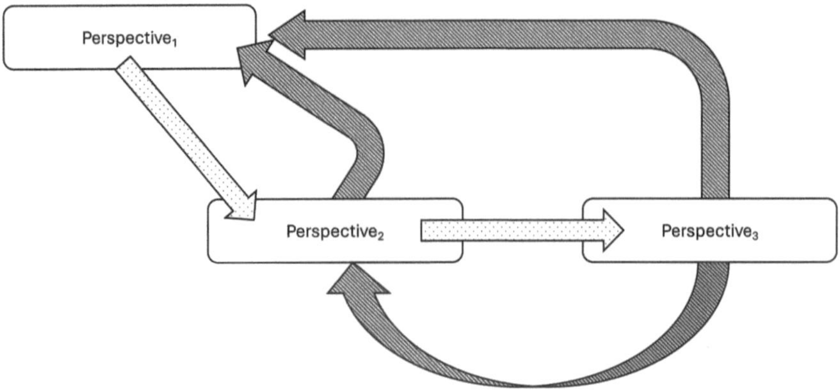

FIGURE 5.3 Possible Lines of Iteration between Three Perspectives; Light Arrows Feed Forward Changes, Dark Arrows Feed Backward

as small populations. These hypotheses then lead to various tests—appeals to simulations, for instance—before finally being brought together in a recontexualized explanation of the Avalonian Ediacaran. Iteration goes beyond enumerative induction because what is learned from one perspective can, under the right conditions, be transported to another perspective and back again. This is due, I take it, to the lack of perspectival isolation as discussed in the last chapter. Under the right conditions, as we'll see, this brings epistemic benefits.

Iteration in this context, then, occurs when (1) some information is generated via some perspective (that is, a characterization and analysis with some perspectival tools), which (2) is then applied to another perspective, which results in (3) that information being transformed and/or (4) that perspective being transformed. This then leads to (5) that information, transformed, being returned to the original perspective and/or (6) the information being brought to a third perspective. Figure 5.3 captures the basic idea.

With iterativity on the table, the task now is to show how, in combination with strategic perspectivalism, loss and putative loss can be overcome. That is, how historical scientists discover lost worlds.

Recall a problem raised in our discussion of de-extinction: if we attempt to clone an extinct animal, how would we be sure that it, in fact, behaved and looked the way that its past world cousins did? We only know mammoths from their carcasses, their elephant relatives, and their genomes. So, if we had a cloned "mammoth," how would we be sure that elements of its phenotype (especially its behaviour) were relevantly similar to past mammoths and not due to our cloning process or other sources of noise?

This is a general problem when dealing with loss. The past could always have been different in surprising ways, which undermines uniformitarian assumptions. Historical scientists always labour under the possibility of loss: it could just be that dinosaur bones worked differently from rabbit bones, that the microbial mats of the Ediacaran affected decay differently from today's bacterial films under experimental conditions, and so on. We need to be very careful about how we articulate this challenge. Simply pointing out the possibility of loss and demanding a general answer, if we set the bar too high, more or less amounts to demanding an answer to the problem of induction. That's above my paygrade, seems to be a problem with any inductive process, let alone those involving lost worlds, and frankly, isn't something I find productive or enormously interesting (perhaps it's a little like a philosophical version of asking whether dinosaurs feel pain: if you're not careful, it doesn't lead to a very productive discussion). What is significant for our purposes is how iteration plays a critical role in identifying loss when it occurs and partially overcoming the challenges posed by loss and potential loss.

The strategies I've outlined all have potential failure conditions. In artificing, one might build something that isolates a property the target denizen lacks; in grafting, you might over-emphasize the continuity and similarity between denizen and putative members of the same historical trajectory (or misidentify a historical individual); in anchoring, the apparent similarities between instances of types might be just that, only apparent. Each strategy more or less involves taking a perspective—using tools to isolate some set of properties—which can go wrong in multiple ways. As with any inductive inference, attributions of loss can involve various kinds of errors. We might commit a false negative: attribute loss when there has not been any; we might commit a false positive: think there isn't loss when it has occurred.

I argue that iteration mitigates against these errors. How? I won't claim that iteration and strategic perspectivalism guarantee success—we're working with induction after all—but rather that, under the right conditions, iteration constitutes a self-correcting process that gives us reason to think the investigation at hand is either on the right track or, if we're barking up the wrong tree, this will be discovered.

Let's start by going abstract.

Take some characterization of a denizen, call it p, and some characterization of an anchor, call it p^*. That is, there is a perspectival tool such that we can conceive of p and p^* as being of the same kind. Thus, we've taken a perspective such that p and p^* are putatively of the same kind. How do we determine whether, in fact, $p = p^*$? That is, that p and p^* are, in fact, the same kind in the relevant way, and thus that loss hasn't occurred (at least pertaining to p)?

To answer such a worry, we might conduct some investigation of p^*, maybe model it, measure it, etc. And then see whether we get a similar result by performing an analogous investigation of p (or vice versa). Notice that whether we get a positive result or not, we still learn something important about p, at least in relation to p^*. If we get a positive result, then presumably we should be more confident that $p = p^*$. We might then turn to another anchor, measuring or modelling that. However, if we get a negative result, we will seek an explanation for why the results differed. This might involve adopting a different perspective on p or p^*, for instance, or shifting to another putative anchor. Maybe we would try to use the discrepancy between p and p^* to hypothesize a more general model that explains the differences, or we might hunt for further instances that also show that discrepancy. Failures, then, also act as potential scaffolds for further perspectives.

What we have, I hope, is a kind of "self-correcting" procedure. The bet is that if the isolation has gone wrong, then recontextualizing will backfire; if you've contextualized incorrectly, this will be revealed through iterative tests. There are no guarantees or perfect solutions here; chains of unlikely coincidences can happen, but we have good reason here to think that errors are at the very least minimized.

As we move forward, navigating between different perspectives, we identify false positives and negatives in loss-attribution, gain new knowledge about p, while also gaining new knowledge about p^*. As we saw in the last chapter's discussion of living fossils, our investigation of the past and our investigation of the present are coupled. Thus, I think, iteration both helps us discover and understand the properties of past worlds even if they are lost, in part by making sure that we've characterized them adequately.

You might complain about the abstractness of this argument. You shouldn't. Consider the relationship between Mitchell et al.'s strategy and all of those ps and p^*s. Mitchell et al.'s investigations more or less follow the abstract pattern. With the relevant changes, you could describe most of the cases we've examined thus far in these terms (I'll leave this as an exercise for zealous readers). The power of iterativity becomes clearer, I think, when we connect it explicitly with arguments I've made over the last few chapters about perspectives themselves and how they relate. I'll particularly emphasize the role of recontextualization.

Massimi urges us to imagine perspectives "as opening up a 'window on reality' that extends well beyond the boundaries of the representation itself and where the depth, angle, and scale of the representation leave enough room to make inferences about the space and what's in it" (Massimi 2022, 40). In Charlotte Kenchington's palaeoartistic reconstruction of the Ediacaran Avalonian (Figure 2.3), the vanishing point, created by a horizontal line and differently darkened or lightened rangeomorphs, provides a sense of space that reaches well beyond what is literally depicted in the image. Massimi wants to draw an analogy between that effect and scientific perspectives. My interpretation of this idea is fairly simple: a perspective on a target involves characterizing it as some kind, and this has further implications for that kind and other things that are included within it.

Recall the disagreement between Gold et al. (2015) and Hoekzema et al. (2017) from chapter 4. There, we saw that although both employed different perspectives (Gold et al. characterized rangeomorphs as exhibiting terminal-end growth within a phylogenetic bracket, while Hoekzema et al. characterized them in terms of a developmental sequence abstracted from a set of fossils), these were not independent from one another, as Hoekzema et al.'s approach involved empirically testing an assumption underwriting Gold et al.'s work. Further, both perspectives overlapped insofar as they took *Dickinsonia* to grow via serial repetition. Perhaps unlike perspectives in visual art, scientific perspectives can sometimes be superimposed: *Dickinsonia* is simultaneously both a phylogenetic animal and grows via serial repetition.

Such overlaps, links, and extensions from and between perspectives enable a widening scope of rich knowledge. They extend the range of alternatives available, and thus the investigative options and opportunities for developing knowledge. It turns out, so long as you're clever with perspective, problems of loss are often surmountable. Just don't let N equal zero.

Do we have good reason to think these iterative processes are in fact self-correcting, and that those corrections will lead to knowledge of lost worlds? I think—under the right conditions—we do, and to see this, I'll turn to discussion of iterativity in the measurement literature that follows from Hasok Chang's influential work (Chang 2004).

Here are two related challenges to developing a measurement of some quantity or phenomenon. First, generally, how do we validate a measurement and know when one is "correct"? If you wish to test your measurement, that is, check whether it gives the correct result, presumably, you need to already know what the correct result is, that is, already have a measurement. Many of the things we wish to measure are complex, only partially known to us, and often highly ambiguous prior to developing a measurement. So, what do we check our devices against, if not their own results (Chang 2004 calls this "the problem of nomic measurement")?

Second, how do you develop a measure for something when, at the outset of the investigation, you don't quite know what that thing is? Chang highlights a difference between iterative proofs in mathematics, where "the true function we are trying to approximate is already known or at least knowable" (Chang 2004, 216), and empirical measurement cases, where the motivating phenomenon is often highly unsystematic and unstable. Temperature, as we experience it, for instance, is highly variable, context-sensitive, and messy, witnessed by our various attempts to incorporate "wind chill," "dry heat," and other bids to make sense of the difference between temperature as measured by sensible Celsius (and I guess Fahrenheit if you must) and how we experience variation in phenomenal temperature.

Chang's answer to both challenges appeals to the iterative testing and tweaking of measurement across various, related apparatuses, thus generating "a process in which the successive stages of knowledge, each building on the previous one, are created in order to enhance the achievement of certain epistemic goals. For each stage, the later stage is based on the earlier stage, but cannot be deduced from it in any straightforward sense" (Chang 2004, 226; see also Tal 2016 and 2017). What is self-corrected in Chang's cases, then? Correction involves (1) the convergence of iterations toward a fixed point—similar in a sense to consilience of evidence—and (2) "agreement between the concrete image of the abstract concept and the actual operations that we adopt for an empirical engagement with the concept (including its measurement)" (Chang 2004, 216). That is, we have some idea of what we're trying to measure, and through mutual adjustment, this and our measurements come into agreement. Our experiences of temperature provide a starting point and reference point for the development of increasingly sophisticated apparatuses for measuring temperature and theorizing about it. This then leads us to identify useful fixed points, such as the boiling point of water, to build further apparatuses and measurements for.

What might license the results of such iterative processes? As Valde and Scarffe have recently put it, "Rather than resting on a non-social or objective foundation, their processes get their legitimacy from reflexive interactions that are, at least in part, governed by social norms" (2025).[1] Insofar as such processes are "objective," it is in some form of coherentist, socially-mediated way. But this needn't involve vicious circularity, nor does it involve losing empirical contact. To see why, let's use the term *agreement* to do a little schematization of iterative accounts of measurements.

We can distinguish between the *target*, the *operational constraints*, and the *criteria* of agreement. Chang is interested in measuring temperature; his target phenomenon is (say) differences in heat and cold. The operational constraints come from the apparatus used and the derivable data—the perspectival tools, if you will, and the theories of temperature developed. Chang's criteria of agreement are coherentist in nature, but coherence understood in a broad sense: it concerns not merely the relationship between propositions, but how the components of the system contribute to the success of that system's aims. So, what it is to *agree* for Chang is for the target and operational constraints to converge coherently. Other philosophers interested in measurement, looking at differing forms of measurement, have different accounts of agreement.

Alisa Bokulich, for instance, appeals to iterative coherence tests across radiometric data, arguing that these increase accuracy. The points of these tests are not "to arrive at a specific age for the event but rather to pinpoint potential sources of error in the initial methods and assess the magnitude of their effects" (Bokulich 2020, 433). Here, an iterative process in the development of a measurement leads to its increased accuracy in measuring age. So, the target is geological time, the operational constraints involve the apparatus and theory of carbon decay and other radiometric phenomena, and the criterion is increased accuracy. Here, increased accuracy concerns the ability to identify and control for error.

A third example is Sharon Crasnow's (2021) discussion of political-scientific measures of democracy. These measures, at base, take a bunch of measurable proxies as an indication of how 'democratic" a country is. She develops a form of "objectivity" that holds when a measure "is subject to theoretical, empirical and pragmatic constraints" (Crasnow 2021,1209). She particularly emphasizes how the normative aims of the measure interact with empirical results in an iterative manner: "While conceptualizing democracies can be

1 Valde and Scarffe (2025) draw a convincing analogy between the socially mediated coherence of iteration in science and in law.

a starting point for developing measures, concepts may alter in response to changes in regimes around the world, the results of empirical research on the causes and effects of democracy, debates about the values democracies exemplify, the effects on policies of different ways of conceiving of and measuring democracy, and on the goals of research" (Crasnow 2021, 1224).

So, here agreement occurs across multiple elements: the target (democracy and its proxies), the operational constraints, the means of generating data about those proxies, the practical effects of those decisions on policy (depending on how we carve things up), and a set of criteria involving values and goals. Crasnow argues that agreement here (she uses the language of objectivity) occurs when these come together to succeed across these various criteria.

I think the message we should take from the work on iteration in measurement is clear: in the good case, we can argue that an iterative procedure is self-correcting when we can point to a set of constraints emerging from the target, operational constraints, and criteria, which plausibly lead to the relevant kind of agreement.[2] Let's now turn from measurement to consider lost worlds.

Attempts to measure temperature had to grapple with the unsystematic and ambiguous initial phenomena of temperature as we experience it. Historical scientists tackling lost worlds face a similar challenge: is the past world in fact lost and, if so, what properties in particular are lost? When delivering their initial neutral result, Mitchell et al. were unsure whether this could be understood via our usual ecological models, or whether new models needed to be developed specific to the Ediacaran. That is, is the Avalonian Ediacaran lost insofar as it is an unusual edge-case of usual ecological dynamics that we don't see today, or is it lost due to being a completely different kind of ecosystem, demanding a new set of dynamics? To resolve this ambiguity, a complex investigation was required, which involved iterating between various anchors, models, and what we know of the Avalonian systems themselves. Agreement, here, was between these various resources and, I think profoundly, *toward* the past world.

So, the agreement we want is toward a past world, and thus knowledge of what is lost within it. So, what resources do we get from strategic

2 If you find this kind of coherentism distasteful, a view like Khalifa and Goldberg's (2022) might be more attractive to you. Although in section 4, I'll provide some reason to doubt that the epistemic phenomena that interest them are common or central in the science of lost worlds.

perspectivalism? I'll suggest that the role of recontextualization, and particularly that grounded in knowledge derived from traces, puts historical scientists in a relatively strong position regarding whether their processes are self-correcting toward how the past world in fact was.

In the cases I'm interested in, we're not developing a measurement (although, as mentioned in chapter 1, there is a lot for philosophers to think about concerning measuring the deep past!) but aiming to empirically discover a past world. Here, the agreement is not between an "abstract object" ("temperature" in Chang's work), various phenomena related to that object, and various apparatuses and procedures of measurement. Instead, it is between things like fossils and their interpretations, models of general dynamics, currently existing analogues, and so forth. Here, recontextualizing plays a critical role. It is a grounding in traces that plays a large role in ensuring self-correction throughout the procedure.

Derek and Michelle Turner have recently highlighted the role of "contextualized research" to explain the difference between "ancient-alien" theories and legitimate archaeological theories of past human structures and cultures (Turner and Turner 2021). Ancient-alien theories typically hold that some achievements of our ancestors couldn't have been accomplished by such "primitive" folks unaided, instead positing that some extraterrestrial intelligence was to blame. They often involve highly fine-grained analyses of particular artifacts (for instance, von Däniken's interpretation of a Mayan fresco as depicting an astronaut), combined with a form of analogous reasoning that bears a superficial similarity to anchoring. For instance, ancient alien theorists will point to multiple examples across the globe and history to underwrite their claims. However, this similarity between anchoring and conspiratorial analogous reasoning is indeed only superficial, and this is due to the role of recontextualization. Turner and Turner's discussion is worth quoting at length:

> When archaeologists are investigating a particular question about a particular site, they do so in a rich informational context that includes established investigative practices and lots of relevant background knowledge, as well as multiple lines of evidence. This is a way of leveraging previous epistemic successes. Understanding the imagery on a Mayan ruler's tomb does not involve just looking at it, as von Däniken did with Lord Pakal's sarcophagus lid before declaring it to be an

> image of an ancient astronaut. Instead it requires understanding context such as the identity of the dead king and his place in Mayan history as revealed by past research on Mayan texts. Understanding how Mayans adorned their bodies might be relevant to understanding the object near Pakal's nose. Also highly relevant are Mayan cosmology, symbolism and artistic conventions as understood from texts, inscriptions, thousands of objects of art, and architecture at many other sites. One might want to consider the social and political nature of the city where Pakal ruled, as understood from texts and excavations, as well as the layout and social meanings of the pyramid in which the king was buried. Studies of the bones found inside the tomb and the tools and materials used to make the sarcophagus might also be relevant . . . When archaeologists seek to interpret something like Lord Pakal's sarcophagus lid, they rely heavily on a supporting structure of previous research, even if very little of that previous work actually focused on the artifact in question [references removed]. (Turner and Turner 2021, 20)

We've seen similar recontextualization in the palaeobiological examples that have energized this book. Analyses of *Dickinsonia*'s development are recontextualized into later Ediacaran environments as well as phylogenetic hypotheses; trackways putatively demonstrating limping dinosaurs are contextualized with dinosaur physiology and possible environmental variation; Mitchell et al.'s neutral ecological result is made sense of by considering further features of Avalonian assemblages: their small size, short duration, and so on. This shows the crucial role that records—traces—play in overcoming challenges from loss. Downstream information that hasn't been erased provides a crucial lynchpin tying analyses built from artificed experiments and models, and various anchors, to the actual past world we're interested in. There is continual empirical contact—the world "pushes back"—and for historical scientists the agreement is to a past world revealed through traces.

This emphasis on iteration brings together, and I think makes sense of, a bunch of philosophical claims about the nature of historical inference. Both Chapman and Wylie (2018) and Thomas Bonnin (2019) have used Toulmin schemas to understand palaeoscientific arguments. These schemas are extremely helpful, eschewing hypothetico-deductive or probabilistic reconstructions of arguments in favour of various claims with various relata such as "warrants," "rebuttals," etc. I take their work to be a description of the kind of iterativity I've tried to highlight here. It also makes sense of the claims Kim Sterelny and I have made about the epistemic benefits of narratives (Currie and Sterelny 2017; Currie 2017a). We argue that one advantage narratives have over more abstract theoretical explanations of historical episodes is that the narrative, stretched over time as it is, has more opportunities for empirical testing. This can be understood as a claim about grafting: taking an episode as part of a token trajectory enables multiple locations for iterative testing.

Another of Chapman and Wylie's claims about archaeology is that it lacks epistemic foundations. As they say of archaeological data: "Neither these data nor the evidential claims based on them constitute a self-warranting empirical foundation, and yet they can powerfully challenge and constrain the reconstructive and explanatory claims we project onto the cultured past" (Chapman and Wylie 2018, 6).

On this view, the reason that archaeological interpretation is not (always) beholden to archaeological presuppositions is because such presuppositions do not play unquestionable, foundational roles. Rather—to put the point in the terms I've developed here—they consist of sets of perspectival tools that, when judiciously applied, generate various perspectives and provide the basis of iterating studies (we'll return to this thought in section 4).

Thus, iteration and strategic perspectivalism allow historical scientists to uncover lost worlds.

3. Bringing It All Together: How the Metazoan World Arose

"Finally, I want to ask: what sealed Broadfoot's fate?" The next transparency shows a series of black lines stretched across a horizontal axis, each ending with the outlined representation of some lineage. Most lines associated with mammals, including Broadfoot, reach about halfway across, while many graced with reptiles stretch across to the right of the image, where a red line indicates 'present day'.

"What pushed these animals aside and made space for their replacement? That is, what heralded the so-called 'Age of Reptiles'?" Prof. Ichthy indicates everyone around them with a sweep of a flipper. "This is a difficult question. Understanding a lost world is one thing; understanding how it became lost is another entirely. Happily, understanding both lost things and how they became lost are not independent projects. That is, examining Broadfoot and their relatives, and trying to understand how their world fell, and ours arose, are complementary scientific endeavours.

"The iterativity of knowledge just keeps paying dividends, it seems."

Let's briefly characterize the conceptual resources I've been collecting thus far (take a deep breath!).

I've introduced the notion of a *past world*, a subsection of actuality indexed to some set of *denizens*, that is, the entities, processes, properties, relations, and dynamics that populate that world. *Loss* occurs when denizens of some past world do not exist in the present, *erasure* when the traces of those denizens are not retained. A world's being lost is, then, a derivative notion: a world is lost when some of its significant denizens are lost. *Perspectives* (in this context) are ways of characterizing denizens that are enabled by *perspectival tools* such as fossils, models, comparisons, experiments, observations, and so on. In trying to understand putatively lost worlds, historical scientists make creative and opportunistic use of perspectives ("strategic" perspectivalism), developing iterative investigations between and within them. These investigations involve some combination of artificing (building simulacra of denizens), grafting (unifying denizens with the present and other times via causal trajectories), and anchoring (characterizing denizens as tokens of a type and comparing them to other examples). Because perspectives reach beyond the target that we've taken the perspective on, under the right circumstances (where there are sufficient perspectival resources!), historical scientists can solve challenges from loss by adopting self-correcting processes of comparison and testing across perspectives, thus uncovering lost worlds.

Phew, it's been a busy few chapters.

Most of the case studies I've drawn on use some combination of the strategic package I've developed. However, my descriptions aimed to bring out a particular aspect (artificing, grafting, anchoring). This was somewhat artificial, downplaying how strategic perspectivalism involves weaving variations of these strategies together. So, in this section, I will provide an example that highlights this pluralistic aspect, thus demonstrating the epistemic prowess of the bag of tricks I've been describing. To drive the argument home, then, let's turn to another set of questions arising from the Ediacaran: why did the complex metazoans of the later Ediacaran and Mesozoic turn up when they did? Why did the Avalonian world fall? We'll look at Nick Butterfield's controversial (see Cole et al. 2020) answer. As we'll see, it involves a pretty remarkable example of strategic perspectivalism that prominently involves grafting, artificing, and anchoring together.

Let's begin with some very big-picture and very deep-past context.

At some level of description, life as we biochemically know it arose in the "Great Oxygenation Event" around 2.4 billion years ago (Ligrone 2019; Lyons et al. 2021). Before that, the Earth's atmosphere was dominated by methane, and from around 3.5 billion years ago, the biosphere consisted of single-celled critters sporting anaerobic metabolisms. There was a steady supply of both oxygen and hydrogen, as sunlight split water vapour in the atmosphere into its constituent elements, but it wasn't until cyanobacteria arose that oxygen began fixing in the atmosphere. The major metabolic innovation of cyanobacteria is photosynthesis, which uses sunlight to oxidize water, thus producing excess oxygen. As cyanobacteria steadily did their work, an oxygen pump was created, and the atmosphere became the nitrogen- and oxygen-dominated beast we know and love today (and the anaerobic organisms, for whom excess oxygen is poison, sought refuge in the deep oceans and, eventually, our guts).

Fossil evidence shows that eukaryotes arose around 800 million years after the Great Oxygenation Event. Complex multicellular eukaryotes—animals and so on—took another billion years (Mukherjee et al. 2018). What gives? Why would it take a boring billion years (one thousand million) for complex metazoans to arrive on the scene? A common answer appeals to the biochemical requirements of eukaryotes, especially their being oxygen-greedy. Most eukaryotes[3] must absorb oxygen to function, so oxygen-availability is a constraint on eukaryotic complexity and size. Oceanic oxygen levels fluctuated heterogeneously throughout the Phanerozoic, not reaching high levels until the dawn of the Cambrian, and only stabilizing around modern levels in the late Jurassic (Lu et al. 2018). That this correlates with the Cambrian explosion—complex metazoans finally getting their act together—has not gone unnoticed. We have, then, what seems like a neat explanation for what has been called the "boring billion," the eternity between eukaryotes arriving on the scene and metazoans arising. Perhaps our ancient ancestors were awaiting the right oxygenic conditions.

3 For examples of non-aerobic eukaryotes, see Fenchel (2011).

Butterfield argues that this oxygen-driven explanation gets things backwards: it was the eukaryotes themselves that generated the oxygen-rich conditions that underwrote their evolution. Let's see why he thinks this, and along the way, see him graft and anchor (with a dash of artificing).

Butterfield makes a few arguments against the oxygen-driven explanation. I'll mention two. First, he points to monophyly. All evidence suggests that animals form a "well-behaved" clade, that is, they share a common ancestor. That is odd: if the only thing holding a bunch of microbes back from complex multicellularity is oxygen availability, then wouldn't separate lineages take advantage of those conditions once they turned up? So, Butterfield argues we should expect polyphyly under such conditions. If so, then the monophyly of complex animals speaks against an oxygen-driven trajectory. Second, he highlights "the fundamentally non-uniformitarian nature of the pre-metazoan biological pump (precluding any simple extrapolation of atmospheric composition from marine redox signatures)" (Butterfield 2009, 3). It's worth pausing to understand this, not only because (as we'll see) it is crucial to Butterfield's positive account, but because it involves an appeal to loss.

We can understand the biological pump as the process by which carbon is transported from the sea's surface to the ocean floor (de La Rocha and Passow 2006). Some proportion of this occurs by the passive sinking of corpses and excreta (sometimes called "marine snow"). But another proportion is "active." Basically, large numbers of organisms travel from the ocean surface at night to the lower oceans by day. A great migration occurs wherein herds of smaller aquatic life, krill and the like, head up to feed at night on phyto- and zooplankton, followed by a vast ecological web of fish, cephalopods, and cetaceans (oh my!). As the sun rises, krill lead a caravan of trophic hangers-on back to the lightless depths. It seems basically accepted that this process is driven by predation. Krill (and similar organisms) feed during the night because darkness provides cover, making them trickier prey, and return to the safety of the ocean's "twilight zone" during the day (Lambert 1993). As Pinti et al. have put it, we can understand the migration as "the product of a co-adaptive 'game' where many animals seek to optimize their migration patterns relative to the migration patterns of their respective prey, predators, and conspecifics" (Pinti et al. 2023, 997).

What is the relationship between all this and lost worlds? These days—and by that I mean more or less since the Cambrian—the relationship between oceanic and atmospheric oxygen has been basically stable, meaning we can infer from information about deep-water geochemistry to atmospheric oxygen and carbon (Mackensen and Schmiedl 2019). Butterfield has argued against "the assumption that ancient oceanic structure was comparable to today's" (Butterfield 2009, 3) on the basis that, well, it couldn't have been: without complex metazoans, you don't get the biological pump, without the biological pump, the biochemical cycling of oxygen and carbon is different, and it is that cycling that ensures continuity between atmospheric and oceanic carbon.

Butterfield's argument here is a nice example of Page's notion of the past being the key to the present. Nowadays, there is a stable relationship between oceanic and atmospheric carbon and oxygen (well, sort of stable—Butterfield 2009 points to a bunch of exceptions), which is handy because it lets us infer from proxies concerning oceanic carbon to the composition of atmospheric carbon over time (which is important for establishing fluctuations in the relationship between global temperature and atmospheric carbon). But how stable is that regularity, and what does it depend on? By examining a time before the biological pump, Butterfield is able to examine that relationship.

This is a classic example of anchoring: a regularity's robustness and limits are tested by linking them to various instances across time that are relevant to them. We can then iteratively shift between instances and models of that regularity. This allows us to potentially identify cases of continuity and of loss: if Butterfield is right, then pre-Cambrian biochemistry played by different rules than those obeyed from the Cambrian onwards. And he thinks he has a reason for the posited discontinuity, specifically, the organisms of the Ediacaran themselves influenced patterns of oceanic oxidation: "Multicellularity and large size introduce an entirely novel range of fluid-dynamic properties and ecophysiological opportunities for organisms, not least an unparalleled capacity to circulate and swim through the water" (Butterfield 2009, 3). In short, metazoans didn't need to wait until oxygen levels were right. They got there themselves through some nifty ecological engineering.

Early (non-complex) multicellular eukaryotes were single-celled organisms knitted together without any particular architectural or structural flourishes (compared to, say, the elegance of the rangeomorphs), and—as we saw in chapter 3—the dynamics of obligate osmotrophy makes getting bigger tricky. As energy dispersal relies on passive diffusion, and as surface-area-to-volume ratios get less efficient as size increases, the traditional osmotrophic system breaks down as you get larger. However, Butterfield points out that the common ancestor of eukaryotes were single-celled organisms with flagella. Flagella are the basic means of bacterial locomotion: probably most well-known as the "tails" of sperm cells. Butterfield argues that the effect of these tiny wriggling organelles "fundamentally changes the nature of trans-membrane gas exchange" (Butterfield 2018, 3). They do this more or less by creating a current that enhances oxygen diffusion, pulsing it through their bodies. By banding together, they amp up this effect: "Simply by arranging themselves into colonies, flagellated cells gain access to fundamentally greater levels of gas exchange than would be possible in isolation. Far from being frustrated by low levels of oxygen, then, multicellularity presents a unique opportunity to escape the 'tyranny of diffusion'" (Butterfield 2018, 4).

"Simple" metazoans, such as coral, create significant vortices that structure oxygen supply, especially flow, in their local environments. This increases the availability of oxygen and the efficiency of oxygen absorption. This underwrites Butterfield's arguments against an oxygen-driven explanation of the timing and nature of complex metazoan evolution. Oxygen isn't a constraint on getting bigger: getting bigger or banding together in communities is a solution to a lack of oxygen.

So, why the long wait for complexity? What explains the boring billion? Butterfield argues that the answer lies not in an external constraint like oxygen availability, but in the sheer difficulty of evolving the phenotypic and developmental components themselves, what we might call the internal infrastructure constitutive of complex metazoan life: "The challenge appears to lie in the design and assembly of the machine itself" (Butterfield 2018, 5). Developmentally speaking, simple multicellularity only requires iteration. But tissue-grade multicellularity, where various cells come together as a tissue to perform a function, is significantly more complex. Butterfield's explanation needn't commit to the idea that complex metazoan traits have low fitness—that they are selected against—rather, it could also be that their evolvability turns on highly-specific and unlikely developmental trajectories (that is, they are highly "peculiar" in the parlance we'll introduce in chapter 8).

So, Butterfield's negative argument relies on adopting a perspective enabled by a set of tools: examination of currently living analogues of simple metazoan life; metabolic and fluid-dynamic analyses of flagella and other cellular organelles; and various artifactual models of those dynamics, all contextualized to a lost world that existed before complex metazoans and the biological carbon and oxygen pumps. How then does Butterfield account for the emergence of those "modern" systems? Here, he takes on a further set of perspectives, and for our purposes, is quite explicit about this: "Animals are large aerobic heterotrophic organisms, so it makes sense to view them

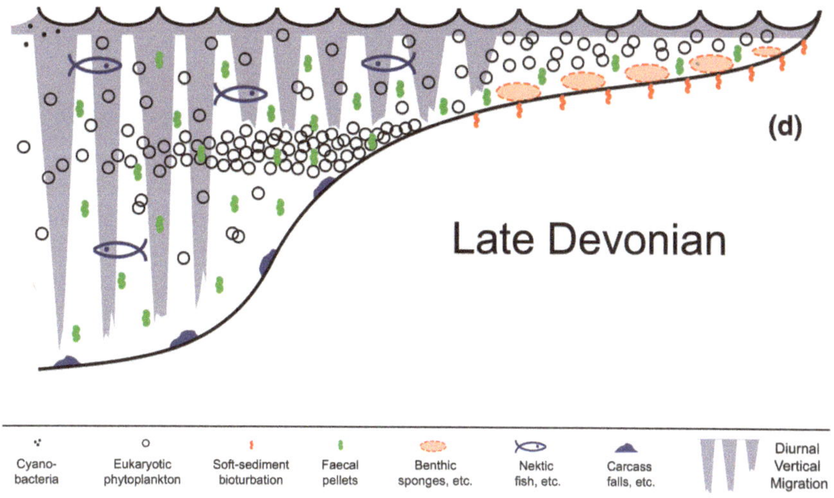

FIGURE 5.4 Modern Oceanic System (Detail from Butterfield 2018, Fig. 3). Reprinted with Permission from *Geobiology*

as oxygen sinks. But as swimmers, pumpers and tethered sources of drag, they also contribute importantly to environmental mixing. And as collectors, transformers and translocators of organic carbon, they fundamentally alter rates and styles of biological oxygen demand" (Butterfield 2018, 7).

Let's unpack this. First, animals consume oxygen (are aerobic) and do not generate energy from the sun, but get it by consuming others, using their oxygen supply (they are heterotrophic). Thus, they are oxygen sinks. Second, animals shift water around with all their swimming and suchlike, thus acting as environmental mixers: they shift the biochemical nature of the ocean, in particular, leading to well-oxygenated surface waters. Third, they change the carbon game by absorbing and excreting large amounts of carbon. For instance, krill interact with oxygen systems differently from phytoplankton: they don't produce their own, for one thing. Thus, by consuming large amounts of phytoplankton, krill convert one kind of carbon-oxygen system into another. Each of these perspectives enables the isolation of particular properties, their comparison with relevant instances, and their subsequent recontextualization. For instance, add these aspects together with the diurnal migration from the surface to the depths, and we have the remarkable biological pump represented in Figure 5.4.

Figure 5.4 represents the oceans as we know them: the daily migration ("diurnal vertical migration") transports carbon and oxygen to the sea floor in a great pump, while swimming biomasses mix the oceans, generating fairly homogenous oxygenation, although with higher concentrations in the surface waters. The representation is highly schematic, focusing on biological interactions and feedback while ignoring physical processes such as tides.

But how did we get here? Butterfield might have convinced us that the oceans of the deep, deep past are lost—that our own oceans work by fundamentally different dynamics—but what were those previous oceans like and how were they transformed into our own? Here, we find grafting coming to the fore: there is a story to tell connecting the unfamiliar past to the familiar, and by linking these through a series of trajectories, they become evidentially relevant not because they are the same, but in part because they are different.

Let's begin in the pre-Metazoan oceans, seen in Figure 5.5. In the pre-Cryogenian (earlier than 720 million years ago), primary production in the oceans was dominated by cyanobacteria. Reliant on sunlight, these spread throughout the surface waters and shallower seas. The evolution of sponges and other filter feeders in shallow benthic environments cleared cyanobacteria from those contexts (shifting them to the surfaces of deeper oceans). This sets the stage for the evolution of algae (Brocks 2018) and phytoplankton: larger eukaryotic photosynthesizers. By the Avalonian period, rangeomorphs, sponges, and other filter-feeding metazoans create an increasingly well-mixed and concentrated oxygenic and carbonic environment for shallower benthic communities.

Butterfield's narrative continues. By the Cambrian, the late Ediacaran evolution of moving metazoans (worms and our friend *Dickinsonia)* had begun processes of "soft sediment bioturbation": basically, the seafloor was transformed as various critters burrow through it, creating richer carbon sequestering and oxygenic exchange (Teal et al. 2008). Further, the evolution of early muscled and skeleton-based metazoans enabled mobile filter feeding and eventually the evolution of predation, and thus the modest beginning of something like the modern biological pump, as in those parts of the ocean deep enough to escape light, eukaryotes begin their daily plunge and nightly surfacing, as we see in Figure 5.6.

With the emergence of large fish in the late Devonian, we get the architecture of the modern oceans. In the last chapter, we saw how Ediacaran fauna might not be failed experiments in evolution after all, but rather include stem-metazoans capable of helping us make sense of the evolution

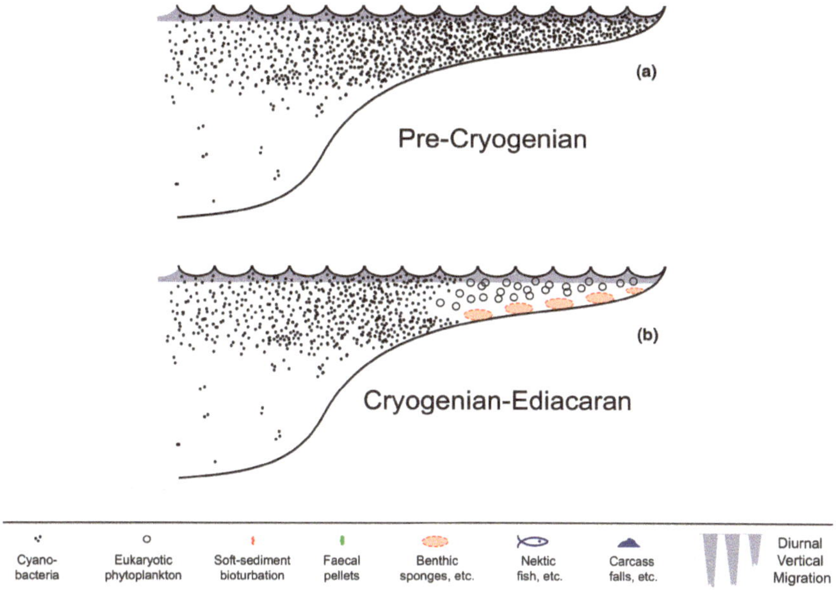

FIGURE 5.5 Pre-Metazoan Oceans (Detail from Butterfield 2018, Fig. 3). Reprinted with Permission from *Geobiology*

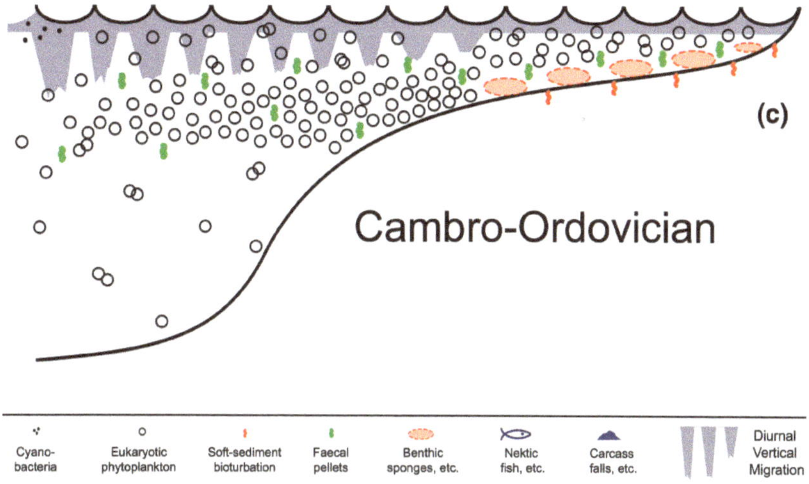

FIGURE 5.6 The Cambrian Ocean (Detail from Butterfield 2018, Fig. 3). Reprinted with Permission from *Geobiology*

of, say, bilaterans in the case of *Dickinsonia*. Here, we see another example of grafting: the fauna of the Ediacaran, not only sponges but presumably rangeomorphs and their ilk, began processes of filtering and structuring their aquatic environments, and concentrating, mixing, and distributing oxygen and carbon, in ways that in part created various conditions required for the eventual evolution of the oceans we know today.

By adopting multiple perspectives, Butterfield iteratively builds a rich picture of a lost world, but not only that, he also provides a picture of how the world became lost and its role in creating the conditions leading to more familiar times. Crucially, the model is continually re-contextualized with features of the past worlds we're interested in. This continual return to past contexts makes it more likely that mistakes, when they arise, will be discovered. The more alignment we find between accounts like Butterfield's and context concerning the past world, the more confident we should be that the lost world has been found.

It's worth noting that I'm not here committed to Butterfield's model being correct. For instance, Austin Booth and Ford Doolittle have provided empirical and conceptual reasons to question the uniqueness of both eukaryogenesis and the special powers of eukaryotes (Booth and Doolittle 2015). Although their ideas don't conflict directly with Butterfield's, they could undermine part of his picture. My claim is that the process of iteration engaged in should make us optimistic that if Butterfield is wrong, we'll have the tools to find this out.

Thus, we can see how iterative strategic perspectivalism generates a self-correcting investigative process that can uncover past worlds in spite of loss. Before turning to more philosophical discussion, I want to finish with a brief discussion of scientific laws.

4. Coda: Scientific Laws

In chapter 8, I'll discuss the "peculiarity" of the past—approximately, its contingency—but our discussion of anchoring raises questions about laws. If various anchors can be understood as tokens of historical or ahistorical types governed by the same set of rules, then shouldn't we think of those rules as putative laws of nature? In this coda, I'll briefly discuss the *epistemic* features of laws as they play out in strategic perspectivalism, leaving metaphysical discussion for that later chapter.

A fairly common refrain among historical scientists is that their inferences about the deep past are underwritten by invariant scientific laws, particularly those revealed by chemistry and physics. Might these provide foundationalist anchors, playing necessary or in some sense privileged roles in the science of lost worlds? I don't think so, and will try to articulate why here. Note that the central point of this discussion doesn't turn on whether laws generally hold, but rather where and how they are applicable.

Philosophers often understand scientific or natural "laws" as *non-accidental, universal* regularities (Carroll 2020). That is, a type-level process (rangeomorphs grow via a three-branch structure, not *this* rangeomorph grows via a three-branch structure) that ranges over all relevant systems (fractal growth leads to self-similarity, not self-similarity is found in some fractally-growing things), that possess those properties for some appropriate reason (geometric relationships, say, ensure the relationship between fractal growth and self-similarity, rather than it just happening to be the case that surviving fractally-growing things exhibit self-similarity). Such regularities, if only we could find them, were taken to play an important role in science because—at base—they enable prediction (if we find a rangeomorph, it will grow fractally) and they enable explanation (rangeomorphs have self-similarity because they grow fractally).

One philosophically exciting idea concerning loss connects it with laws of nature. If a law can be "lost" or a law can "arise," then doesn't this imply that the laws of nature could be (in fact have been) different? If a law of nature is some kind of metaphysical necessity—true in all metaphysically possible worlds—then laws cannot be lost and cannot be found. It might be that various perimeters within the laws differ across worlds or even times, or vary across "initial conditions." Marc Lange distinguishes between *temporary laws* and *eternal but time-dependent laws* (Lange 2008). The former are laws that only hold for some period, while others are laws that come "online" once the right conditions arise. The practice-oriented discussion here cannot distinguish between these, but I consider this a feature, not a bug.

We don't need full-blown laws to predict or explain (Mitchell 1997), so we shouldn't get too concerned about their availability or otherwise in science. However, I want to consider a fairly common refrain from scientists seeking to understand putatively lost worlds: that appeals to scientific laws will ground their discoveries and justify their investigations. Here's an illustrative example from the introduction to Maynard-Smith and Szathmary's *Major Transitions*:

> There are obvious difficulties in discussing unique events that happened a long time ago . . . [however] we have agreed theories both of chemistry and of the mechanism of evolutionary change. We can therefore insist that our explanations be plausible both chemically, and in terms of natural selection. This places a severe constraint on possible theories . . . Further, theories are often testable by looking at existing organisms. (Maynard-Smith and Szathmary 1997, 3)

Maynard-Smith and Szathmary are proposing a general strategy for understanding a particular set of important changes in the history of life, "major transitions." These changes involve shifts in the transmission of ancestral information or shifts in biological individuality. A classic example is the evolution of multicellularity (see chapters collected in Calcott and Sterelny 2011). The investigative strategy proposed could be understood as a kind of combination of anchoring and artificing. Maynard-Smith and Szathmary's proposal is to understand major transitions in terms of the abstract mechanics of population genetics and similar modelling approaches. Roughly, build an agent-based model that explains under what evolutionary conditions "de-Darwinization" occurs, for instance, an evolutionary process that transitions from a bunch of "selfish," competing single-celled organisms to an organism consisting of cooperating parts. The models must be consistent with evolutionary and biochemical laws and tested on extant model organisms.

Such a model would then be applied across those major transitions. This is an ambitious project, no doubt, but what matters here is what Maynard-Smith and Szathmary appeal to in constraining their project: it must fit within the theories of chemistry and evolution. That is, they must appeal to fairly generally applicable laws. One way of understanding the putative role of laws here is through what Khalifa and Goldberg call *epistemic outsourcing*: "The phenomenon whereby results taken as basic or given in one scientific field are justified owing to the intellectual efforts, evidence, and methods in another scientific field" (Khalifa and Goldberg 2022, 383). Epistemic outsourcing is *social* and *foundational*. The former insofar as it relies on (something like) the testimony of the outsourced scientists, and the latter insofar as it is taken as "basic," that is, unquestioned, by the outsourcing scientists. If laws from physical (and other) sciences are epistemically outsourced by historical scientists, then we should see them playing foundational, taken-for-granted roles in their reasoning. They might be applied, for instance, in an "off-the-shelf" manner to their particular purposes.

No doubt, well-verified robust regularities from physics and chemistry (but also from evolutionary biology, ecology, physiology, etc.) play critical roles in uncovering lost worlds. Mitchell et al., for instance, use fairly straightforward models of spatial niche-differentiation; for another example, the geometric growth models we've seen (of *Dickinsonia* and rangeomorphs) follow extremely general mathematical rules; and for another example, we've just seen how crucial various basic biochemical dynamics are for Butterfield's account of the arrival of complex metazoans and the biological pump. But to think that, say, Butterfield's investigation is a primarily biochemical one is to miss the epistemic trees for the forest. As he puts it: "All biological exchange ultimately depends on chemical diffusion, but it is the associated fluid-dynamic context that determines physical and ecological properties" (Butterfield 2017, 10).

Models of chemical diffusion are crucial, but these must be put in the context of an aquatic medium, which itself—as Butterfield argues—is deeply affected by physiological and anatomical features of the organisms living within those environments. So, no doubt, what we might call "laws," or at least models of them, are important perspectival tools, but I don't see them as playing a privileged or necessary role. That is, there might be some level of epistemic outsourcing going on, but the deployment of that knowledge is not foundational.[4]

Further, we might imagine Mitchell et al.'s work as being about the "laws" of ecology. That is, there is a set of "rules" that range over all living ecosystems. And in a sense, their work supports this idea: the same model that should lead us to expect niche-driven spatial differentiation in most ecosystems should also lead us to expect neutral-driven differentiation in cases where competition is limited, assemblages are frequently disrupted, and so on. But notice the role of "laws" here. While Mitchell et al. certainly framed the discussion—set up an interesting challenge—it was via an anchoring strategy, the close attention to contextual knowledge about the Avalonian, as well as

4 Khalifa and Goldberg's account of "foundational" is about the epistemic status of propositions, approximately whether they are questioned within the relevant epistemic practice. I'm here less interested in the epistemic status than I am in the epistemic *function* of foundations. Even if some scientific propositions are unquestioned by historical scientists (so they are "basic" in Khalifa and Goldberg's sense), they do not—or at least do not typically—play a basic or foundational role in their reasoning.

various analogous cases, that progress was made (see Sánchez-Dorado 2024a for an emphasis on iteration between various "idiographic" and "nomothetic" models in the Earth sciences).

The lesson here is that models of highly abstract, robust, and general regularities (or "laws," if you must) can play important roles in both identifying and solving challenges from loss. But it is worth noting that their very abstractness can be limiting. They afford a perspective that allows unifying very many targets—as we saw in Mitchell et al.'s case—but if opportunities for iteration between perspectives are limited, especially without a rich trace record and investigation of diverse analogies, this in itself is not particularly productive. This leads me to be somewhat sceptical of the kind of investigative strategy that Maynard-Smith and Szathmary suggest. Although laws and extant animals can teach us much, without an actual anchor into the deep past (provided by, say, fossils), I'm not confident that we'll see as much substantive progress (Currie 2019d).

So, I want to suggest it isn't laws per se, but the adoption of strategic perspectives and iteration that allow us to discover much about lost worlds. Our models of highly general regularities, to be employable, must often be made bespoke: carefully tailored to fit the local, idiosyncratic contexts that must be grappled with if we are to understand the deep past.

Prof. Ichthy takes a deep breath. "So," they begin, "earlier I was worrying, well, at least feigning worry, about whether my work on Broadfoot amounts to mere speculation. I hope now to have convinced you—or at least to have convinced myself!" They give a self-deprecating chuckle. "That the worry is unjustified. Even though Broadfoot is a strange creature, unlike anything we've seen before, we are nonetheless able to piece together this ancient lineage. So, how do we do it?" Once again, Prof. Ichthy's use of 'we' is a self-conscious nod to colleagues and students.

"I've emphasized how important taking a *perspective* on Broadfoot is. If something looks unique, then, think of ways in which it might not be unique. Broadfoot is a unique species, that's for sure. But they aren't the only mammal, they aren't the only swimmer, they aren't the only venomous creature, they aren't the only vertebrate, and so on. These perspectives let us link Broadfoot with an array of other things, be they living analogues, their surviving relatives, the things we build, and contextual factors from the environment Broadfoot lived within. And with each of these new perspectives and links, we get more opportunities for bringing our knowledge into conversation, iteratively working our way closer and closer to a rich and, I think, plausible picture of this weird, wonderful little animal." Ichthy pauses, perhaps considering how exactly to finish the lecture.

"I mentioned earlier that one reason to care about Broadfoot and the science of Broadfoot is that thinking about the past can be transformative of how we think about the present and the future. And I can say I've learned a lot thinking about Broadfoot and the other mammals they shared that past world with. Although I only see that past world in parts, it provides such a broad perspective—a grand vista—from which to think about our world and how changeable and how fragile and, yes, how valuable it is."

Ichthy pauses again, their flipper resting fondly on the case containing the skull.

"Were it not for this specimen, I could never have imagined a world with Broadfoot. And I'm so glad that I got to, because the world I live in is so much the richer for it." Ichthy smiles a little bashfully as they indicate the end of their lecture with a small nod.

6

Possible Worlds

The applause closing Prof. Ichthy's lecture ranges from polite (various academics) to appreciative (closer colleagues and grad students) to almost inappropriately rambunctious (Ichthy's family, who whoop and whistle). Ichthy beckons their students onto the stage to share the applause, which they do somewhat bashfully. As the applause dies down, a second expectant hush descends, and Ichthy indicates they'd be happy to take questions.

Audience members glance about through the hush, waiting for someone to break the silence. Finally, a younger plesiosaur breaks the awkwardness by raising a flipper and, at Icthy's nod, bends their serpentine neck upward to speak.

"Uh, so, you've said a lot about the things we know about Broadfoot, that it was, like, a mammal and stuff. And you said some stuff about these big things that happened in prehistory, like, how the age of the mammals ended. But have we also learned anything about how things could be? Like, did these events have to happen? Could the mammals have survived? If they did survive, what would that be like? Would mammals have developed reptile-like intelligence—could they have had professors and universities and sports, and the stuff that, like, we have? I know it sounds kind of silly . . ."

The questioner trails off in some embarrassment before Ichthy comes to a magnanimous rescue: "Could all this knowledge of the actual past also be knowledge of what is possible? That is, about how things could be? An excellent question!"

ILLUSTRATION 6.1 Prof. Ichthy's Audience

Historical scientists generate knowledge despite erasure and loss. How should we characterize that knowledge? Should we call it knowledge, or understanding, or know-how, or know-that? Further, what is this knowledge *of*? Do they know facts about states of affairs in the past, or *merely* what was possible, or only some consistent set of things going on in their minds? What kind of epistemic good or goods are we talking about, and what kinds of things do these goods pertain to?

Heady, deep, philosophical questions indeed.

My immediate answer, you'll likely not be surprised to hear, is pluralistic: given the diversity of evidence historical scientists have at their disposal, the complexity of the world they are trying to understand, and the strategically perspectival, iterative tactics they employ, it would be surprising if there was a monistic epistemic story to be had. Their knowledge comes in many forms and is of many things. While this pluralism is plausible, it doesn't necessarily make for an enlightening discussion. Instead, I want to focus on a particular aspect of the knowledge historical scientists generate: the *modal* element. Above, I made telltale use of the term "merely" in relation to knowledge of what is "possible," even italicizing it for emphasis. In this chapter, I'll argue we should dispense with the diminutive "merely." A major—sometimes the main—dividend of historical science is knowledge of possibility, and it is my aim in this chapter to defend this claim and the legitimacy of modal knowledge. It might be useful to have a slogan:

> The science of lost worlds is a science of objective possibility.

What does this amount to? Although sciences of the deep past tell us about the actual deep past—no doubt—they also tell us about possibility in rich, complex ways. Moreover, this modal knowledge is not merely of "epistemic" possibility, that is, what could have happened in the past given what we know. It is also of "objective" possibility. Which is to say, when the science succeeds, it provides true modal information, that is, beyond what actually happened, and concerns what could have, or might have, or must have, or would have happened had something else occurred (Wirling and Grüne-Yanoff 2024). And, as we'll see, this modal knowledge is complex and structured. It concerns not so much what is possible *tout court*, but rather concerns modal spaces consisting of sets of complexly interacting conditionals.

I'll further argue that the best way to understand the justification of this modal knowledge is not, as others have suggested, through the workings of scientific imagination, but instead through an "artifactualist" account of perspectival tools. In brief, historical scientists construct and examine various objects—perspectival tools—and demonstrate their having various dispositions, that is, sets of modal properties. These dispositions are projected onto past targets as well as onto more general tendencies, patterns, and regularities. As we'll see in the next chapter, the scientific imagination is expanded through the investigation of lost worlds. But I don't think imagination, as understood by most philosophers, plays any special role in underwriting or justifying modal knowledge. There's no special pleading required to understand how scientific representational strategies generate modal, as opposed to non-modal, knowledge.

My strategy is to begin with the old philosophical chestnut of "how possibly" explanations. I'll carry out some disambiguation and identify a set of "how-possibles" that fit the phenomena at hand. I'll then argue for a minimally realist account of this objective modal knowledge. What I don't mean by this is that scientists get it right all the time; rather, I have in mind the humbler idea that the reference of their claims is often modal structures, not actuality. This will involve a commitment to artifactualism.

With that in place, I'll turn to a case study: this time, astrobiology. Astrobiology is as much a science of the deep past as palaeontology. I'll argue that construing astrobiology as being in the business of identifying extraterrestrial life is, surprisingly, missing a large part of the story. As we'll see, astrobiology (or at least the corner I'll explore) provides dividends in the form of understanding the relationships and dependencies—the rich modal structures—holding between various biotic and abiotic processes and potential atmospheres. That is, astrobiologists map out how various planetary compositions provide the conditions of possibility of differing atmospheric compositions, given various other biotic and abiotic factors.

Much of the philosophical discussions I'll be drawing on herein concern modelling, especially the (surprisingly small) literature on the modality of models. I'm not sure whether anything much hangs on any of the practices I'll cover being called "modelling" or not. The last twenty years or so have seen an enormous amount of work in the philosophy of modelling, and I don't really see much unity in how philosophers of science conceive of models. Joe Wilson (2024) has recently pointed to various similarities between how philosophers have characterized model-based and historical science, using this to put pressure on attempts to characterize sciences in terms of "prototypical" methods. In this context, attempting to delineate between what's a model and what's not a model seems at best a little boring and at worst obscurantist. Presumably, a table (of contents, or of data) isn't a model—though perhaps it sometimes counts as a "data model"—these are nonetheless perspectival tools that scientists make use of.

So, suffice to say, I'll be talking about "perspectival tools" in the sense I introduced in chapter 3: these isolate and instantiate various properties to be explored by providing a space for comparison (as in, say, phylogenetics) or direct intervention (as in experimental taphonomy), or via inference (as in spatial point analysis). We've seen a wide variety of perspectival tools, from Woodruff and Varricchio's (2011) scale model of a dinosaur burrow to various geometric representations of the growth of ancient organisms to phylogenetic and morphological trees, tables, and morphospaces. It is tempting to think of these practices as situating a phenomenon or a set of phenomena within a "logical space" (Elgin 2017, 155); that is, the perspectival tools delineate a set of constraints within which the target is situated. This is about right, but as we'll see, I'll deny we should necessarily think of these spaces in "logical," or abstract terms: perspectival tools are artifacts, that is, physical objects with their own dispositions. These dispositions themselves set a space of possibility.

The science of lost worlds is a science of objective possibility, then, in two senses. First, it provides modal knowledge of more than epistemic possibility; second, this knowledge is "objectual," that is, grounded in the dispositions of particular objects.

A quick note before diving in. A discussion of modality in palaeoscientific contexts might immediately call to mind Stephen Jay Gould's famous "life's tape" thought experiment (Gould 1989). To illustrate his view on life's contingency, Gould asks us to imagine "rewinding" life to some earlier, presumably crucial, stage. He asks, were the tape then allowed to run forward, whether we'd see the same history unfold: for many events in life's history, Gould bets that we wouldn't. The thought experiment has been used to frame everything from philosophical analysis of contingency to experimental laboratory work. Here, however, I'll not be discussing the modal nature of Gould's thought experiments. This would distract us from the significantly more grounded modal knowledge produced by the cases we've been examining. Regardless of what we might think about replaying metaphorical tapes (and, indeed, we might think imagination plays a role in these thought experiments), the bona fide modal knowledge apparent throughout the more workaday investigations we've been concerned with, from Avalonian ecology to dinosaur burrows and pain behaviour, to the phylogeny and development of *Dickinsonia*, is deserving of its own philosophical accounting.

1. How Possibly: At Least Three Ways

Let's frame this chapter's discussion by distinguishing between different kinds of how-possible explanations. The term "how-possibly explanation" is traceable to William Dray's work in the philosophy of history (1957), where he argued that historical explanation often proceeds according to a distinctive form (one differing from scientific explanation, as understood by Hempel). Dray's notion of "how-possibly" explanation was picked up in various areas of philosophy, particularly in the philosophy of biology's discussions of adaptationism (Brandon 1990; Forber 2010), and along the way, it shifted somewhat from its original meaning. Alisa Bokulich (2014) has provided a nice nutshell of this history, and, along with Thomas Reydon (2023), provides much-needed clarity that I'll draw on. I'm interested in disambiguating forms of "how-possibly" to zero in on various ways that modality might feature in epistemic practices like explanation. As we'll see, many discussions of how-possibly take modal knowledge to be *in service of the actual*, but I'll argue this strategy misses some of what is crucial about understanding lost worlds. Instead, we should (at least sometimes) take the "how-possibly" of historical science to refer to objective possibility, and to be targeted at that objective possibility.

I'll call our first form of "how-possible" *mere-how-possibly*. These are, I think, what philosophers of science often have in mind when they appeal to "how-possibly explanations." To show something is possible is to demonstrate that there is at least one way that it could happen. Typically, mere-how-possibly explanations take a set of constraints and demonstrate that some outcome is consistent with those constraints. Consider challenges to neo-Darwinian adaptationism from appeals to "irreducible complexity" (remember those?). It is denied that some complex trait could be generated via an incremental evolutionary process of selection (therefore, through some argumentative magic, said complex trait must have been designed by . . . someone . . .). Typically, the response is to provide some possible evolutionary routes through which said trait could have evolved incrementally, thus defusing the challenge. Here, the adaptationist need not show how the trait in fact evolved, but rather show how it possibly could (Weber and Depew 2004).

In some scientific disputes, demonstrating that a hypothesis is merely possible is crucial. For instance, "Kon-Tiki experiments" aim to show that some hypothesis is possible via some kind of physical demonstration (Novick et al. 2020). Thor Heyerdahl drift-voyaged from South America to the Pacific to show that a Pacific settlement from that direction was possible with those technologies, and further Kon-Tiki experiments using Polynesian craft and navigation techniques played a decisive role in showing that non-drift voyaging was possible in the other direction. These demonstrations of mere possibility removed major objections to the relevant hypotheses and were crucial to the eventual acceptance of a non-drift, westward model of Polynesian settlement. Further, because mere-how-possibly explanations concretize at least one possibility within a set of constraints, they can play crucial roles in exploring those constraints and various other forms of scaffolding (see Walsh's 2019 discussion of scaffolding). Mere-how-possibly will return in chapter 7's discussion of palaeoart.

Another form of how-possibly captures Dray's original meaning, so let's call it *Dray-how-possibly.* Here, instead of providing some general modal information—whether some outcome is consistent with some set of constraints—how-possibly explanations, as originally envisioned, involve providing *actual* information, which demonstrates that some *actual* outcome is possible. As Dray put it: "The explanation consists in showing that, in spite of all appearances to the contrary, the event was not impossible after all" (Dray 1968, 391). Explanations of magic tricks often take this form. The magician was able to pull a rabbit out of their empty hat—the trick was possible—because of a combination of misdirection and a secret compartment in the hat that gave it the appearance of being empty. We're given some actual information (the secret compartment), which explains how what was actual was possible.

Consider explanations of global biogeographical dispersal prior to the development of tectonic plate theory. How were various animals able to cross the oceans between, for instance, South America and Africa? In a mere-how-possibly context, the detractor might be happy with an appeal to land-bridges, independent of whether they actually occurred. For a Dray-how-possible explanation, we point to the actual spread of tectonic plates to explain how this could have occurred.

Demonstrations of possibility can play both *mere* and *Dray* roles. If one was informed that Polynesia was settled from Southeast Asia, and expressed bafflement as to how this could be achieved, an explanation of the highly sophisticated navigational techniques and technologies that Polynesian ancestors possessed might remove that bafflement, thus forming a Dray-how-possibly explanation. Instead, one might simply be interested in whether it is possible to get from Southeast Asia to Polynesia via some kind of craft, and would accept *one possible story* (regardless of whether it is the actual story). As Reydon summarizes: "Dray's how-possibly explanations tell the actual historical story to show how an improbable event was possible, whereas Brandon's and Resnik's how-possibly explanations only speculate about history" (Reydon 2023, 166).

That is, because mere-how-possible explanations are in the business of demonstrating that some outcome was possible, as opposed to showing that some actual event made an outcome possible, insofar as they're interested in the actual, it is in a merely "speculative" fashion. Bokulich (2014) demonstrates the complex, context-dependent relationship between different how-possible and how-actual explanations.

For my purposes, there is a further similarity between these how-possible explanations. They are typically about, I think, epistemic rather than objective possibility. In these cases, *the possible is in service to the actual.* That is, we care about what is possible because we ultimately care about what is actual: we're interested in further constraining the set of worlds that could be the actual world. This is necessarily the case for Dray-how-possibly, and typically the case for mere-how-possibly. To get the flavour of this, it's helpful to work through an example from Massimi. "As new pieces of experimental evidence e1 . . . e3 are brought in, they refine the how-possible inferences by excluding more and more of the live objective possibilities x1, . . . xn. This is because new evidence gradually and increasingly constrains what is physically conceivable" (Massimi 2019, 877).

Roughly, Massimi is describing how the scope of what is "physically conceivable" (see below) shrinks as a clearer picture of the actual emerges through incoming evidence. Despite her use of the term "objective," this is epistemic possibility: some proposition *might* be actual insofar as it accords with the other things we know. In principle (and perhaps with some wiggle room), as knowledge approaches completeness, physical conceivability and actuality approach overlap (we'll return to physical conceivability in the next section). This is what I mean by saying the "possible is in service of the actual." We care about mere-how-possibly explanations because we want to know whether some actual outcome could have occurred given such and such a surprising circumstance; we care about Dray-how-possibly explanations because we want to know how some apparently improbable event occurred. These claims are all of the form "such and such happenings are all consistent with what we know." For Dray-how-possibly, we say "this event was possible after all because of how the past was"; for mere-how-possibly, we say "it could have happened like this, for all we know." Objective modality claims are more of the form "if such and such were to occur, then the following would happen," that is, they aim for true claims about possibility itself. This is the kind of thing I'll have in mind when I discuss the dispositions of artifacts below.[1]

We needn't think of possibility as playing a mere in-service role. As Daniel Swaim has put it, epistemological accounts of modality do "not fit well with a great deal of scientific practice, wherein the point is to lay bare (to the extent possible) what the structure of physical possibilities is really like, and subsequently, why things turned out to be the way they are" (Swaim 2021, 264). Thomas Brandstetter characterizes models in geology as "how-possibly experiments," aiming to test hypotheses about what actually happened by determining whether they could have, possibility in service of the actual. But he goes further to claim that some models were "miniature theatres for dramatizing the action of forces on matter over time. In this sense, the model experiments were time machines: they made it possible to visualize geological time itself" (Brandstetter 2011, 135). Beyond being in service of the actual, then, they also provide routes to examine geological processes—and potentially objective modal knowledge.

Let's consider some discussions I take to indicate a third set of how-possibly: explanations that attempt to map out true objective possibilities, sometimes explaining via situating a target within that modal space, thus providing information about how and why that target could, would, and does occur. This involves appeal to objective rather than epistemic possibility, as—when we have a successful explanation—the space of the possible doesn't shrink as more information is included. As we'll see, often but not always, these modal spaces are generated by perspectival tools.

A view about how models provide insight into possibility is associated with Batterman and Rice (Batterman and Rice 2014; Rice 2018). The basic idea is that various models and targets are part of the same "universality class." I interpret "universality class" in a particular way: to be in the same

1 The term "disposition" looms large both in analytic metaphysics and scientific metaphysics. Here, I'm using it as a term to pick out the various modal properties, often conditionals, that constitute the knowledge generated from perspectival tools. I don't take myself to be committed to any particular claim about the nature of dispositions generally, although (1) I do take dispositions to be necessarily modal, and (2) I don't think all modal claims are dispositional.

universality class is to share dispositional properties. For instance, the Avalonian ecologies that concern Mitchell et al. are, if their analyses are right, in the same universality class as contemporary ecosystems insofar as when they share the same sets of properties—small populations, fast generational turnover, lack of nutrient constraint—they're disposed to exhibit neutral behaviour. The same might be said of the highly abstract models ecologists use to explain these. There are, I think, similarities between the argument I will provide in the next section and aspects of Batterman and Rice's solution. However, they are focused on how minimal "toy" models capture some fairly universal aspects of real-world targets, while I'm interested in a broader array of perspectival tools, as well as their combinatorial features. The important thing about their account, however, is that possibility is not in service of the actual: in addition to learning some dispositional properties about our target (say, the rangeomorph-dominated ecosystems of the Avalonian), we *also* learn modal information about the universality class (say, the properties of neutrally-driven ecosystems). It is in virtue of two systems sharing membership in the same universality class that the same modal claims apply to them.

Focused on the historical sciences, Meghan Page has argued that one benefit of studying the deep past is probing and testing the fragilities exhibited by regularities established under contemporary conditions. As she puts it:

> Historical investigations often play an important and ineliminable role in advancing our knowledge of causal structure and stable regularities. When confronted with complex geological processes that occur over large regions of space and time, historical science acts as an experimental context to test the stability of currently observed regularities. As a result, historical investigations enhance our understanding of the complex causal webs that produce natural processes and help predict how such processes might evolve or shift in the future. (Page 2021, 462)

Sometimes palaeontologists are in the business of using contemporarily established regularities to make inferences about the past, most blatantly in trace-based reasoning. But sometimes they are in the business of probing the stability of contemporarily established regularities by applying them to the past. Consider again the question Mitchell et al. put to the surprising neutrally driven spatial ecologies of the Avalonian. They ask: do these show that our usual theories about ecosystems simply *fail* under those conditions? By anchoring their analyses across a set of other ecosystems and models, Mitchell et al. argued that when properly understood, our ecological models can, in fact, predict neutrality. This represented an explanation of Avalonian neutrality (it was due to short generation time, low dispersal, and frequent disruptions). But further, it deepened our modal understanding of the factors shaping niche and neutrally structured ecosystems. We now know better under what circumstances ecosystems might exhibit those structures. Often, the pluripotency of perspectival tools is fundamentally modal.

Mitchell et al. thus investigate what, in 2018 (chapter 7), I called *fragile systems*. Here, the target is not some actual event, or even a set of events, but rather the conditions under which certain dynamics hold. In the more extreme cases, such as Mitchell et al.'s, we could characterize these as testing for certain kinds of loss, for instance, of the conditions for the applicability of certain kinds of dynamics.

Expanding from this kind of thought, we can characterize much work across the historical sciences as grappling with various fragile systems; this is particularly true of artificing strategies. Woodruff and Varricchio (2011) empirically investigate a set of dependencies between burrow infill and bone position; Laflamme and Narbonne (2008) model dependencies between rangeomorph morphology and optimization in osmotrophy; Cuthill and Conway Morris (2014) examine osmotrophy in light of fractal morphologies. Each of these artifacts instantiates a possibility space, and examining them generates knowledge of that space (Currie 2020, 2022). Further, the results of artificing are drawn on, integrated, and contextualized with historical reconstruction to provide complex trajectories, such as Butterfield's explanation of the long road from Ediacaran-style oxygenation to that which emerged through and beyond the Cambrian.[2]

Daniel Swaim's account of *narrative possibility* makes a similar point (Swaim 2021). Focusing on narrative explanations (roughly, explanations that take the form of a token sequence of causal events, culminating in the *explanandum*), he develops an "ecological" approach. This involves, first, central subjects—the token historical individual the story is wrapped around—and their capacities; second, these capacities take on various *affordances* given

---

2 In Currie (2021b and 2024a), I develop a notion of "profiles" that further fleshes out this kind of notion.

the environments into which those historical individuals are embedded. In Butterfield's explanation of the diurnal pump, he pointed to the oxygen-structuring capacities of simple eukaryotes and showed how their being embedded in pre-Cambrian environments afforded more concentrated oxygen and thus a scaffold for the emergence of larger and more complex eukaryotes.

In these circumstances, scientists of the deep past are not only concerned with token denizens in the past—particular events and processes—they are concerned with the dynamics, dispositions, and fragilities of various system-types, whether they are token-historical, historical, or ahistorical. In these circumstances, as our knowledge grows, possibility does not shrink to actuality, although it might become more complex and structured.

These increases in modal complexity and structure were what I had in mind earlier when I appealed to "sets of complexly interacting conditionals." Consider Mitchell et al.'s work. They shift from modelling Ediacaran ecosystems via a dichotomy between neutrally and niche-structured spatial locations, to introducing various conditionals regarding small populations, dispersal distance, and more. These conditionals interact: if an ecosystem has small populations, it will tend toward neutrality; if an ecosystem is often disrupted, it will tend toward neutrality. These conditionals work together in explaining their result. So, in Mitchell et al.'s work, modal complexity is added as various conditionals concerning the factors associated with neutrality are brought online. The product of the research is a modal space that tracks the conditions under which neutrality and niche differentiation are expected, and a how-actually explanation[3] when the Avalonian past world is situated within that space.

I think these extremely common circumstances across investigations of past worlds underwrite a prima facie motivation for taking this scientific work to be uncovering objective possibilities.

Fragile systems are explored through the use of perspectival tools. Not all perspectival tools aim to capture objective possibility directly. As we saw in chapter 4, cladistics and other grafting techniques make use of perspectival tools, but the aimed outcome (at least sometimes) is actual ancestral relationships. When we have several possible clades, this should be understood in terms of epistemic possibility. However, these are often used as the basis of developing modal relationships. Knowing ancestry, for instance, is necessary

3 The explanation, as I've characterized it, is potentially a Dray-how-possibly explanation.

for identifying patterns of convergence across taxa, thus underwriting claims about the conditions under which various adaptations might arise (Griffiths 1996; Currie 2013). Recall that perspectival tools are typically pluripotent: applying not only to single instances but drawing out similarities between instances and modelling them more generally. This last function is plausibly understood in terms of objective modal properties.

So, no doubt epistemic conceptions of possibility matter to historical investigation—indeed, the exploration of epistemic possibility is central to chapter 7's discussion of the scientific imagination—but they don't exhaust what is modally relevant for the investigation of lost worlds. Historical science is concerned not simply with the actual and with epistemic possibility, but with modality and objective possibility, and, following accounts like Swaim's, we can understand that possibility as "baked into the world's structure" (Swaim 2021, 265). But which epistemology should follow this ontology? That is, how do historical scientists gain knowledge of objective possibility?

2. Imagination Bad—Artifactualism Good

"But how might we know about these possibilities? You might start by considering our imaginations." Ichthy taps the top of their head, their spectacles bouncing a little comically. "I can entertain the possibility that the mammals survived, and that they evolved taxa with lives much like ours. I can imagine, surreal as it may be, some sentient mammal giving a lecture much like this one. And maybe my being able to imagine this gives us some reason to think that sentient, intelligent mammals are, indeed, possible."

Ichthy pauses dramatically. In the audience, one graduate student nudges another with a slight smirk, indicating that they've heard what's coming before.

Ichthy puffs their chest out. "But I am a scientist, and just picturing or considering things in the cage of my mind should carry very little weight indeed. Instead," Ichthy gestures to the toy Broadfoot still held by one of the students, "we build things. And the things we build tell us about what could be."

I've thus far laid out a few options regarding the kinds of modal knowledge sciences of the deep past might generate and argued that, at least sometimes, objective possibility better captures the dividends. This leaves a set of crucial epistemic questions open, especially what Wilring and Grüne-Yanoff call "the epistemic question" concerning modal models. They ask: "What must models or modelling practices be like in order to provide good reasons for modal claims (*under what conditions*) and why does fulfilling those criteria make models epistemically valuable with respect to modal truth (*in virtue of what?*)? [italics in original]" (Wilring and Grüne-Yanoff 2021, 4).

Multiple questions, then. In this section, I'll make some steps toward answering them. I'll start with a set of views that appeal in some sense to the cognitive, imaginative capacities of scientists, and argue that these don't capture what is going on in the cases we've discussed. I'll then turn to, I think, a better kind of view—artifactualism—and argue that it does.

The argument in this section is narrow in several senses. First, if you'll permit me instrumental use of a much-maligned distinction, I'm stepping into the *context of justification* regarding the role of imagination. That is to say, I'm not here interested in the role imagination plays in the generation and development of ideas (discovery and pursuit) but in claims that point to imagination playing a special role in justifying scientific modal claims. Second, I'm interested in the justification of modal knowledge derived from perspectival tools. As we'll see, some of the prominent views on the table accord imagination a special justificatory role for that modal knowledge, and I deny this. So, imagination might well play justificatory roles throughout science, but it doesn't play a special one in this case. Third, I'll be taking imagination in an "imagistic" way, that is, as a mental faculty that involves sensory-like representations. As we'll see in the next chapter, there are other ways we can understand imagination that are more useful for our purposes. Indeed, in next chapter, I'll argue for an artifactualist account of scientific imagination generally.

I'd advise not getting too caught up in what I mean by "knowledge" here—I'm using it loosely, in a way that could include close by epistemic notions, such as understanding or grasping. I also don't think my usage needs to distinguish between propositional and non-propositional ("know-how" or ability-based) conceptions of knowledge. Although I'll speak of dispositions and express counterfactuals propositionally, I suspect this is a matter of presentation.

2.1 Imagination Bad

The idea that models are valuable because they provide modal knowledge is a move often made by philosophers trying to make sense of highly abstract models that bear little or no resemblance to their target systems—sometimes called "minimal" or "toy" models (Reutlinger et al. 2018; Gelfert 2019). Given that such models surely can't give us true information about actual targets per se, perhaps they can give us true information about what is possible. Such models idealize in order to isolate a system's essential dynamics. As Michael Weisberg has put it, a minimal model "contains only those factors that make a difference to the occurrence and essential character of the phenomenon in question" (Weisberg 2007b, 642). The epistemic questions ask, then, what reason do we have to think that such modal information is justified? And, if models provide knowledge of objective possibility, in virtue of what do they do this?

Some answers to these questions appeal to scientific imagination. Following Wilring and Grüne-Yanoff's characterizations, there are at least two versions of the proposal. One takes models to describe fictional worlds and rests their credibility on the imagination-driven credibility of fictions. Meanwhile, another set is interested in the "conceivability" of worlds, that is, the capacity of relevant experts to imagine them. What unifies these, more or less, is that the justifying circumstances under which the models do their modal work are—so far as I can tell—inside scientists' heads. For Michaela Massimi, for instance, modal claims derived from scientific tools are justified when the relevant scientists can *physically conceive* of those claims. That is, they can imagine it under the constraints of their background knowledge and the laws of nature they know. Till Grüne-Yanoff, by contrast, takes the fiction described by models to be judged as *credible* in ways similar to how we judge the credibility of fictions: although fictions depart from the actual world, we still seem able to judge them as more or less credible. "A model is credible—and hence a good guide to justified possibility claims—when the model world is internally coherent and a competent user of the model would judge the development in the model world to be intuitively plausible, conditional on the model set up" (Wirling and Grüne-Yanoff 2021, 6).

So, in addition to requiring coherence, such views take a scientist's *judgement*, a judgement underwritten by some imaginative act, to be necessary for the justification of the modal knowledge generated. Reflecting on the modal claims we've seen historical scientists make, I find such appeals implausible.

Let's run imagination-views through a simple case. In chapter 3, one of our examples of artificing involved Woodruff and Varricchio constructing a burrow, filling it with rabbit bones, and probing the relationship between bone deposition, initial bone position, different kinds of burrow inflow, and different compositions of flow. They primarily did this to test hypotheses concerning a token past world: whether the critters found in the fossilized Jurassic burrow should be interpreted as starting within the burrow or having been deposited from the outside. They also intended the model results to have more general applications, applying to fossil burrows across a wide range of actual and non-actual instances. As I argued in chapter 3, their experiment is pluripotent. In addition to testing hypotheses concerning that fossilized burrow, they are reasonably understood as aiming at objective modal knowledge: namely, of the dependencies between those variables in burrows. They thus tell us about the dispositions of skeleton deposition in burrows. And knowledge of these dispositions underwrites counterfactuals and other modal claims. Their argument depended on this being the case: we should interpret the dinosaurs as having been in the burrow at the time of death because if they had not been in the burrow, the bones would have been distributed differently.

Let's focus on the counterfactual: *if the critters had been outside the burrow, then they wouldn't have left that pattern of bones*. By *credibility*, this modal information is justified by the relevant scientists considering the relationship between the fictional world where the critters didn't start in the burrow—imagining it, using their background knowledge—and then, given the model setup and their own expertise, concluding that the counterfactual is "intuitively plausible." By Massimi's "physical conceivability" view, similarly, we should accept the counterfactual because it is physically conceivable to the scientists in question, that is, given the relevant constraints and relevant "laws," they can imagine this outcome.

I don't doubt that Woodruff and Varricchio did imagine and think through the results of their studies in the way described by these views, but I don't find this a convincing way of accounting for the epistemic question. Appeals to imagination explain neither the conditions under which their experiments generate modal knowledge nor what justifies it.

Scientists' imaginings are extraneous to the justification of the modal knowledge they generate. By setting up an experimental apparatus and performing various interventions, Woodruff and Varricchio empirically generate and explore a modal space. They are able to map, in the experimental system, the conditional relationship between initial and final bone position against various differences in inflow, bone density, etc. The justification for this conditional knowledge turns on features of the experiment, their data-recording practices, and statistically relevant properties concerning how representative the experiment is. The addition of the scientists doing some imagining doesn't seem to bring any obvious justificatory weight: after all, there is a physical object that generated the data. No doubt this requires interpretation, but I don't see any particular reason to think this interpretation would require imaginative capacities, especially. If their empirical arguments hold water, they have established a set of dispositions possessed by the experimental apparatus they built.

However, generating modal knowledge pertaining to the experiment isn't the end of the justificatory story. We must also consider whether that set of dispositions can be projected onto both the Jurassic burrow that it is intended to probe and onto more general characteristics of burrow infills (perhaps concerning the "universality class" at hand). I find imagination extraneous here as well. As we saw in chapter 3, Woodruff and Varricchio carefully point out various differences and discrepancies between the object of intervention and the eventual target, and possible idiosyncrasies about their experimental apparatus that could undermine its representativeness. For instance, their experiments involve a single, complete rabbit skeleton, as opposed to the three incomplete skeletons of the target. In their discussion, they note that "[t]he single rabbit skeleton consists of 230 elements compared to the observed 122 in the Blackleaf Formation specimen. The number of total elements likely would impact the depositional process to some extent, but it is not clear a priori, whether the rabbit assemblage should be increased or decreased in number for proper modelling" (Woodruff and Varricchio 2011, 144). Although this appears somewhat inconclusive, I take the point to be that it is unknown what kinds of effects having more or fewer elements would have, but that, regardless, it is unlikely to affect the main thrust of the experiment, which is to determine the relationship between initial skeleton position and position in deposition.

It is this careful consideration of possible differences, similarities, and confounders that underwrites the justification of projecting the modal conditionals, not the capacities to conceive of the conditionals holding in those other cases, nor judging them as credible. The experiment involved isolating a set of properties thought to be possessed by the denizen, determining the objective modal features of those properties, and then arguing that we should expect similar properties with similar modal features in the lost denizen.

Philosophers familiar with the literature on projectability and experimentation would, I hope, have found this argument somewhat obvious: of *course*, the projectability of experimental results turns on their internal and external validity (Weber 2004; Currie and Levy 2019). I agree, and this is precisely my point. I don't see any special difference between projecting modal information about an experimental setup versus projecting other kinds of information. That is, where imagination-based views make a special appeal to the role of imagination in justifying modal knowledge, I think the same basic structure holds for both modal and non-modal knowledge.

Perhaps a thought experiment will help (perhaps . . .). Let's imagine that Woodruff and Varricchio simply *cannot* conceive of the dinosaur bones ending up in the positions they were found in if they started within the burrow. A failure to conceive, in this instance, involves not finding the result "intuitively plausible." We might envision this as the scientists running a kind of mental simulation, and the patterns of bones just not behaving as they ought. Or the scientists being unable to figure out, through their own reasoning, what follows from their results. I can imagine this being very frustrating for the scientists, and I can imagine this motivating them to do further exploration. And indeed, I can imagine them complaining that they *don't understand* their results. But, crucially, would this undermine the counterfactual claim: "if the critters had been outside the burrow, then they wouldn't have left that pattern of bones"? I don't think so; after all, why should a scientist's capacity to run mental simulations make any kind of difference to *empirically demonstrated* modal knowledge? It is that empirical demonstration, I want to say, that justifies the counterfactual, not the capacity of some scientist or other to imagine it.

Following Grüne-Yanoff's definition of credibility (see above), you might respond that the scientists in the thought experiment are irrelevant because either the model isn't credible or the user is not competent. But I don't think this kind of response works. It seems to require claiming that credibility or competence turns on whether the scientist understands the system they are engaging with. Another set of scientists, reading Woodruff and Varricchio's results, would primarily judge their credibility just as we usually do: Is their evidence good? Was the experiment set up well? Has external validity been sufficiently established? Is the lab capable of this kind of work? When a scientist asks after another's credibility, they ask these kinds of questions; they don't ask after scientists' capacity to imagine. Again, I don't mean to claim here that imagination plays no role in science, or even in justification; rather, I'm arguing that it doesn't seem to play any kind of special role in establishing the modal knowledge generated by studies like the one at hand.

You might complain that I'm taking "imagination" or "conceivability" too literally—to say that something is "conceivable" isn't to say that some actual agent needs in fact conceive it, but rather that some sufficiently idealized agent would do so (what has been called a "non-epistemic" account of conceivability; Massimi 2022, 148). It is sometimes a little ambiguous how psychologized these terms should be taken to be. By Massimi's account at least, "imagining" appears to be a practice both attributed to a particular epistemic subject but also an epistemic community, suggesting that conceivability is at least to some extent de-psychologized (depending on how happy you are with communities having minds). However, I think ultimately we should read Massimi's and other appeals to "imagination" in this context to place psychologistic capacities in a crucial position for justifying the modal deliveries of models. The constraints of background knowledge, laws, etc., work for Massimi to hold imagination to some kind of empirical account: "Scientific imagining has to respond to the tribunal of experience" (Massimi 2022, 150).

You might further complain about my choice of case: it is an experiment, not the kind of toy model that imagination-folks are concerned with! Fair enough: let's consider a much more "modely" case: Emily Mitchell and company's attempt to understand the rangeomorph-dominated ecosystems of the Avalonian Ediacaran. As we saw, ecological models of the spatial and taxonomic relationships of those fossil beds turned up a unique, neutral-dominated pattern. In response, Mitchell et al. adopted a strategy I've called *anchoring*, navigating among various natural systems toward informing a model of those systemic dynamics. They considered various properties of those Avalonian systems—their relative immaturity due to frequent disruption, small sizes, lack of resource constraint—and then compared these with contemporary systems that also exhibited these properties, demonstrating that they too exhibit the telltale signs of neutrally structured ecosystems. This involved modal claims, such as *if the Avalonian ecosystems had more resource competition, then they would be more niche-structured,* and *if the Avalonian ecosystems were more resilient, they would be more niche-structured.* The justification of these counterfactuals doesn't at all seem to me to involve the scientists imagining or conceiving of them, regardless of their expertise and how much constraint the "tribunal of experience" provides. It comes from the specific arguments they provide and the demonstrations they perform. A thought experiment à la the one above can be rerun: even if Mitchell or her colleagues were unable themselves to imagine the counterfactuals, if they failed to physically conceive of them, or failed to judge them as credible, the justificatory work would nonetheless be carried by the dispositional properties they establish in their models, and the links they empirically draw between models and real-world targets.

A final problem: I'm not sure whether imagination-based answers to the epistemic question of modality are well-placed to capture objective modal knowledge. Recall that epistemic possibility is, basically, what is left open given the constraints of some individual or epistemic community's knowledge. As Knuuttila has put it, epistemic possibility "depends on scientific knowledge of the actual state of the world. Any scientific claim that is not ruled out by scientific knowledge is epistemically possible" (Knuuttila 2024, 65). By contrast, *objective* possibility is not simply a claim about what is consistent with our knowledge of the actual, but involves modal claims beyond actuality. The counterfactual statement *If there were more competition for resources in Avalonian systems, then they would be less neutrally structured* is epistemically possible if the statement is consistent with what scientists know. It is objectively possible if, regardless of scientific knowledge, the counterfactual is true (or sufficiently veridical, or justified, or whatever). As we saw, one way of differentiating between these is whether the space of possibility shrinks as more information comes in. The epistemically possible is geared toward the actual, and so—when we are lucky—these will converge. Not so of the objectively possible. It might turn out that Avalonia's systems are not neutrally structured after all, but the above counterfactual conditional is not thereby ruled out. My argument is that sciences of the deep past provide us with knowledge of objective possibility, not merely epistemic possibility. Now, consider the imagination-based accounts on the table.

Both families of imagination accounts—those relying on "conceivability" and those relying on "credibility"—involve a criterion of consistency and a psychologistic criterion: roughly, the relevant representations must be consistent with what else the agent knows, and the agent must grasp them in the right way. This is intended as an answer to the *justification* of the modal knowledge. The consistency aspect appears to simply track epistemic possibility. To be consistent with a set of scientific knowledge *just is* to be epistemically possible vis-à-vis that knowledge. It no doubt, then, confirms epistemic possibility. But does it give us any particular guide to objective possibility? It might do: if we think our scientific knowledge is really well-grounded, if we think it tracks the structure of the world, then perhaps it does. But in that instance, the justification isn't to do with imagination, but concerns whether we have good scientific knowledge. How about the psychologistic criteria? Does an expert being able to grasp, or run a mental simulation, or draw the right inferences, or in some sense "get" the result, add epistemic weight to a conclusion being about objective possibility? In some cases, perhaps, as a matter of testimony, the intuitive judgements of experts under the right conditions might be a good reason for us to take a hypothesis seriously. But given how epistemic agents are situated within their knowledge, we must ask: what are they running mental simulations *of*? Well, it would be their understanding of the relevant phenomena under the constraints provided by their background knowledge. In other words, epistemic possibility. So, only when epistemic and objective possibility track one another should we expect imagination to track objective possibility. And, as we've just seen, that appears to only be a matter of the consistency of our background beliefs.

You might worry that my arguing that imagination enjoys no special role in underwriting modal knowledge doesn't undermine the role of imagination if it underwrites knowledge, or at least knowledge generated from models, more broadly, as is often the case for some views, especially those sometimes called "fictionalist" (see Toon 2016; Levy and Godfrey-Smith 2020; Salis 2020). The fictionalist, I think, can happily agree with the negative claim of this section. Typically, appeals to imagination by fictionalists are not appeals intended to directly justify claims derived from models. Rather, imagination matters for fixing a model's *content*. For a fictionalist, claims about models (that, say, Cuthill and Conway Morris' geometric programs represent rangeomorph development via some fictionalist semantics) should be understood as invitations to imagine: "We can regard a model's equations (or even a verbal description of the model) as a prop that, given suitable (scientific) principles of generation, imply what the model's content is" (Levy 2024, 102). Fixing a model's content is crucial for identifying claims derived from the model, but this doesn't say anything in particular about what justifies that content or propositions derived from it. As such, one might entirely go along with my arguments that imagination plays no special role in answering the epistemic question about modal models, and still adopt a fictionalist view (in chapter 7, however, I will disagree with fictionalism about the scientific imagination more generally).

So, at this stage at least, my argument isn't that imagination isn't crucial for knowledge generally, but simply that it plays no special role in justifying claims to objective modal knowledge.

So, answers to the epistemic question about modality that appeal to imagination won't work because (1) imagination is extraneous to justification, (2) they put too much weight on scientist's grasping compared to the work put into designing, tweaking, and making arguments concerning perspectival tools, and (3) they appear to give us epistemic, not objective, possibility.

Happily, there's another view within the modelling tradition that provides, I think, an answer much more aligned with the modal knowledge we've discussed: *artifactualism*.

2.2 Artifactualism Good

Emphasis on imagination—mental simulations, "grasping," or other psychologistic accounts of the modal power of perspectival tools—doesn't sit well with the *bespoke* nature of many perspectival tools. Palaeontologists do not simply grab well-known or well-used models and imagine them in new scenarios. Palaeontologists build, adapt, and tweak model design, construction, and deployment toward particular idiosyncratic features of the systems they're trying to understand. As I'll argue in chapter 7, imagination is deeply coupled with perspectival tools. As such, I think an artifactualist account has a better chance of answering questions regarding the justification of modal knowledge.

Artifactualism is a set of views concerning models, elements of which are "implicit, in many practice-oriented approaches to modelling" (Knuuttila 2024, 122). Artifactualists emphasize models' materiality in the context of their use and, in various senses, emphasize the non-representational aspects of models. The view has been largely developed by Tarja Knuuttila (2005, 2011, 2021) with the central contrast being representational, particularly fictional, accounts of models (see also my attempt at a similar view, Currie 2017b, and de Oliveira's more radical form, 2022).

Most accounts of models are *representation-first*. That is, they think of models as primarily standing in some relationship to some actual target system, and ask in virtue of what that relationship is representational. On such views, Cuthill and Conway Morris' model is fundamentally a model *of* rangeomorph growth. Such accounts start with what Knuuttila has called "external representation": "the relationship of a model to a real-world target system" (Knuuttila 2024, 65). Artifactualist accounts begin instead with the model itself. A model is not fundamentally a representation of another system, but a system *itself*—a set of interdependencies, expressed through some (to use Knuuttila's (2011) language) *mode* and implemented in some *media*. The former is the symbolic devices at play—the language, if you want—and the latter is the material substrate at hand (in 2017, I made approximately the same distinction by distinguishing "vehicle" and "content," terms that are too representationalist in retrospect). To use a model, scientists must *interact* with it in some way, as Guilherme Sanches de Oliveira puts it:

> In order to be used by scientists, even [mathematical] models must be implemented in some way that enables interaction and manipulation, including, for instance, as markings and inscriptions with pen on paper or chalk on a blackboard, or as a programming code typed up and displayed on a computer screen. It might be tempting to think of "the model" as transcending, or being independent from, any particular physical implementation—still it's precisely as some physical implementation or other that the model enables scientists to intervene in some way (e.g., changing parameters, settings or variable values) and to visualize or otherwise measure the effects of those interventions. (de Oliveira 2022, 4)

Let's briefly work through an example. A mathematical model will follow some mode. Cuthill and Conway Morris' (2014) model of rangeomorph development consists of a procedure for generating geometric patterns based on the serial repetition of self-same fractals. This procedure is the *mode*. However, this mode must be realized in some *media*. Cuthill and Conway Morris used a computer, but in principle they could have used pen-and-paper, or straws, or imagined it in their heads. In a certain sense, the various media are different ways of "instantiating" or "realizing" the mode, but each medium has very different properties that are often highly relevant to its epistemic possibilities. Computational media are constrained by hardware and software and are often able to generate a large number of runs on comparatively short timescales. Pen-and-paper media leave much more open to human calculation error and take much more time—however, in computational cases, human error in the form of coding mistakes is often much, much harder to discover.

When it comes to modal knowledge, artifactualist accounts—whether "hybrid" versions that make concessions to representation (as favoured by myself and Knuuttila) or the more "radical" forms defended by de Oliveira—have a crucial advantage over views that appeal to imagination. The various artifacts scientists build are not considered representations in the first instance, but are instead considered "systems-in-themselves."[4] As systems-in-themselves that scientists have constructed and examined, their use is primarily intended to establish the dispositions of those objects. That is, they generate knowledge of objective possibility concerning those objects. The trick, then, is to provide good reason to think the relevantly same dispositions will occur in the objects they ultimately want to understand, or in the "fragile systems" they're targeting.

The artifactualist perspective also provides some clues as to under what conditions modal knowledge might be justified. Knuuttila does appeal to imagination-based accounts in her artifactualist justification of modal models: "The question of conceivability is crucial for modal inferences" (Knuuttila 2024, 9). However, the way this plays out in her discussion is less about mental simulations and more about serious study of "whether the possible systems studied by modelling could be actualized" (Knuuttila 2024, 11). As she points out, often synthetic biologists attempt to determine whether theoretical models can be built using organic components, explaining how the various media through which the structures are realized enable different kinds of knowledge. I have made a similar point about how ecologists navigate between highly abstract models, field data, and simple "bottle" experimental systems:

> A coupled differential equation is one way of representing a set of assumptions about the causal interactions between trophic levels. Positing such an equation sets a particular modal domain, and allows us to explore it. That is, it underwrites our learning how to navigate the domain specified by the equations. We see the same in bottle experiments. But—crucially—the mediums are different, the way we specify our assumptions are different, and both the modal domains, and the way we

4 This term—"systems-in-themselves"—might be somewhat misleading, as it might imply that they do not have functions. As artifacts, they *do* have functions. However, these are often open-ended, in flux, and, crucially, they are not restricted to *representational* function (Currie 2017b; Knuuttila 2024).

> explore and navigate these domains are different: thus, I argue, different understandings are generated. (Currie 2020, 924)

For artifactualists, modelling practices often involve using different media to realize "the same" mode. This is done to establish the more robust dispositions across those systems of interest. In light of the book thus far, we already have the means of describing this: *anchoring*. When we're trying to understand lost worlds (and I bet in many other scientific circumstances), toy models and other scientific artifacts are other anchors across which we iterate.

A useful way of making sense of artifactualist epistemology is Catherine Elgin's notion of "exemplification" (Elgin 2017). Something *exemplifies* a property when it possesses that property. In virtue of having that property, examining that something can be a way of demonstrating features and dispositions associated with that property. I might represent a *wing* in tap-dancing verbally: the dancer scrapes their foot out to one side, then quickly hits the floor before stepping down. Alternatively, Kirsten might exemplify a wing by, well, putting on her tap shoes and performing one. Similarly, Cuthill and Conway Morris' geometric model exemplifies some properties of rangeomorph development in virtue of in fact possessing those properties. Exemplifying and examinations of exemplifications can act as demonstrations of, for instance, modal properties of the relevant objects.

So, here is the basic strategy for generating modal knowledge according to an artifactualist account. In standard, say, experimental settings, we ask after the projectability of results, the representativeness of the specimen, and so on. The basic answer to these questions involves finding some warrant, a set of reasons, to think that the specimen and results are sufficiently similar to our ultimate target. Steel (2008), for instance, relies on what he calls *comparative process tracing*. This involves identifying the mechanisms responsible for experimental behaviour and tracking whether the same mechanisms are present in the target. The artificed dinosaur burrow and careful comparison between it and reconstructions of the Jurassic burrow are an example. Now, having identified a disposition in the system we are examining—mapping out a modal structure for it—we then, in just the same way, ask if that modal structure is representative (and *how* representative it is) of the non-experimental structures we're interested in. If we're thinking of exemplification, we'd here ask whether the exemplified properties are in fact the ones we're interested in, in the system of interest.

You might wonder what account of justification I have in mind. I don't take myself to have committed to any particular position between, say, internalist accounts that take justification to lie in the beliefs of agents or externalist accounts that incorporate aspects of the world into justification. I think that either internalists or externalists can make sense of what I've said, but I suspect my externalist leanings are probably quite obvious throughout the discussion.

One way of reconciling appeal to epistemic concepts like imagination and artifactualism is to tie imagination closely to "understanding," but rather than appeal to "theoretical" conceptions of understanding (e.g., De Regt 2017), appeal to more embodied, situated, or "know-how" conceptions of understanding. On these views, what it is for a scientist to understand (and, let's claim, "imagine") is to have certain abilities or capacities (see Leonelli 2009; Le Behain 2016; Currie 2020). As Leonelli puts it: "Understanding can only be qualified as 'scientific' when obtained through the skilful and consistent use of tools, instruments, methods, theories, and/or models: these are the means through which researchers can effectively understand a phenomenon as well as communicate their understanding to others" (Leonelli 2009, 190).

Embodied forms of understanding marry naturally with artifactualism. Scientific understanding acts through the objects scientists construct and interact with. It is the capacity to do things with these objects: to draw inferences, make predictions about their behaviour, manipulate them in predictable ways, and so on . . . that constitutes understanding. If imagination is similarly embodied, then it too folds into an artifactualist account of modal knowledge. In chapter 7, we'll return to these ideas, but for now I'm left to draw some general lessons and apply them to a final case.

3. The Science of Lost Worlds Is a Science of Objective Possibility

Scientific justification, at least of the kind of modal knowledge that concerns us here, doesn't happen inside scientists' skulls. Justification lies in intimate connection with artifacts ranging from material specimens to theories to experiments and other kinds of apparatus. In light of loss or the possibility of loss, historical scientists construct perspectival tools: traces, epistemic artifacts, material objects—media—that realize a mode. Demonstrations of these artifacts' dispositions ground scientists' knowledge of the dispositions of the systems that interest them. These artifacts have genuine modal properties, and in virtue of deep understanding—when we have it—of the relationships between those artifacts and the systems they model, those modal properties can be projected via inference onto the past. Thus, when the conditions are right, historical scientists don't simply tell us what actually happened, but can tell us what might, could, or would have happened under different circumstances. Further, those artifacts often instantiate complex, multi-layered modal spaces, which not only enable testing various hypotheses about the actual past, but in themselves represent genuine epistemic achievements.

The argument from the last section, at base, claimed that imagination-based accounts of the justification of modal models—even if they work for the toy models they target—can't make sense of the practices of scientists interested in uncovering lost worlds. Modal claims are justified not based on some scientific capacity to conceive of possibilities, but via demonstration in an artifact that exemplifies the properties of interest. The results of said demonstration are then transferred to the target system or system type.

On this view, no special answer or pleading, nor an appeal to imagination, is required to justify modal knowledge.

I've argued that to understand the sciences of the deep past, you should understand them as tackling two interwoven challenges. First, erasure: information from the past is often fragile and incomplete, typically requiring interpretation through complex chains of reasoning and rich background theory. Second, loss: past worlds are sometimes deeply unfamiliar due to containing unseen, mysterious denizens. Historical scientists always navigate challenges from at least the possibility of loss. Both loss and erasure contribute to the heavily modal nature of scientists' work. Tackling erasure requires understanding the dependencies between traces and the past denizens they are sourced from, as well as various forces and sources of bias and information loss; at base, it requires understanding the dynamics of information loss and decay. Tackling loss (and its possibility) involves understanding the fragility of systems. The well-mixed oxygenation of the oceans is crucial for supporting life as we know it, but that turns on the biological pump generated by the daily migration of all of those aquatic creatures, and that pump itself has a history.

Faced with loss and erasure, then, scientists of the deep past construct artifacts that—they hope—exemplify the kinds of properties they need to understand to make sense of past worlds. These artifacts are (1) applications of strategic perspectivalism: they isolate particular properties; (2) bespoke: they are coopted and geared toward local and often highly particular questions;

(3) pluripotent: while informing us about various instances, they also explore and probe fragile systems, dynamics, and regularities (that is, possibility); and (4) iterative: they are in dialogue with other artifacts, with traces, and often with observational data from our contemporary world. These iterative explorations grant modal knowledge about the conditions under which phenomena occur, their persistence, and so on. This modal knowledge is not merely epistemic, but objective.

This set of claims underwrites another regarding the nature of success in the historical sciences. As Derek Turner emphasizes, we will likely get many details concerning past worlds wrong (Turner 2004, 2005, 2007). The dinosaurs may have been located outside of the burrow, rangeomorphs may not have been obligate osmotrophs, Avalonian ecosystems may not have been neutrally structured, and the emergence of metazoa may not have turned on the emergence of key innovations that structured their oxygen environments. But this is just to say we didn't get the *actual* world right.

We may still, nonetheless, have gained crucial insights into modality. We may come away knowing more about the relationship between burrow infill and bone placement, the properties and limits of obligate osmotrophy, the dynamics and conditions under which neutral ecosystems occur, and how organisms can influence oxygen flow.

So, the sciences of past worlds are sciences of possible worlds, and that possibility is objective. It is objective both in the sense of going beyond epistemic possibility and in the sense of being objectual, that is, grounded in the dispositions of objects.

4. Astrobiology and the Actual in Service of the Possible

In chapter 2, I introduced problems generated from loss via an analogy to astrobiology. Many astrobiologists examine the atmospheric composition of exoplanets (inferred from traces in the form of wavelengths detected by telescopes) in an attempt to identify "biosignatures" (that is, atmospheric compositions that require feedback from a biosphere). There is a sense in which they have a single N of life: that found on Earth. This raises a question: whether or not life on our planet is representative of life generally. If we get that wrong, our characterization of biosignatures could be too narrow (hence generating a bunch of false negatives) or too broad (generating false positives). As I argued, only having 1 N is not so much a problem concerning the amount of specimens or data, but a problem of being able to generate sufficiently rich background knowledge required to make sense of data. If astrobiologists had a bunch of living planets to examine, they would be much better positioned to develop the background theory required to detect unambiguous biosignatures (if such a thing can be had).

Astrobiology is often characterized as a multidisciplinary (or at least a multiple-lines-of-evidence) effort to discover extraterrestrial life. This is typically tied to understanding the origins of life on our own planet. NASA emphasizes this aspect, for instance, characterizing astrobiology as "the broad-based study of the origins of life here and the search for life beyond Earth" (Kaufman 2022). It is, then, at least doubly historical: by understanding various processes that operated in Earth's *past* astrobiologists hope to develop ways of detecting life from exoplanet *traces*; they're further interested in the pasts of neighbouring planets and the possibility of their having housed life or having once had conditions relevant to it (hence the importance of the "warm & wet" early Mars hypothesis).

So, we can more or less divide astrobiology into two potentially quite different but mutually informing tasks. First, what kinds of processes lead to life? That is, under what conditions might life arise? Second, what signals might there be of extraterrestrial life from exoplanets, that is, can we develop an understanding of biosignatures? I'll focus on this latter question.

As I've indicated, much philosophical discussion of astrobiology, and epistemic worries from astrobiologists themselves, turns on the extent to which we can use life on Earth as something like a representative sample when understanding life in the universe. The only specimen we understand in any detail is Earth and Earth's history. So, how can we know whether biosignatures developed based on our understanding of Earth have the right scope? If they're too narrow, we'll potentially miss a lot of life; if they're too broad, we'll potentially misidentify it. Worse, atmospheric composition underdetermines the processes that maintain or generate them. Earth's oxygen-richness is due in part to biotic processes, but on other planets, oxygen-richness could be due to abiotic processes (Meadows et al. 2018). Further, as various sinks could mask oxygen production, a *lack* of oxygen is also no indicator that earth-like life is not present (indeed, it's only been in the last 400 million years or so, a tenth of its life, that Earth would have produced a stable oxygenated signal).

The basic shape of this problem should be familiar to taphonomists. Many mineralogical forms could be produced by biological or purely geological origins. Thus, telling whether a weirdly-shaped rock's properties are indicative of biological or geological signals turns on having a rich understanding of the various ways that various processes could generate those properties. In most circumstances, trace-based reasoning doesn't occur independently of analysis of specific details of context. The same holds for biosignatures: biosignature interpretation relies on "understanding a planetary environment to identify characteristics that may potentially reduce our ability to see a biosignature, even if a strong biological source exists" (Meadows et al. 2018, 632). But this shouldn't be a surprise: it is another way in which recontextualization is crucial for successful recovery of lost or putatively lost worlds.

Notice that I have presented biosignature research as involving modality, but primarily in the service of the actual: we're interested in the dependency relationships between exoplanet atmospheric composition and possible biospheres because we want to know whether that planet has life. But we needn't view astrobiology in these terms. We might be interested in how different atmospheric compositions might come about, in mapping kinds of possible atmospheres, and the kinds of biotic and abiotic processes that could generate and maintain them. To illustrate this and see why it matters, let's turn to a recent example.

Nicholson et al. (2022) are interested in understanding biosignatures in the context of "nutrient-limited" biospheres. Their study focuses on conditions of the early Earth during the Archean, before the "Great Oxygenation Event," when our planetary biota was largely dominated by methanogens. "Given that Earth spent roughly a third of its lifetime in the Archean, it is natural to begin our study of the vast possibilities for biosignatures with this long–lived and comparatively simple biosphere [references omitted]" (Nicholson et al. 2022, 222). The aim of their modelling is to provide environmentally relevant biosignatures for planets in that kind of state. The approach is to use a relatively simple model of methane exchange, but let the biological aspects act as free parameters, evolving and interacting with the atmosphere. As such, they hope to map out a modal space consisting of possible arrangements of atmospheric composition and biological states.

In a nice example of artificing (and very characteristic of astrobiology), Nicholson et al.'s model consists of an atmosphere and an ocean, with three chemicals circulating between them (hydrogen, carbon dioxide, and methane). This simplicity is important for their goal; they characterize their target abstractly to highlight—exemplify—the relevant factors: "As we are interested in the overall behaviour of the simple life-environment coupled system and are not trying to recreate the climate of a real planet, we use these simplifications to keep the abiotic environment simple" (Nicholson et al. 2022, 225). It's also another example of how even abstract models are rendered highly bespoke in strategically perspectival contexts, to facilitate the particular kinds of isolation and eventual recontextualization crucial for the investigation's aims.

Microbes are modelled as a single species that metabolizes carbon dioxide and hydrogen into methane and water. This generates adenosine triphosphate (ATP), which is used to generate biomass. Microbe population dynamics are determined by ATP availability: at each timestep, some proportion starve due to a lack of ATP, and some proportion reproduce due to sufficient ATP (and some proportion do neither). The biological and abiological chemical composition is coupled in various ways: most obviously by the availability of ATP, but also by, for example, the "burial rate," wherein microbe deaths act as a hydrogen sink as they are buried under the ocean floor. The basic experimental setup is to run the abiotic system until it reaches an equilibrium, and then introduce the microbes, varying factors like death rate, and how ATP-greedy the microbe is.

The major result of the various studies is that atmospheric methane remains largely insensitive to biological action, but is highly sensitive to oceanic hydrogen. Thus, variation in methane in such a planetary system can be chalked up to hydrogen variation regardless of the processes of microbes. The upshot of this is that population dynamics are less important than nutrient-availability. On planets like the early Earth, it is oxygen-availability that dampens the system: "When considering a nutrient-limited biosphere it is more important to accurately model the processes that regulate the availability of the limiting nutrient, and determine the limit to which life can exploit this nutrient, than it is to model any specific population dynamics for that biosphere" (Nicholson et al. 2022, 237). For the direct purposes of developing an atmospheric biosignature for this type of planet, this isn't great news: detecting oceanic oxygen and factors affecting it (volcanism, for instance) on exoplanets seems like a big ask, to say the least.

How should we characterize the epistemic dividends of work like Nicholson et al.? Their aim is certainly to help develop biosignatures: that makes sense of the various decisions made in constructing their simulation. But what we learn doesn't only pertain to detecting biosignatures. We're also learning, at least potentially, about the kinds of dynamics that might rule atmospheric composition and various other global systems. We're mapping out possible planets. And this mapping isn't merely in service of the actual:

astrobiologists make sense of complex dynamics and dependencies between global systems. It is, at least in principle, a science of objective possibility.

There is no doubt, then, that astrobiology can be characterized in part as the search for extraterrestrial life. But to leave our description there is to do it an epistemic disservice. Astrobiology doesn't simply ask whether we can detect life on other planets. At its best, astrobiology is interested in the space of possible planets: what kinds of atmospheres could there be, and how could those atmospheres be generated and maintained? But this exploration isn't empirically unrooted—it is not a mere modelling exercise—but is anchored by empirical examples, from the signatures of distant exoplanets to the current and past Earth.

There's a more general lesson here concerning the dividends of historical science. Typically, the explicit aim is to answer questions about the actual past—for example, did this dinosaur burrow? Were rangeomorphs governed by ecological dynamics different from those of contemporary ecosystems? Was there life on distant planets? But the dividends of the study are modal spaces: dependencies between bone position and burrow infill, ecosystem properties that determine neutral spatial distribution, possible constellations of atmospheric composition and underlying biotic and abiotic processes. Perhaps ironically, the focus on the actual provides anchors for the exploration of possibility.

You might complain here: surely the scientists take themselves to be trying to uncover biosignatures, not trying to explore conditionals concerning atmospheres and possible biospheres? And indeed, this being an aim of research like Nicholson et al.'s needs to be taken seriously to understand why they undertake the research in the way that they do, why various aspects are salient, and so on. I agree with this. But I distinguish between the *aim* of an investigation and the *dividends* of an investigation. Dividends can be byproducts, unintended upshots, or unexpected windfalls. Scientists needn't be aiming for dividends that they nonetheless achieve.

In a certain sense, then, we might think of these cases as not only involving the possible in service of the actual, but the actual in service of the possible.

7

Imagined Worlds

> A member of the public raises a claw: "Uh, you painted Broadfoot a kind of browny colour—do we know Broadfoot was brown?"
>
> Prof. Ichthy laughs, "Well, maybe we don't know Broadfoot was brown, but we can certainly imagine Broadfoot being brown."
>
> Noting the suitably nonplussed reaction from the crowd, Ichthy becomes a little more serious: "Let me explain. There are a lot of things we're not yet sure of, such as Broadfoot's colouration. However, we are in a position to make pretty educated guesses. I'm sure some things in the past were surprising. Some mammals, I'm sure, sported wonderful and weird colouration. After all, a bunch of the animals around us have surprising colours. However, the browny colour we've speculated for Broadfoot is both the kind of colour you might expect an animal who lived like that to have, and it tracks the colour of most mammals we know of today. So, although we're not certain, there are constraints on how we imagine. And while I probably wouldn't stretch that to count as bona fide knowledge, I think these constraints matter."

Scientists confronted with putatively lost worlds are confronted both with the limits of their imagination and, at the same time, the tools needed to expand it.

In the introduction, I boldly claimed that this book is about the scientific imagination, but at this stage, you'd be forgiven for scepticism on that score. Thus far, all I've said about the imagination is to argue that imagination fails to play a special epistemic role in grounding modal knowledge in science. It's time to give a positive characterization of imagination in science, or at least the imagination involved in uncovering knowledge of lost worlds.

My account of scientific imagination will be rooted in artifactualism: I'll argue that perspectival tools underwrite, expand, and extend scientific imagination; indeed, those tools partly constitute scientific imagination. I'll not deny that the psychologistic aspects of some scientists' imagining play important roles, but will claim that for the majority of the time, imagining is best understood as carried by perspectival tools. The tool is often required for the imaginative act, often carries the imaginative act's content, and is necessary for the act's spread and development among scientific communities. In arguing for an artifactualist account of the scientific imagination, then, I'll soften the focus on scientists' individual psychologies.

Imagination so understood, then, does play a justificatory role in our knowledge of the past; however, imagination is significantly de-psychologized as compared to the views discussed in chapter 6, and I retain the position that imagination doesn't play a *special* role in uncovering modal knowledge: as we'll see, it is just as necessary for knowledge of the actual past.

Before presenting my view on imagination, I'll discuss the role of imagination and modality in palaeoart, that is, attempts to provide artistic representations of the past worlds that historical scientists hypothesize about. I'll focus on a case from recent history: "shrink-wrapping." I understand this kind of palaeoartistic practice as the combination of two different "modes": both representing extinct taxa *as* living organisms, as associated with natural history illustrations, and being epistemically responsible via being constrained by contemporary palaeobiology. The demand to represent them as living organisms requires the "concretization" of various hitherto abstract ideas. And this requirement sometimes clashes with the expectation that palaeoart should be epistemically responsible. I will, then, articulate a dilemma for palaeoartists—a clash between the demand to represent extinct taxa as natural-history organisms and the demand to be epistemically responsible—and I'll then suggest a particular way out of that dilemma, one that is reflected in how palaeoart is often practiced now. In particular, I'll suggest that palaeoartistic representation ought to be pluralistic and take diversity in representation, thus exploring epistemic possibility as a central motivator.

This chapter, then, has two major parts. First, a side path into palaeoart. There, I'll both make some direct arguments about palaeoartistic practice and provide the contours for the more general account of imagination in the second part. Before any of that, however, we need some further preliminaries about imagination.

1. The Poverty of Imagination

> "But why care about what I can imagine?" Ichthy asks, reflectively. "Prior to discovering Broadfoot, such a weird little creature existing wouldn't have occurred to me. My imagination is terribly, terribly restricted. But also, if I had imagined Broadfoot, such imagining would be rightly dismissed as so much fancy. After all," Ichthy shrugs, "I can imagine whatever I want, so of what scientific import is me happening to imagine something?"
>
> "As a scientist, my imagination is in a somewhat paradoxical bind. It is both too restricted and too free."

Uncovering lost worlds requires imagination: at a minimum we must conceive of denizens differing from those that are familiar to us. Beyond mere conception, we often attempt to represent them in various ways: with linguistic description, various abstract representations, attempts to generate various of their properties, such as what they sounded like and looked like. The imagination involved in uncovering lost worlds is deeply driven by empirical discovery: Prof. Ichthy is fond of emphasizing that it is only in virtue of the fortuitous discovery of Broadfoot's fossil that they even conceived of such an organism. As James McAllister has put it: "In empirical discoveries . . . scientists are repeatedly confronted with phenomena that cannot be comprehended in terms of the familiar. Far from outstripping the range of empirical experience . . . imagination is broadened by empirical input: Scientists learn after having made discoveries how limited their imaginative powers have been" (McAllister 2012, 15).

My aim is to provide an account of scientific imagination geared toward understanding this feature of the science of lost worlds. To get there, I need to situate myself within the voluminous discussion of imagination.

As you'd imagine, there is a *lot* of work across the academy on imagination. And this work approaches the imagination, both conceptually and empirically, in a variety of different ways (e.g., Forgeard and Kaufman 2015). One advantage of this diversity is that, in tackling imagination, one is fairly free to stipulate which aspect interests you in the relevant context. What, then, is the appropriate conception of imagination for considering the science of lost worlds?

Arnon Levy has distinguished between *thin* and *thick* conceptions of imagination. Where thicker notions involve some kind of mental imagery, "a kind of off-line sensory experience" (Levy 2024, 100), a thinner account of imagination "is to merely entertain, to take under consideration without presuming the truth" (Levy 2024, 100). Similarly, James McAllister distinguishes between notions of imagination that focus on *mental imagery* and another—thinner—usage that cross-cuts it. "Mental imagery" isn't restricted to images but includes other sensory imaginings, such as imagined tactile feel, smell, or taste. The focus on mental imagery is highly psychologized: imagination is that thing that makes use of sensory phenomena in your head. The other sense of imagination, which McAllister and I are broadly interested in, is concerned less with mental images but instead with the unknown. He understands "imagination" as involving "a mental capacity for conceiving entities, states of affairs, events, and phenomena that have not previously been observed. In a generic sense, imagination plays a role whenever someone originates novel output of any kind (McAllister 2012, 11).

This is a fairly minimal conception of the imagination: the mental capacity for originating novel output. This differs from mental imagery in obvious ways. First, mental imagery need not be novel—consider visual memory, for instance (this is one way that McAllister's conception departs from Levy's). Second, novelty need not involve mental imagery. No doubt, conceiving of lost worlds sometimes includes mental imagery, as it does the production of images. However, sometimes discovery takes more strictly symbolic forms, from narrative to mathematics. McAllister's notion is further constrained by being about scientists' use of mental capacities "when they conceive possible features of the world that they have not previously encountered empirically" (McAllister 2012, 11). This combination of mental capacity and the unknown is common in accounts of imagination. Both Sheredos and Bechtel (2020) and Mike Stuart (2020) define imagination in terms of creativity and imagistically. As I'll provide reasons against tying imagination—*creative imagination*—to mental imagery per se, the account is somewhat at odds with these. However, this doesn't mean that notions of the cognitive are left out. As we'll see, I'll understand mental capacities in an extended or situated (de-psychologized) way suitable to an artifactualist account.

So, when I'm interested in the *scientific imagination*, like McAllister, I'm interested in the capacity to generate novel conceptualizations of the unfamiliar, specifically lost worlds. This concept of imagination is closely related to work on speculation and creativity, and I take myself to be expanding on previous work on those topics (Boden 2004; Hills and Bird 2019; Stuart 2022; also Currie 2019e, 2021c, 2022).

It is worth noting that my focus departs a fair bit from much of the work on imagination in the philosophy of science, which tends to target scientific thought experiments (Brown 1991; Meynell 2014; Stuart 2016; Murphy 2020). Surprisingly often, particularly when introducing novel theories, scientists employ fictional—often impossible—scenarios to illustrate or make sense of the ideas they explore. Philosophers disagree on whether the mental imaginings involved in such thought experiments play a role in justification properly speaking. Nancy Nersessian (2007), for instance, argues that sensory imagination is necessary for the persuasive force (and indeed the justification) of scientific thought experiments, while John Norton (1996) has it that every scientific thought experiment can be stripped of its imagery and presented as a formal or semi-formal argument (see Stuart 2020 for more). This dispute turns on whether the act of imagining itself is necessary for the thought experiment's persuasive force.

The importance of this argument expands dramatically if, like some in the debate, we take there to be an equivalence between the kind of imagination employed in thought experiments and in scientific modelling. Salis and Frigg (2020) are especially explicit about this: "TEs [thought experiments] and SMs [scientific models] involve the same kind of imagination. The imaginative activities involved in SMs and TEs can be analysed together" (Salis and Frigg 2020, 22). When I get to sketching an artifactualist account of imagination, I'll agree to some extent with this: insofar as mental imagery and, say, computational systems can be media that accommodate modes, they can do imaginative work in similar ways. However, "perspectival tool" is a broader category than "model," so it doesn't follow that the same treatment will work across all cases that interest us. And, indeed, when we're interested in McAllister's conception of imagination, thought experiments and models are just a subset of the varied ways in which we might conceive of the unfamiliar (see Skaf and Stuart 2025 for further discussion of the differences between thought experiments and scientific models).

McAllister describes two contrasting objections to the use of scientific imagination. The first is the *arbitrariness objection*. We are, in some sense at least, free to imagine what we like. Not in the sense of being able to imagine anything at all, but in the sense of our imaginative capacities being insufficiently constrained for us to have good reason to think imagining tracks what there is, or even what there relevantly might have been. As McAllister puts it, "An entity or state of affairs conceived by an act of imagination is just one among many that could be imagined: None of these has any more claim to reality than any other" (McAllister 2012,13). The second is the *poverty objection*: "The fact that it is impossible or difficult to rule out that our imagination is insufficiently creative to apprehend reality" (McAllister 2012, 15). There are a lot of weird things out there, so why think that we've imagined all of them?

Arbitrariness and poverty together are especially troubling for uncovering lost worlds. Consider the plethora of explanations posited for pizza-disk fossils. They have been imagined as jellyfish, arthropods, microbial colonies, and false taxa. By the arbitrariness objection, our merely being able to conceive of these interpretations as such doesn't seem to carry much weight, given the other ways in which we might imagine. We can imagine them in a bunch of ways, so why think any particular imagining is important? By the poverty objection, our *inability* to conceive of the range of options undermines the importance of our imagining of pizza disks as they could have been. Considering how surprising the past often turns out to be, why think that the space of what we do imagine maps in any interesting way onto how things were?

Both arbitrariness and poverty have a common source—one that has been prominent in recent philosophical discussions of the epistemic power of imagination—a lack of *constraint*. As Amy Kind has put it, there are roughly three reasons folks give for thinking that imagination cannot give us knowledge: "First, imagining is typically under our voluntary control. Second, imagining is not world-sensitive—the content of an imagining is determined by the imaginer, not by the world. Third, imagining is uninformative—an act of imagining can provide us with no new information" (Kind 2016, 145).

Each of these problems, especially linked with arbitrariness and poverty, lends support to the idea that imagination needs to be constrained. *Voluntariness* feeds into arbitrariness: if we can imagine what we want, then imagination isn't sufficiently constrained to be epistemically relevant. *World-sensitivity* can relate to both poverty and arbitrariness: our imagination's partial freedom from the world makes it insensitive to it, but also, what we *can* imagine is apparently set by our mental prowess, not the potentially much weirder world. Finally, *uninformativeness* is due to imagination requiring us to reshuffle knowledge we already have: the poverty of our current information thus restricts what we might imagine. Together, these form a challenge regarding *epistemic relevance* (Kind 2018; Rucińska and Gallagher 2021). We can imagine a bunch of things, but why think the possibilities we imagine relate to past worlds? Our imaginations are extremely limited, so why think what we do imagine relates at all to what the past was like?

The challenge of epistemic relevance asks how we might constrain imagination in the right way, that is, find epistemic principles that, if imagination followed, would solve this set of objections. As Kinberg and Levy put it: "To the extent that one's imagination abides by appropriate principles one can use it to generate knowledge—including, in particular, knowledge of concrete actual, worldly facts" (Kinberg and Levy 2023, 327). Philosophers tend to be somewhat circumspect about what these constraints are supposed to be (although we'll see a suggestion appealing to embodiment later in the chapter), but are much clearer on what the constraints are supposed to do. First, the imaginer begins with a setup that sufficiently captures the real-world scenario that interests them. Second, the imaginative scenario should then unfold as it would were the scenario to occur in the real world. At base, then, whatever the constraints might be, they should ensure that the imaginative exercise tracks worldly events.

However, there are two recent arguments suggesting that appeals to constraint—or at least systematic appeals that are supposed to underwrite the particular power of imagination—cannot be had. Mike Stuart (2020) has argued from something like a pessimistic meta-induction that, for any attempted general claim about the proper constraints on imagination such that it can be epistemically viable, there will be examples of scientists breaking those rules in epistemically valuable ways (and, presumably, cases of scientists *obeying* those rules in ways that are disvaluable). For Stuart, then, scientific imagination is anarchic in the Feyerabendian sense. That is, there is no systematic story to tell about the successful use of imagination in science. Second, Kinberg and Levy point out that the epistemic stories we might give in constraining imagination are never particular to imagination; rather, they pick out general epistemic rules of thumb, in particular, hypothetical reasoning more generally. As such, they cannot be solutions to problems particular to imagination. I'll return to these challenges later in the chapter.

In previous work, I've characterized some palaeoscience as *phenomena-driven*, that is, their models and explanations are built, examined, and interpreted in continual conversation with the past as revealed by traces (Currie 2019d; Dresow 2021). This feature of the palaeosciences provides initial answers to how scientific imagination might be constrained in epistemically relevant ways. The scientific imagination of palaeoscientists is not so arbitrary, but directed and underwritten by real traces. The scientific imagination is not so impoverished, but boosted by unexpected empirical discoveries and extended by perspectival tools. I think this character is best understood by an *artifactualist* account of scientific imagination. To get there, I want to first tackle a broad case study from the recent history of palaeoart.

2. The Palaeoartist's Dilemma

> "Can we go back to that illustration of Broadfoot, please?" Ichthy waits patiently while the grad student at the OHP fumbles through the slides, dropping a few on the floor, before finding the right one and slipping it into the projector. Ichthy considers the image, peering at it through their spectacles, before continuing. "Some of this reconstruction we're quite sure of, particularly the anatomical details, that is, the bone and muscle. Others we're less sure of. The colour, as we've discussed, is speculative. Not wholly so, of course: again, the mammals around these days tend towards rather drab browns, and that makes sense for camouflage."
>
> One of the keener grad students jumps into the conversation: "This troubles me, though, in illustrations like this, both what we know really well and things we—or the artist—basically made up are presented with the same confidence. This makes it seem like we know a lot more about them than we really do. It would be better to represent them less realistically, I think."
>
> "Maybe," Prof. Ichthy responds, a slight grimace across the jaw indicating some discomfort at the interruption, "but without the realistic interpretation, people would find it harder to imagine Broadfoot as a real, living creature, and I think that's important too . . ."

Lost worlds are times in the past that played by different rules, that contained differing denizens than now, and that are eerily unfamiliar. Learning about them, then, both extends and challenges the scientific imagination. In this section, I want to explore how lost worlds might be represented in vivid, powerful ways in what we might call *science-oriented art*, that is, art that is in the business of seriously representing scientific knowledge, such as we might see illustrating scientific books, sculpted in museums, and in animations accompanied by David Attenborough's voice-over. I'll thus dip my toe into what Victoria Cain has called "the ongoing dilemma of picturing what can never be witnessed" (Cain 2010, 285).

"Dip my toe," indeed. Work on the relationship between artistic illustration and science (to say nothing of the relationship between art and science generally) is an enormous topic (Ambrosio 2014; Daston and Galison

2021; Ambrosio and Sánchez-Dorado 2024) that I can't hope to do justice to here. Similarly, the historical, social, and political elements underlying representations of the historical and Earth sciences are rich (Rudwick 1992; Cain 2010; Rieppel 2012, 2019; Jones 2019; Tamborini 2024). I'll restrict myself to a recent historical episode that highlights a particular epistemic and aesthetic challenge for palaeoart, a dilemma that emerges from the requirement to represent extinct taxa as concrete organisms (in characteristic poses, in characteristic environments) on the one hand, and the requirement that palaeoart be epistemically responsible on the other. This will involve articulating what this kind of palaeoartistic representation affords, which will set us up for a more general defence of an artifactualist account of scientific imagination.

As we saw in chapter 1, a crucial part of 19th-century naturalists' arguments for the importance of palaeontology and the Earth sciences was the discovery of deep time. Throughout the history of palaeontology, the spectacle of various forms of visualizations—from Henry De la Beche's *Duria Antiqiour*, to the Crystal Palace sculptures, to Charles Knight's murals, to *Walking with Dinosaurs*—has played an important role in popularizing and shaping these discussions. Understood broadly, "palaeoart" has been produced across multiple institutional and professional contexts, from formal relationships between professional artists and museums in the early 20th century, to today's expansive online fan communities (Romano and Manucci 2023). However, the term itself was only introduced in 1986 by Mark Hallett (Ansón, Fernández, and Ramos 2015), and has a somewhat narrower use, restricted to what I'll call "science-oriented art," representing extinct organisms and other elements of past worlds. Summarizing Hallett's view, Romano and Manucci put it as follows: "[Palaeoart] would be linked to the reconstruction process starting from available material, a process for which an entirely hypothetical component is essential and unavoidable. A 'palaeoartist' must be able to merge congruently aesthetic and scientific knowledge" (Romano and Manucci 2023, 7).

Debus and Debus (2002) helpfully distinguish between "palaeoart" proper and *palaeoimagery*, a broader category that includes any representation inspired by palaeontological knowledge. I'm interested in that former, more restricted, category.

I'm here interested in approximately the period from the 1980s to today. The repositioning of palaeontology into palaeo*biology* in the 1970s and '80s also reinvigorated palaeoart toward representing the lost worlds palaeobiologists sought to understand, especially in the "Dinosaur Renaissance" where the public and palaeontological perspective of dinosaurs went from, as palaeontologist Robert Bakker said in the 1970s, "the epitome of extinctness, the prototype of an animal so maladapted that it dies out . . ." to ". . . more interesting creatures, better adapted to a wide range of environments and immensely more sophisticated in their bioenergetic machinery than had been thought" (Bakker 1975, 58; see also Ansón, Hernández, and Saura Ramos 2015; Monnin 2023). Bakker's and others' palaeontological illustrations drove the Dinosaur Renaissance by representing dinosaurs in the dynamic ways they hypothesized.

Considering how prominent representations of extinct animals are in the popular imagination, palaeoart is a surprisingly idiosyncratic "discipline." These days, palaeoartists are very rarely palaeontologists and are typically self-trained with input via informal advice through informal channels. The relationship between palaeoartists and palaeontologists could perhaps be understood analogously to how Caitlin Wylie describes the relationship between palaeontologists and fossil preparators: they play different roles in knowledge production and follow differing incentive structures (Wylie 2009, 2015). Palaeoartists are fairly disconnected, then, from the kinds of institutional structures, training, and activities we might identify with "science." And yet, palaeoartists are often deeply informed by science: palaeoart is as much an epistemic activity as it is an aesthetic one.

I will focus on palaeoart that (1) attempts to represent extinct taxa *as concrete organisms* and (2) intends primarily to represent the best current scientific knowledge (I'll explain what is meant by these in due course). This excludes a lot of palaeoart. Karolina Twardosz, for instance, represents extinct taxa *as cartoons*, that is, using the symbolic forms, tropes, and techniques of cartoon drawing as opposed to those of natural history. Jim Robins' illustrations of Prof. Ichthy are also partly cartoonish and non-natural-historical in mode. They thus adopt different "modes," and this makes a difference to their representational capacities. To foreground future arguments and jargon: although both Twardosz and Robins communicate scientific details in part, the "cartoony" mode affords significantly more "implicit non-committal" representation; they are significantly less powerful as objects of "virtual witnessing" in light of this.

Many representations of dinosaurs are purely for entertainment, or at least willing to sacrifice accuracy for other values (Michael Crichton swapping the more accurate "Deinonychus" for the cooler-sounding "Velociraptor" being an infamous example). Regardless, I'm interested in the kinds of illustrations that often accompany journal articles, press releases, museum representations, and palaeobiological books: those that primarily aim to communicate scientific knowledge of the deep past.

My discussion in this section will somewhat isolate palaeoart from its interaction with palaeontology. Especially in light of my emphasis on iterativity, this is unfortunate: no doubt there are many instances of palaeontology influencing palaeoart and palaeoart in turn influencing palaeontology. And in addition to the bountiful historical work already on this topic, I think there is a lot of philosophy to be done in this arena as well. But for my purposes, I'm interested in palaeoart for its own sake because, as we'll see, it provides a rich inroads to thinking about how artifactual modes enable and constrain various epistemic activities.

I'll argue that for science-oriented palaeoart, popular in the wake of the Dinosaur Renaissance, a dilemma arises between the two requirements listed above: first, that extinct taxa be represented as organisms, that is, as concrete three-dimensional living creatures; second, that palaeoartists be epistemically responsible, that is, not misrepresent scientific knowledge. The dilemma coalesces around the practice of "shrink-wrapping."

2.1 Shrink-Wrapping: What Went Wrong?

Taking inspiration from the scientifically-informed palaeoart common in the United States in the early 20th century, especially Charles Knight's work, palaeoartists in the 1970s onward began shifting from what pioneering palaeoartist Greg Paul called "a damaging laissez-faire attitude about constructing dinosaurs" (Paul 1987, 6) toward a more scientifically-informed approach. Part of this new seriousness was due to progress in science: "In some cases dinosaurs can be restored with remarkably high fidelity, almost as accurately as some recently extinct animals" (Paul 1987, 40).

Paul and his fellow travellers (such as Mauricio Antón) have a somewhat ambiguous legacy. As we'll see, on the one hand, they have been hugely influential in demonstrating just how powerful and how scientifically informed images of lost denizens can be, but on the other hand, they brought to prominence a particular style of representation, often called "shrink-wrapping,' which carried some epistemically damaging consequences.

To get a sense of how scientifically-informed palaeoart proceeds, it is useful to start with the information typically emphasized in how-to guides.

Instead of beginning with how to draw, these often dive into rigorous anatomical detail. Sticking with Paul, here he is discussing the skulls of archosaurs:

> Archosaur skulls are like those of reptiles and birds, and unlike those of mammals, in lacking facial muscles and having the skin directly appressed to the skull . . . These features make them much easier to restore. Easily the most prominent skull muscle is the powerful posterior pterygoideus that wraps around and deepens the back third of the mandible. Jaw muscles bulge gently out of the skull openings behind the orbit and in thecodonts and saurischians also out of the preorbital opening and depression. (Paul 1987, 6)

Scientifically-informed palaeoart starts from a *musculoskeletal* reconstruction, which represents the basic skeletal anatomy and musculature of the taxa. The artist then layers *integument* upon their reconstruction: skin, feathers, fat, colour, and so on. The skeletal reconstruction is, more or less, taken as given, and the integument is as informed by modern science as can be.

The style that Paul popularized, "shrink-wrapping," is "the colloquial term for paleoart that squeezes the bones and muscles under a gaunt layer of skin" (McDermott 2020). Illustration 7.1 shows a shrink-wrapped Prof. Ichthy.

Shrink-wrapped Ichthy reveals significantly more skeletal detail than standard Ichthy. The structure of the fins, ribs, and backbone is plainly visible. Why represent extinct organisms like this? The aim is betrayed in Paul's own advice: "It is more important to execute the distinctive profile of each dinosaur than to meticulously add hypothetical details" (Paul 1987, 42). That is, the palaeoartist should emphasize the well-understood skeletal and muscular anatomy, not the highly open integument. This was motivated by an effort to be *epistemically responsible*, that is, to not speculate too far beyond what is scientifically established. Science writer Amy McDermott reports this in her interviews with palaeoartists: "Shrink-wrapping is the conservative approach to imagining an entire animal based on fossil bones rather than speculating about soft tissue as well, says New Mexico-based freelance paleoartist Matt Celeskey; Ugueto believes that shrink-wrapping caught on simply because artists and palaeontologists have been so focused on the form suggested by bones" (McDermott 2020, 2729).

The thinking here is straightforward. To be epistemically responsible, palaeoartists ought to emphasize the parts of their representations that are

ILLUSTRATION 7.1 Prof. Ichthy, Shrink-Wrapped

scientifically well-established. And with some exceptions, integument is highly underdetermined by scientific knowledge: "Integument is the great unknown for many fossil animals; its preservation is rare and infrequent, and even when it's preserved, it may be massively distorted or rearranged relative to position in life" (Conway, Kosemen, Naish, and Hartman 2013, 11; see also Eriksson 2014, 100–101). It's worth noting that the shrink-wrap style was influenced by more than attempts to avoid speculation. First, there were (and are!) some empirical reasons to think that the anatomy of some dinosaurs, particularly the head, is fairly shrink-wrapped. Second, the clear representation of muscles and anatomy harkens to the idealized, heroic human forms of classical sculpture. It is a style that lends itself to the more dynamic vision of dinosaurs pushed by Bakker, Paul, and other Renaissance palaeontologists and palaeoartists. As Victor Monnin has put it, Bakker's style, in particular, involved:

> A powerful visual rhetoric speaking to the reader's imagination learning about dinosaurs' physiology and behaviors. If the dramatic scenes and acrobatic poses can easily be criticized as exaggerated, it is mainly because they do not serve as plain decoration to the written argumentation. On the contrary, their apparent exaggeration serves to create an even clearer contrast between a "traditional" dinosaur iconography and the author's own vision of dinosaurs as warm-blooded animals. (Monnin 2023, 9)

So, aspects of the style played a rhetorical function, but nonetheless, avoiding speculation about integument is the most prominent justification for the style.

There is an interesting similarity between shrink-wrapping and styles of visual representation from the 18th and early 19th century, which Daston and Galison (2021) characterize as following the virtue of "truth to nature." In these cases, illustrations strive not to represent specimens as they genuinely looked, but rather in a "perfected" way that brings out the underlying truth among nature's frustrating variability. Linnean description, for instance, "singled out those features common to the entire species (the *description*) as well as those which differentiated this species from all others in the genus (the *differentia*) but at all costs avoided features peculiar to this or that individual member of the species" (Daston and Galison 2021, 60). These are "reasoned images": not faithful representations of how the plant literally looked, but images that capture the morphological features that distinguish Linnean categories.[1] As in Linnean representation, shrink-wrappers highlight particular features of their subjects, namely, the bone and muscular anatomy. But they do so for different reasons: instead of representing the "underlying truth" revealed by careful study of nature's variability, their primary epistemic aim is to avoid speculating about that variability.

So, here is the first constraint on palaeoart, which led to the influential "shrink-wrapped" style: only represent properties that are sufficiently scientifically established (or, perhaps, *emphasize* well-evidenced properties). This is an instance of the second component of palaeoart, above: represent the best scientific knowledge, that is, be *epistemically responsible*. However, being informed by palaeontological science is not the only aim of this kind of palaeoart.

In his palaeoartist's handbook, Mark Witton emphasizes the necessity of "the skills of a natural history artist, such as being able to produce realistic scenes of the natural world, to arrange interesting compositions, and predict how light and shadow would play out on alien, ancient forms" (Witton 2018, 8). Palaeoartists are expected to represent extinct taxa like dinosaurs *as if they were living organisms*, specifically, using the same techniques used to represent extant organisms. This is one way in which the palaeoartistic cases I'm interested in depart from other public-facing scientific representations and science-oriented art (think of representations of galaxies, atoms, genes, viruses, and so on). While these sometimes employ techniques that imply realism, palaeoartists explicitly make use of natural history techniques that are also used for extant—even familiar—animals. This has a proud and long tradition. The first real attempt at a pictorial representation of a palaeontological scene, De la Beche's *Another Dorset*, "belongs squarely in a well-established pictorial tradition, namely that of natural history illustrations" (Rudwick 1992, 46). This, then, is the first requirement above: represent extinct organisms as concrete, living organisms.

The term "abstract" and its converse "concrete" are, to say the least, vexing across the philosophies of both science and art. In practice, processes of

1 Fascinatingly, Jim Robin's depiction of a theropod graduate student with a toy Broadfoot from chapter 3 isn't a representation of any particular theropod taxa, but rather of a generic theropod, including the paradigmatic features of the group, rather than of a particular species.

abstraction should not be understood as simple acts of omission (Leonelli 2008) but can be actively generative and creative (Stuart and Kozlov 2024; Sánchez-Dorado 2024a). As such, "abstract" and "concrete" are not necessarily mutually exclusive terms. Instead, one useful way of understanding abstraction is in terms of *framing*. This refers to what is included, excluded, and emphasized in the representation, especially when combined with conventions concerning the function of the representation. Abstraction is one way of providing a "prescriptive frame" that encourages certain kinds of viewing, interaction, and interpretation (Walsh, Roberts, and Currie, forthcoming). Consider Catherine Elgin's example: "A fabric swatch exemplifies its pattern, color, texture, and weave. It marginalizes other properties of the cloth—its shape, location, age, and so on" (Elgin 2017, 20). A fabric swatch is certainly concrete—it's a physical object, after all. But there are conventions about what swatches are *for* (giving one a sense of the fabric's aesthetic properties for design purposes), which, in combination with its physical properties, lead it to be taken as representing the fabric more generally. Below, I'll provide an artifactualist gloss on these elements. For now, the point is that palaeoartists represent extinct organisms using the genre conventions and techniques that are used for natural history illustrations of extant animals. Part of the prescriptive frame is an assumption of *realism* by the audience.

One useful way of understanding the function of the realism implied by the prescriptive frame of natural history illustration draws on Shapin and Schaffer's notion of *virtual witnessing* (1986). They developed this idea to make sense of the literary devices used by early modern natural philosophers, who attempted to describe their experiments in ways that ideally lead readers to have an experience approximating having been there—hence, to "virtually witness" the experiment. Historians of palaeontology have deployed the notion in visual contexts (Rudwick 1992; Cain 2010). In the case of post-palaeobiological-revolution palaeoart, extinct taxa are presented *as if* they are subjects of natural history portraits of living taxa. The animals are presented *as concrete living animals*. We might think the same of computer-animated visual media like *Walking with Dinosaurs*, which make use of the techniques and tone of natural history documentaries (although these sometimes have a somewhat tongue-in-cheek parodic undertone). To see how natural history illustration might act like virtual witnessing, consider this portrait of Misty ShadowFang.

Robins' image of Misty ShadowFang isn't photo-realistic, but it is *concrete* in the sense that it represents Misty as a particular, individual organism. It is a *portrait* and, as such, aims at both sufficient accuracy and to capture her "essential air" (Freeland 2007).[2] There is a sense in which the portrait could serve as a stand-in for seeing Misty herself, and hence lead to *virtual witnessing*. This is achieved by various artistic tricks and techniques, as well as by the genre conventions—and thus the prescriptive frames—of portraitists and natural history illustrators.

ILLUSTRATION 7.2 Misty ShadowFang

Presenting extinct taxa similarly to illustrations of concrete living organisms leads to similar implications. That is, similarly to how seeing this image—or even a photo—of Misty could approximate truly witnessing her, so also examining palaeoart could approximate virtually "seeing" extinct taxa. Virtual witnessing is afforded by the organism not being presented abstractly, but as a concrete object. Thus, in creating natural-historical concrete images of extinct taxa, palaeoartists deploy a combination of deep anatomical knowledge with a set of artistic techniques to display the organism vividly.

To get to the palaeoartist's dilemma, and to see what went wrong with shrink-wrapping, we can follow Catherine Elgin in distinguishing between two ways in which a representation might be *non-committal*.

> A representation is implicitly noncommittal with respect to a property if it makes no commitment as to whether the represented object has that property. A stick figure is implicitly noncommittal with respect to its subject's girth. It simply does not go into the matter. A representation is explicitly noncommittal with respect to a given property if its representing the

2 Freeland argues there is an essential tension in portraiture between *revelation* (being faithful to the topic) and the *creativity* of artistic freedom. I take this to be a different tension than the one I'm interested in.

having of one feature precludes its taking a stand on the having or lacking of another. A representation of a man wearing a hat is explicitly noncommittal with respect to whether he is bald, because representing him as hatted makes it impossible for the picture to commit itself on the question of his baldness [italics in original]. (Elgin 2017, 154)

Explicit non-commitment involves masking some properties via the representation of other properties. An example from palaeoart is a common trick in representing *Tiktaalik*, a late Devonian fish that is often considered a transitional fossil from fish to terrestrial animals. We only have fossils of the organism's front half. By representing it as coming out of the water, half-submerged, palaeoartists are both able to highlight its land-and-sea transitional status, and make an explicit non-commitment about the critter's unknown back half. Figure 7.1 is a sculptural example from the Royal Tyrrell Museum.

So, with dim lighting, half-submersion, foliage-afforded obscurity, and so forth, palaeoartists can use explicit non-commitment to avoid speculation.

Implicit non-commitment involves forms of representation that make no commitment to some properties in virtue of the mode of representation deployed. Under usual circumstances, a stick figure is just not in the business of representing girth.[3] Skeletal-muscular reconstructions are similarly abstract: they don't make any claims about integument. The "cartoon" approach Robins adopts in representing Prof. Ichthy, similarly, affords implicit non-commitment (otherwise we'd get our knickers in a twist over Ichthy's bowtie). However, implicit non-commitment clashes directly with the requirement that palaeoartists represent extinct taxa as concrete natural-history objects.

Representing an organism in the shrink-wrapped style is an attempt to be (something like) implicitly non-committal about integument. But, as palaeoart, it also involves representing extinct taxa as whole, concrete organisms. As such, it represents the extinct taxa as in fact being a shrink-wrapped organism. The commonality of this style led to the common perception of dinosaurs and other extinct taxa as somewhat ghastly, monstrous, skeletal beasts. And it did so apparently with the backing of up-to-date scientific evidence.

Worse, shrink-wrapped organisms are also implausible.

---

3 Although, as Rose Trappes points out to me, they are tacitly committed in various ways vis-à-vis our expectations of "default" girth, gender, and so on.

FIGURE 7.1 Royal Tyrrell Museum's *Tiktaalik* Sculpture (from r/NatureWasMetal)

In 2013, Conway, Kosemen, Naish, and Hartman published *All Yesterdays: Unique and Speculative Views of Dinosaurs and Other Prehistoric Animals*, a fascinating book that both argues against the then-prevalent shrink-wrapped style and provides examples of more diverse palaeoart. We'll turn to that positive project in the next subsection. A major aspect of the book is an attempt to represent living animals as shrink-wrapped, including domesticated cats, swans, and so forth. The results are very different from the morphology of these extant animals. And they are profoundly cursed.

So, in attempting to be epistemically responsible, palaeoartists in the shrink-wrapped style inadvertently generated misleading representations of extinct taxa. We have, then, all the pieces needed for a dilemma:

1. Palaeoartists are expected to depict extinct taxa *as whole organisms.*

2. Palaeoartists are expected to be *epistemically responsible.*

3. Depicting taxa as whole organisms is necessarily speculative.
4. Speculation is inconsistent with epistemic responsibility.
5. Therefore, palaeoartists face a dilemma between two requirements: holistic depiction and epistemic responsibility.

By "whole organism," I'm referring to the techniques of natural history illustration, which imply a realist framing. "Speculation" is a tricky term in science with many meanings (Currie 2021c). In this instance, the idea is that in shrink-wrapping, palaeoartists present at best speculative, and likely false, elements of their subjects in a realist mode. It is that element, representing extinct organisms as truly shrink-wrapped, that amounts to an epistemically irresponsible representation.

In the next subsection, I'll argue against premise 4: done right, palaeoart can be speculative but remain epistemically responsible.

2.2 Diversity of Representation as an Aesthetic/Epistemic Virtue

The answer to the palaeoartist's dilemma is to speculate more, not less.

Recall the artifactualist machinery of "medium" and "mode." Palaeoart, as I've described it, works across a range of media: from drawings to paintings to three-dimensional sculptures to moving images. The palaeoart I'm interested in does more or less have a unified mode: representing extinct taxa as living organisms (following the practices of natural history representation) epistemically responsibly (informed by palaeobiology). However, there is an apparent clash within this mode: the concreteness required to represent extinct taxa as living organisms seems to undermine attempts to be epistemically responsible. I suspect that Greg Paul was well aware of this:

> I disagree with photographic-image restorations. Extinct animals cannot be restored with complete confidence or be photographed. Art is the use of visual cues and creative licence to convince people that they are seeing a version of reality. Paintings and other restorations should be just what they are, aesthetic and realistic illusions of how the subjects may have appeared in life . . . [artistic techniques can] create a work that may even verge on the surrealistic but that is based on facts

and possibilities. Such works not only inform, they challenge the viewer. (Paul 1987, 44)

I doubt Paul's comments about art apply generally, but for this version of palaeoart, the techniques of the visual arts use various tricks to "convince people they are seeing a version of reality." They are tools of virtual witnessing. However, I've argued these tricks undermine implicit non-commitment. This led shrink-wrapping to fail in its aim of epistemic responsibility.

As Paul implies, however, we should not give up on virtual witnessing outright, that is, get around the dilemma by denying the first premise. Although less realist representations of extinct lineages are welcome and often powerful, there are both epistemic and aesthetic reasons to sometimes represent extinct taxa as concrete organisms. Aesthetically, at least some of the power of palaeoart is in its concretely revealing the past to us. Visual representation makes lost worlds salient, both to practitioners and the public. It is the sense of virtual witnessing which, plausibly I'd think, leads to much of the awe and other aesthetic attitudes and affect that palaeoart provides us. More epistemically, concretizing our ideas can aid in exploring these relationships, finding consistency or otherwise among them. In other words, exploring epistemic possibility. Given some musculoskeletal reconstruction and other known contextual features of an organism, what behaviours and integument are possible? Similarly to how Butterfield's narrative grafting pulled together the various perspectives involved in explaining the emergence of the metazoan-filled, well-oxygenated Mesozoic, a good piece of palaeoart can do the same for anatomical, ecological, phylogenetic, and energetic traits interpreted from fossils.

This kind of palaeoart, then, has distinctive epistemic and aesthetic power in uniting natural history representation and palaeoscience.[4] Doing this epistemically responsibly, I think, does not involve avoiding speculation, but instead generating as many different images that accord with contemporary science as we can. The answer is not to be conservative but to be diverse. This lesson is well-represented in *All Yesterdays* (Conway, Kosemen, and Naish 2013): Conway, Kosemen, and Naish experiment not only with integument

4 There is a lot more to be said about the aesthetics of palaeoart, and the relationship between the aesthetic and epistemic generally. For recent discussion of aesthetics in scientific practice, see Ivanova and Murphy (2023); for palaeoart-specific aesthetics, see Xhignesse (2024); for discussions of aesthetics geared toward the historical sciences, see Turner's *Palaeoaesthetics* (2019a), Caitlin Wylie's work (especially 2015), and my own contribution (Currie 2023).

but with various behaviours and environmental contexts. A striking example is their reconstruction of aquatic long-necked *Elasmosaurus*, thrusting their necks high out of the water in an ornate and energetically expensive mating display. Their commentary is revealing: "No record of such behavior could possibly exist—it is wholly and unashamedly speculative—and yet things equally spectacular must have happened throughout the history of life" (Conway, Kosemen, and Naish 2013, 20).

In chapter 6, we discussed *mere-how-possibly* explanations, explanations that pick out one way things could have gone. I don't think that Conway, Kosemen, and Naish mean exactly the same thing here. The point of concretizing one possibility among others isn't merely to show one way things could have gone, that is, to demonstrate possibility, but to aid the reader in realizing just how many wild and wonderful denizens could have occupied past worlds. *All Yesterdays* was followed by *All Your Yesterdays*, in 2017, a project that brought together a wide array of differing palaeoartists at a global scale and at various levels of notoriety and professionalism to further represent the possible variation.

Vividly representing some of the diversity of forms and behaviours that could have existed in the past makes salient just how broad epistemic possibility is, and how surprising lost worlds might turn out to be.

So, in addition to (1) whole-organism representation and (2) consistency with science, I'd add a third requirement to palaeoart: (3) palaeoart should strive to collectively generate *a sufficiently diverse* set of representations. Within these constraints, it strikes me, the dilemma is resolved. The dilemma was generated by the idea that concretizing to achieve virtual witnessing was not epistemically responsible. But if we switch our attention to a group-level property—diversity—we see that *variation* in how the integument is filled out comes with epistemic benefits, not costs. Most obviously, it involves exploring epistemic possibility, that is, what is open given current palaeontological knowledge. Thus, observing that variation both communicates what is well established, via what doesn't vary, and what is not, via what does vary.

The upshot of an appeal to diversity is not unbridled speculation. The arguments of *All Yesterdays* don't license, as Darren Naish has put it, "people to speculate away and do whatever the hell they liked, evidence, conservatism and critical thinking be damned" (Naish 2017, 7). Rather, the combination of diversity and coherence with science leads to exploring epistemic possibility. The edges of what is "coherent" with up-to-date science are, I take it, controversial, as is what counts as "up-to-date." But this just makes creative and speculative use of palaeoart an excellent forum for productive discussion of those porous boundaries.

One complaint about whole-organism representation, parodied by Ichthy's graduate student, is that it represents little-known organisms with as much confidence as those for which we have excellent specimens. We know what mammoths look like—we have frozen examples—but integument is much rarer for dinosaurs. Yet whole-organism representation renders both as virtual-witnessing material. Following the second two requirements, above, we should expect representations of mammoths to be significantly less diverse than many of the dinosaurs. As the constraints of scientific consensus tighten, the open diversity decreases. Thus, although any individual reconstruction will be concrete, the relevant information about certainty and open epistemic possibility would be communicated across reconstructions.

An upshot of this meta-variation, wherein well-known critters are represented in narrower sets of ways than less-well-known critters, is that those observing the art will likely come to know about current constraints on scientific knowledge. It further undermines what Witton, Naish, and Conway (2014) decry as the "culture of copying" in palaeoart. Instead of relying on what have in effect become tropes of palaeontological representation that lack scientific grounding, palaeoartists are free to explore epistemic possibility (see Turner's (2019a) discussion of a mounted *T. rex* and *Triceratops* in the Smithsonian).

An odd luxury of many philosophical prescriptions is our getting to ignore the economic, cultural, and social realities of the practices at hand, and I'm fairly blatantly doing that here. There are plenty of realities on the ground that restrict palaeoartistic imagination (see Witton, Naish, and Conway 2014). In the wake of *All Yesterdays*, *All Your Yesterdays*, and other projects, and with the internet affording much further democratization, communication, and dissemination, palaeoart has become, as Chris Manias has put it to me, "a large-scale amateur-dominated field" (personal communication). No doubt, there's a lot to be said both positively and negatively about this, but so far as I can tell palaeoart has taken up something similar to my third requirement.

In that spirit, let's give the last word to Greg Paul: "We need to be daring and bold. Not only are traditional concepts of dinosaur appearance no more intrinsically valid than controversial ones, more often than not they are incorrect. The reasons for illustrating alternative views of dinosaurs are excellent" (Paul 1987, 40).

3. Artifactualist Imagination

> The next question is from one of Prof. Ichthy's relatives: "Where do you get your ideas?" Ichthy only pauses for the briefest moment: "I look at things and build things."

In section 1, I articulated a set of challenges to taking scientific imagination seriously. Given imagination's arbitrariness and poverty, why think that imagination is epistemically relevant? As we saw, one strategy seeks principles by which imagination might be *constrained* and *enabled* such that imaginative acts are not mere fancy and yet are able to extend beyond our meagre resources. However, we also saw arguments against our chances of finding legitimate constraints. As Kinberg and Levy argue, suggested constraints on imagining are not about the powers of imagination per se; as Mike Stuart argued, imagination in science is anarchic: that is, there is no set of informative principles that range systematically across scientific uses of the imagination.

In section 2, we dived into some recent palaeoart history. As we saw, the mode of representing extinct taxa as living organisms and the requirement to cohere with up-to-date science place significant constraints on palaeoartists' imagining. The *whole organism* constraint precludes the ability to be implicitly non-committal about organismic traits: the organism is to be shown as a concrete object. Cohering with science requires that palaeoartists, insofar as they can, restrict their representations to what is epistemically possible. As we saw, this involved intimate knowledge of anatomy derived not only from living animals but also from fossils. In artifactualist terms, the imagination of the palaeoartist is constrained by the mode and medium of their practice. Palaeoart is neither special nor unique in this regard; indeed, mainstream palaeontology involves a diverse array of visual practices, from fossil illustration to digital visual techniques and more (e.g., Cunningham et al. 2014), all of which are similarly constrained by mode and medium. This also isn't limited to visual practices, but includes epistemic practices geared toward lost worlds generally.

In this section, I'll bring these together to develop an *artifactualist* account of the scientific imagination. In answer to Kinberg and Levy's point, I'll argue that restricting ourselves to a *mentalized* conception of the imagination is a mistake; therefore, imagination, as it plays out in the human mind, not having special powers, is a feature, not a bug. To Stuart's complaint, I'll argue that the informative principles cannot be generally constrained because they are constrained by the *material conditions* in which imagination plays out. As material conditions—modes instantiated across various media—are so diverse, so too are the constraints on imagination. As with palaeoart, scientific imagining of lost worlds is best understood in terms of the mode and medium (and *combinations* of modes and media) through which imagining occurs. That is, the perspectival tools—including traces and interpretations of traces—don't merely constrain our imagination. They *are* our imagination.

Rucińska and Gallagher (see also Jones and Schoonen 2024; Savojardo 2024) argue that the imagination can be constrained via a commitment to strong embodied cognition, that is, "the role the extra-neural body plays in cognition—its postures and movements, and the history of bodily interactions that make up a bodily habit, skill or knowhow" (Rucińska and Gallagher 2021, 8144). The idea is that if we think, and thus imagine, through our bodies and through our movements, then large constraints come into play for some imaginative activities. Perhaps a disembodied brain, if such a thing could imagine, imagines whatever it wants, but an embodied agent imagines through the affect, proprioception, and phenomenology of its body as it interacts with the world.

> Imaginings may be closely coupled to the dynamic constraints involved in interacting in the world just so far as bodily movement and action are constrained by how we relate to our environment and engage with others. When we imagine with movement, we can't imagine anything we want, as our movements are constrained not just by the actual parameters of the body and developed body schema, but by ongoing environmental situations. (Rucińska and Gallagher 2021, 8166)

Rucińska and Gallagher are not primarily interested in the scientific imagination: they're interested in both day-to-day imaginings, say, my imagining lifting a heavier kettlebell than I have before, and imaginings that concern analytic metaphysicians (e.g., Kung 2016). Recall that I'm interested in imagination similar to how McAllister had it: the (mental) capacity of conceiving of previously unobserved entities, in particular, as it plays out in the science of lost worlds. Here, the embodied imagining of scientists is potentially very important (when, say, approaching the physicality of a fossil specimen), but to capture how lost worlds are imagined we need to go further, and also consider the varied perspectival tools that are employed. To do this, I think, we best develop an *artifactualist account of the imagination.*

Here it is:

> What can be relevantly imagined scientifically is a function of the constraints and enabling conditions set by the modes and media of the perspectival tools at hand.

Note that this is not a claim about what the imagination *is*, but about what *can* be imagined. However, there's not much distance between a claim about imagination's capacities and imagination's nature. It is further not about imagining per se, but *relevant* imagining, that is, imagining that is epistemically linked to the target. In the instances I care about, these targets are lost worlds and the denizens that comprise them. As we've seen, the palaeoartistic imagination is deeply constrained and enabled by the mode and by the medium. Similarly, the palaeoscientific imagination is constrained and enabled by the various artificing, grafting, and anchoring activities that historical scientists employ. Phylogenetic contexts require that the histories of extinct taxa are rooted in the analysis of fossil and extant characters, various statistical techniques that hypothesize their evolutionary relationships, and are understood in terms of interactions among historical individuals. Mitchell et al.'s imagination is rooted both in the spatial relationships of Avalonian rangeomorph ecologies, as well as information gleaned from the various anchors they bring into play. Cuthill and Conway Morris' imagining plays out in the construction and exploration of geometric models of rangeomorph growth, as well as in the results of re-contextualizing these with rangeomorph fossils and what we've inferred regarding their developmental processes and environments.

In each of these examples, scientists rely not on a single mode or medium, but work across them to both enrich imagining and further ensure and develop relevance to past worlds.

There are at least two related complaints you might have regarding an artifactualist account of imagination. First, isn't imagining primarily a mental activity? One might be happy to say that scientific imagination is situated and scaffolded by artifacts, but still insist that the *real* imagination is in humans conceiving of, or perhaps even engaging in immersive (Kinberg and Levy 2023) mental imagery. Even though McAllister's notion of imagination isn't tied to mental imagery, it is nonetheless tied to mental capacities: is dropping that, as I appear to, a step too far? Second, if we give up on mental conceptions of imagination, what, then, carries imaginative content?

If one wishes to grant our mental activities a special place, perhaps in embodied terms as Rucińska and Gallagher do, one can still say something very much like what I have said by appealing to scaffolding or similar metaphors.[5] Mike Stuart's account of *imaginative tools* is extremely close to, and partly inspirational to, the view I'm trying to develop here. In a very artifactualist turn, he distinguishes between tools of the hand—say, a sewing needle—and tools of the mind: "Tools of the mind (calculators, white boards, computer simulations, artificial intelligence, etc.) are really tools in the same sense that tools of the hand (forceps, pipettes, pumps, particle colliders, etc.) are tools. That is, tools of the mind 'promote or regulate' the 'motion' of the mind, by supplying suggestions and cautions that improve scientific reasoning" (Stuart 2022, 451).

Tools of the imagination are that subset of tools of the mind that "promote or regulate" our imaginings. Stuart is, perhaps responsibly, somewhat non-committal about whether tools of the imagination should be better understood as *aids*, that is, physical objects that increase our natural capacities, or whether they are the *bearers* of those activities. In his discussion of the function of visualizations to aphantasic scientists, he claims that such scientists "use and create visual images . . . to explore ideas in an imaginative way that they cannot do inside their own minds" (Stuart 2022, 451). The artifactualist would phrase this in a slightly stronger way: it is not that the image is being used "in an imaginative way," but rather that the scientist *is* imagining (I have no quibbles with Stuart's further claims about the consequentionalist justification of imaginative tool use).

I think the artifactualist can accommodate the minds of scientists, perhaps even claim they are special. The move is just to say that scientific minds are *themselves* artifacts. The process of becoming a scientist involves both practical and theoretical training over a long term—this is how "paradigms" are built, after all. How scientists think are products of their broader cultural context, idiosyncratic individual differences, and the work they have undertaken as a scientist-in-training and as a scientist. Years of labour shape the practical, epistemic, aesthetic, and moral sensibilities of scientists (Currie 2023b, 2023c). And further, these sensibilities and forms of understanding are often deeply embodied and entangled with material context (see Myers' 2008 on *body work*). A scientist with exceptional capacity for mental imagery,

5 As, for instance, Sterelny does for cognition (2010) and Krueger and Colombetti (2015) do for affective states.

say, might be able to run Cuthill and Conway Morris' geometric models in a purely mental environment. This would just be to say that they use a different medium but the same (or very similar) mode, just as we switch mediums but not necessarily modes, as we switch between a pen-and-paper medium and a computational medium in thinking through an ecological model. And just as switching from paper and computers brings various different affordances and epistemic dangers, so too might switching between a mental perspectival tool and one built of silicon.

In recent work, Kirsten Walsh, Tom Roberts, and I apply so-called "4E" cognition to scientific contexts (Walsh, Roberts, and Currie forthcoming). We distinguish between views whereby scientific apparatuses are considered part of, aids to, or extensions of, the scientists' senses and cognition (e.g., Toon 2014), and views whereby we consider the scientist's cognition as situated or distributed, that is, part of the system that does the science. I take it that an artifactualist would find the latter set of views attractive, and indeed, need not deny scientific agency by doing so.[6] Further, I've been speaking in largely propositional terms about this modal knowledge, but I think less propositional accounts appealing to "pragmatic understanding": knowledge by acquaintance, know-how, or skill (e.g., Stuart forthcoming) fit nicely within a situated framework. I don't take myself to be committed one way or the other to how exactly we characterize the relevant epistemic good here. The crucial element is the relationship between artifacts and imagination in the production of epistemic goods. As Brandstetter says of models in geology, "They liberated the imagination of the scientist, enabling him or her to literally see processes that were essentially outside the grasp of human perception" (Brandstetter 2011, 136). While I'm not sure what to make of the appeal to literal sight, here, the point is that artifacts like models generate bountiful epistemic goods, and some of these involve enabling and carrying out imaginative acts.

In this context, I do think there is reason to resist claims about mental imagination's in principle priority. Consider the *diversity* of scientists' mental

6 There are various ways one might cash out a "situated" view about scientific work. For instance, you might conceive of it in terms of ecological agent-environment dynamics (using something like dynamical systems theory), à la Anthony Chemero's work (2011). A more intentionalist approach could be drawn from my work with Kirsten Walsh and Alice Murphy (Murphy, Currie, and Walsh forthcoming), which considers how experiments instantiate and afford various forms of scientific agency. Hopefully, one's preferred direction here shouldn't make much difference for my purposes.

imaginative abilities. Mike Stuart has examined these across various studies, summarized in Stuart (2024). Scientists vary in their reported and tested abilities in terms of mental imagery (visual and otherwise). I don't think we should say that those with more capacities in this sense are thereby more imaginative scientists. Let's suppose that Cuthill and Conway Morris were utterly bereft of the mental imagery required to perform anything like the geometric transformations described above—let's imagine they are aphantasic. That doesn't seem to block them from generating the model that they did. It strikes me that scientists who generate the same modelling work, where one is aphantastic and the other isn't, should be considered equally imaginative. Further, our scientists with rich mental imagination still must *communicate* their work: science is a fundamentally social activity, and this requires the use of artifacts beyond mental imagery.

None of this denies that, in some contexts, we may want to study mental imagination in science, or that we may wish to argue that highly agential epistemic goods, such as understanding, require according special status to the "internal" mind. Nor is it to say the scientists do not deserve praise for their imaginative work. Rather, an artifactualist account of imagination that considers imaginary content to be carried across various media and modes can make sense of scientific imagination's poverty and constraints, at least in the context of lost worlds.

So, where does imagination *happen*? Should we say that imagination occurs within the scientist's mind—properly scaffolded—or does imaginative content sometimes occur in the objects they construct? Although the artifactualist is led to say that imagination is extended in this sense, I'm not so sure whether answering this question, that is, unambiguously identifying "where" imagination happens, is so important. To see why, I want to provide an analogy with another human practice: making music.

Consider the solo human voice, singing a melody. Here, we're likely to agree that the music comes from the singer. Although the voice was likely trained, the song was written by somebody else, or at least drawn from a musical tradition, it is still pretty clear that it is the singer who generates the music. Now, let's add instruments and other musicians. We'll likely agree that the *musicians* collectively play the *music*, although some of the aspects of the sound come from their scaffolded interactions with each other and their instruments: these afford, to crib from Joel Krueger, acts of "musicking," which generate experiences inaccessible without them (Krueger 2014). Now consider procedurally-generated music, often associated with Brian Eno: this music is generated according to a set of systematic rules. Or consider music developed "in collaboration" with artificial intelligence. Here, we might disagree about whether AI is a collaborator, or, as Roberts and Krueger have argued, might take it as a fiction that the AI truly collaborates (Roberts and Krueger 2022). These questions about *agency*—about who is truly *acting* in the music-making—aside, we can still agree that *music is happening.*

I think we can say the same of the scientific imagination. Shifting from the scientist sitting in a dark room using their rich mental imagery to consider what the world might be like, to scientists as we've seen them in this book, in rich interaction with multitudes of perspectival tools, to—perhaps—the automated generation of ideas and experiments, we might have reasonable disagreements about who or what is doing the imagining. But *imagination is happening.* And, crucially, how the imagination occurs, the mode and medium, makes a difference to what it might do. As Mike Stuart points out, AI-designed experiments are often ugly in the sense that they do not follow the kinds of elegant norms and principles that human-designed experiments at least aim for. This is because their algorithmic mode explores the space of possible design options very differently from how humans do, and their media afford highly precise, high-speed information processing compared to our wetware. As Stuart puts it, "The reason that AI-designed experiments are ugly *and* epistemically good is due to a common cause: the nature of the algorithms involved" (Stuart 2023, 227).

Thus, just as understanding the aesthetic achievements of music requires attending to the various affordances of musicians, their instruments, and the musical tradition they follow, understanding acts of scientific imagination—where their ideas come from—requires careful attention to the modes and media at play. Who is imagining, or where imaginary content lies, are open questions. But in the present discussion, they are secondary.

Artifactualism's main competition links scientific imagination with *make-believe.* This view is developed from a powerful form of fictionalism about models that understands them as props in games along the lines that Kendall Walton has argued (Walton 1993; for modelling views, see Toon 2012; Levy 2015). In brief, make-believe occurs when some combination of props and conventional rules prescribe that we imagine things to be thus-and-so. Say some children have constructed a sandcastle and are together telling a tale of medieval gallantry and the intrigue of castle politics. Here, the sand and its shape, and the conventions the children develop (one of them declaring a shell to be a knight, say) together prescribe their imagining that, say, the seawater crashing into the sandcastle is a spell cast by an evil witch. Similarly, on such views, Mitchell et al. use a set of ecological models as props, which, in combination with the construal of that model, prescribes imaginings such as taking the two-dimensional Euclidean space of the spatial point analysis to represent ecological locations. As with fictions, propositions are generated by a combination of *primary truths*, roughly analogous to what would be explicitly stated in the text, and *principles of generation*, the rules of inference grounded in the conventions and props.

An immediate problem with using this form of fictionalism to analyse imagination is circularity: make-believe is understood as a prescription to imagine, and so trying to analyse imagination in terms of make-believe doesn't seem to get us very far. Fictionalist accounts of modelling explain idealization and other confusing elements of modelling practice by *appeal to* imagination. As Arnon Levy put it recently, "A central motivation for treating models as fictions is their shared imaginative, concrete-hypothetical character" (Levy 2024, 100). But explaining models or fiction with imagination—defining them as acts of imagining, say—requires an independent definition of "imagination" to avoid circularity.

Happily, Salis and Frigg do provide an independent definition of imagination. They commit to *propositional* imagination, that is, imagination that can be understood in terms of some attitude toward a proposition, *p*. For them, some mental act counts as imagining just when the act is (1) *free*, which is to say we're under no obligation to believe what we imagine,[7] (2) *mirrored*, that is, the inferential rules work more or less the same as belief, and (3) *quarantined*, that is, the propositions we imagine and inferences we make on this basis do not directly affect our behaviours or other propositions, except when they are transported beyond the quarantine.

Following Salis and Frigg, the idea is that scientific imagining is an act of make-believe where props in combination with a set of conventional rules prescribe free, mirrored, and quarantined mental propositions. This account has a lot going for it. As Salis and Frigg argue, the view satisfies many desiderata of an account of scientific imagination: it is *constrained*, both by props and by the rules of generation, and it is *social*. "It has an objective content that is normatively characterized in terms of social conventions implicitly or

7 I take it that this doesn't conflict with the poverty of the imagination: we're not free to imagine whatever there is, I assume. This is a little ambiguous in Salis and Frigg: "We can decide freely what to imagine" (Salis and Frigg 2020, 30).

explicitly understood as being in force within the relevant game" (Salis and Frigg 2020, 44). It is also, as Levy (2024) explains, *normative* in the sense that models have content: one might be *wrong* about a model by, say, misconstruing the props or entertaining the wrong mental propositions.

What is the difference between artifactualism as I've described it and fictionalism? I suspect that we can quite happily map talk of *mode* onto social conventions, that is, the rules of the game, and talk of *medium* to props. I'm quite happy to think that a claim like "Mitchell et al. use a computational medium and statistical mode to generate a neutral result" is intertranslatable with "Mitchell et al. use a computer running a set of software as a prop which, in combination with statistical rules, prescribes imagining a neutral result."

Like a fictional account of imagination, artifactualism can provide normativity. This is because artifacts are, well, built and used intentionally. I've previously distinguished between an object's *physical* and *functional* properties (Currie 2017b). Functional properties are those physical properties that contribute to the intended purpose of the artifact. Take the PVC pipe that Woodruff and Varricchio used in their artificed dinosaur burrow. Here are two of its physical properties: it is *rigid*, and it is *white*. The former is a functional property because it is required to deliver the sediment to fill the burrow; the latter is not a functional property insofar as the pipe's colour doesn't make a difference to its purpose. As such, artifacts can have better or worse functional properties, can be misused, and, in contexts where the functions are representational, can be right or wrong. Note that I don't think the functional properties should always be delineated via the intentions of the builder or designer of the artifact, but sometimes also in the context of use: the pluripotency of perspectival tools involves fruitful flexibility as to their use, and hence context-sensitivity as to which physical properties are functional, and how. Thus, functional properties are sometimes set by makers and sometimes set by users (Knuuttila 2024, 118).

Thus, fictionalism and artifactualism are very close. Indeed, Fiora Salis (2021) has incorporated an artifactualist element to a Waltonian account of modelling by emphasizing the material nature of "model descriptions" (approximately, the props). Salis might well be right that perspectival tools function as Waltonian fictions in some contexts, but as we'll see, these don't exhaust our interests here.

One difference between the views, recalling our discussion of artifactualism in chapter 6, is that make-believe is representation-first while artifactualism is not. However, this difference loses much of its import if we buy their intertranslatability, becoming more of a matter of emphasis than anything else. And indeed, the use of props well captures the central motivation of the account I'm pushing: lost worlds reveal the poverty of the imagination, and this is overcome by the use of strategic perspectivalism, which in fictionalist terms would involve the construction and exploration of props and conventions.

However, I think there are two reasons to prefer artifactualism over make-believe fictionalism (see Friend 2020 and Thomasson 2020[8] for other anti-fictionalist accounts of models).

First, and speculatively, on the face of it, make-believe fictionalism takes the propositional content of imaginings to be mental events, and although this might be plausible for scientific thought experiments, this is less obvious in modelling cases. In thought experiments, the props are linguistic descriptions or other representations that we then "play out" in our minds, coming to some judgement or understanding. Modelling cases might well involve a lot of mental action, but it is the model that generates the content on which the judgment is based. When Cuthill and Conway Morris generate rangeomorph morphology out of a computational system following a geometric three-part self-branching system, it is that computational system that generates the results. This process *could,* in principle, be done within someone's head, at which point it would take on the appearance of a thought experiment. But it doesn't. This suggests there is an analogy to the functional equivalence claim that infamously underwrites arguments for extending the mind (Clark and Chalmers 1998). If, in principle, diaries might play the same functional role as memory, and memory is defined functionally, then, in principle, diaries are *part* of memory. By analogy, if *imagining* is prescribed propositional content derived from props and rules of generation, then when the act of making inferences from props is carried out in a model, well, it seems as if the model is doing the imagining.

So, if we are functionalists about aspects of imagination, then we might be tempted to extend the imagination beyond mental acts. But this move appears to break elements of the view: I'm not sure, for instance, what it would mean for freedom and quarantining to apply to a model, as I'm not sure how to make sense of the difference between what a model "believes" and what it merely "entertains." The fictionalist, then, faces a dilemma: either models

8 I leave open whether Thomasson's artifactualism about *fictions* is consistent with artifactualism about *models*.

somehow prescribe that we perform imaginative acts that we do not (or cannot), or imagination is extended in ways that break down our working definition of it. By contrast, an artifactualist, it seems to me, can happily discuss scientists and their models as coupled, embedded systems in which acts of imagination occur.

A more plausible reason to prefer artifactualism, and much more importantly for the discussions in the final two chapters of this book, is the difficulty that make-believe comes to once it is applied to perspectival tools that are not models. Artificing is, at base, modelling with the intention of creating something with lost properties. But neither grafting nor anchoring involves modelling in the sense that fictionalists about models have in mind. When a scientist generates a phylogenetic analysis which, say, places *Dickinsonia* as a stem metazoan, they are not playing make-believe. They are making an inference. The easiest way of seeing this is just to say that quarantining fails: if *Dickinsonia* is a stem metazoan, this has further implications for how we reconstruct that taxon, as well as the order in which synapomorphies came online during the emergence of metazoans, and so on. Salis and Frigg might point out that this doesn't count as imagining at all by their account, which is right, but it certainly counts as a bona fide case of uncovering hitherto unseen or unconceived past worlds. So, that response won't work for my purposes, at least.

More difficult still for the fictionalist is imagining carried out with *specimens*. Not all perspectival tools are representations in the same way as models are. For instance, some perspectival tools are fossils. In chapter 4, we discussed a set of dinosaur trackways that were suggestive of a limp (recall this mattered because it would be indicative of pain-avoidant behaviour). There is no doubt that scientists considering the trackways would use mental imagining in considering how a critter might leave those tracks (as well as other perceptual tools), but it seems inappropriate to describe scientists as entering into a game of make-believe where they pretend that the fossils and trackways are, well, fossils and trackways. The fossil isn't a prop in a game of make-believe in these instances. Not only does quarantining fail as with the grafting case above, but I take it that it isn't at all make-believe that specimens are the specimens they are, with the aetiology they have: we're not *pretending* it is the case, but *inferring* it is the case. There is no doubt that some forms of supposition[9] and other acts of propositional imagining are involved in these cases—fossil interpretation involves interpretation and hypothesizing, of course—but to think that supposition is where things stay is to misunderstand the crucial role that specimens play in establishing the existence of lost denizens.

As I'll further explain in chapter 9, it is worth remembering the central role that specimens—traces of the past—play in the discovery of lost worlds: they are crucial for *re-contextualizing* evidence, information, and hypotheses provided by other perspectival tools. Insofar as traces provide a rich picture of an actual past, they ensure that the possibility we explore is relevant to that actual history. This point might lead us to question the relevance of "historical" studies that lean too heavily on either abstract modelling (Currie 2024a) or studies of contemporary phenomena (Meneganzin and Currie 2025). Without a rich record to be contextualized within, we might well agree that objective modal knowledge is being generated, but we might question whether that modal knowledge is in fact relevant to the historical targets we have in mind.

You might object here: doesn't the artifactualist do similar harm to specimens? Where the fictionalist gets things wrong by considering fossil interpretation as make-believe, perhaps the artifactualist mischaracterizes specimens by calling them *artifacts*. I think this is mistaken. It is certainly the case that applying strategic perspectivalism, as empirical reasoning generally does, requires *construction* in various senses. As Caitlin Wylie has documented and analysed, the use of fossils as data requires extensive *preparation*. But, as she argues, preparation sits somewhere between construction and "bare" empirical data (Wylie 2021). There might be some contexts where we want to distinguish between *constructed* artifacts like models and *prepared* artifacts like specimens, but I don't think epistemic damage is done either by the term "artifact" or the specificities of the view I've developed here.

Recall the distinction between physical and functional properties introduced above. A fossil has a bunch of physical properties—size, shape, composition, etc.—and the fossil preparator manipulates these properties to reveal and hone particular properties functional for palaeontological analysis (as well as storage and display). As such, the artifactualist has the conceptual machinery required to make sense of fossils as perspectival tools, and thus how they constrain and enable scientific imagining.

9 Salis and Frigg categorize *supposition* as a form of imagination that differs somewhat from make-believe.

Finally, the artifactualist is perfectly able to help themselves to the machinery of make-belief when it is appropriate (Currie 2017b). They are free to say that the mode in Cuthill and Conway Morris' models or in the experimental dinosaur burrows is to use those media as props in a game of make-believe. But the artifactualist can also accommodate the variety of other artifact-based interactions required to make sense of past worlds and uncover lost denizens.

Back in chapter 3, I appealed approvingly to philosophical accounts relating to "as if" representation, and this might seem to expunge the kinds of characterizations seen across uses of strategic perspectivalism and make-believe. However, as Salis and Frigg point out, propositional imagination is not restricted to make-believe; it also includes attitudes, such as supposition and counterfactual reasoning. If we isolate modelling cases, then I think make-belief gets us a long way, but it alone doesn't capture the scientific imagination. At its most minimal, to make scientific sense of data derived from observations, scientists must rely on a *frame*: decisions must be made about what to foreground and what not to foreground, what is salient, and what can be dismissed. We've seen palaeontologists artificing burrows frame their work by, say, backgrounding that the bones they use are from rabbits, but foregrounding their being from vertebrates. But this doesn't involve them engaging in make-believe, especially as it comes along with specific arguments suggesting that the particular features of rabbit bones compared to other vertebrates will not make a significant difference to their results.

To wrap up, then, let's consider how the artifactualist makes sense of the constraints on scientific imagination in targeting lost worlds. The various perspectival tools historical scientists construct can be understood as artifacts: they are various modes carried by various media, wherein the medium and use pick out some of the medium's properties as functional in particular contexts. Cuthill and Conway Morris' artificing are geometric modes in a synthetic, computational medium. Many of their model's properties have epistemic functions in virtue of their exemplifying traits that Cuthill and Conway Morris hope are also present in their target system. At the very least, they are functional in virtue of exemplifying traits and dynamics epistemically relevant to the task at hand. Fossils and data generated from fossils, such as Mitchell et al.'s scans and Varricchio and team's analysis of the burrow, are also artifacts. They are prepared in various ways to act as data and evidence; thus, some of their physical properties gain epistemic functions. The mineralogical medium of the fossil is treated as exemplifying an organismic mode. As we saw in chapters 4 and 5, these various artifacts are brought into dialogue in virtue of overlaps in their representational or other functions, and these underwrite iterative, self-correcting investigations that can reveal lost worlds despite how much they diverge from the familiar present.

Thus, relevant scientific imagining is constrained and enacted through interaction with the multitude of perspectival tools at scientists' command. These enable us to imagine lost worlds.

8

Peculiar Worlds

A bespectacled *Allosaurus* takes the next question. They clear their throat and begin speaking in a manner suggestive of someone who, somewhat like Prof. Ichthy, is accustomed to speaking in whole paragraphs.

"Doctor Professor Ichthy, thank you for your interesting lecture and congratulations on your new chair. I must admit, however, that I found much in your lecture deeply troubling. Through the lens of your account of the past, we see a world not of order and stability, but of seemingly endless fragility and chaos. Perhaps better to not say 'chaos,' but 'chaos-*ing.*' *Komplexelieferung*, if I may. We can, I'd say, contrast this chaos with the meaning we find in our immediate experience. From our own perspectives, it seems to me, we form the centre of our own local universes, universes containing and shaped by familiar things. Our friends, family, the day-to-day experiences that make up our lives on scales appropriate to our status as finite beings. This stability is the phenomenology of our immediate experience and provides, I would say, the grounding of the very possibility of our lives' meanings.

Your world vision—'world-making' I'd call it—it seems to me, demands that we abstract from that familiar perspective. This confronts us with the sheer fragility of our own *welterschaffend*, of all that is meaningful to us. It seems our beliefs, our relationships, our very moral worldview, are contingent upon not just some tiny happenstance, but some near-infinite chain of tiny happenstances. Where around us, we see a local reality from which we might find glimpses of true meaning in eternity, you would deny us that eternity as an illusion. We light tiny candles of order around us, about which only chaos swirls.

I ask, Prof. Ichthy, how can we make sense of a world of such unerring hostility, how should we find meaning in such a meaningless world?"

Prof. Ichthy pauses, partly making sure the *Allosaurus* has finished, but also wondering how to proceed. They sigh, "No one warned me philosophers would be attending . . ."

ILLUSTRATION 8.1 Ichthy Through the Looking Glass

In the first chapter, I compared Prof. Ichthy with another fictional Ichthyosaur: that depicted in De la Beche's *Awful Changes*. Prof. Ichthy, through their emphasis on Broadfoot's uniqueness, paints a highly contingent picture of the past. I took Professor Ichthyosaurus to represent the steady-state view of Earth's history as defended by Lyell (or as Lyell was interpreted by his contemporaries). As such, we might imagine Professor Ichthyosaurus being less impressed by the contingency of the past. For him, the patterns uncovered in the past reveal regularities. Loss occurs, no doubt, but that loss is in some sense inconsequential. On this kind of view, if you know the broad-scale pattern, then you know most of what matters. The rest, as Ernest Rutherford might say, is stamp-collecting (Johnson 2007).

For Professor Ichthyosaurus, history matters insofar as examining the past might be necessary for discovering the cycles governing the Earth, but once those cycles are established, it matters not. For Prof. Ichthy, by contrast, history does matter. That is, even if we have a good understanding of the Earth's cycles, we still need to attend to the details of history's particular unfolding.

My aim in this chapter is to consider whether the position developed thus far can play any role in the metaphysical dispute between our fictional ichthyosaurs.

I'll start by developing previous work toward an account of when history matters, building on my notion of "peculiarity." I'll then consider under what conditions scientific results or practices might provide insight into metaphysics. This will underwrite arguments that peculiarity is, at least to some extent, supported by the strategies historical scientists adopt in the face of loss. I'll suggest strategic perspectivalism is a method adapted to, and adopted in light of, a peculiar world. Finally, I'll link history's peculiarity to a set of metaphysical views that emphasize the world's disunified, dappled, or processual nature.

My reference to "stamp collecting" may have got you thinking about laws, a topic I covered somewhat dismissively in section 4 of chapter 5. There, I argued that appeals to general laws are sometimes an important aspect of strategic perspectivalism, but their tendency to abstraction typically makes their epistemic utility limited. Suffice to say, in history we see both fragility and robustness, and robust patterns and mechanisms seem to play law-like roles in scientific explanation and investigation, at least sometimes (Sterelny 2016). In a peculiar world, we'll expect law-like regularities to hold fairly often, but only for a limited scope. Regardless, I'll not directly link my discussion to that of laws here.

As we'll see, I won't be defending a *scale-independent* metaphysics, that is, I'll not maintain that metaphysical views hold regardless of time, place, and level of description. Instead, I'll be interested in what we might call the *palaeobiological scale*. This is an admittedly ambiguous way of putting things, but I'll take it to refer to something like the scales accessible through the fossil record, which includes everything from long-term, millions-of-years-long macroevolutionary trends, through to single-organism properties such as ontogeny, physiology, and so forth. One way of generating the set of scales of interest is to include those captured by the perspectival tools palaeobiologists use. These are certainly broad, but do more or less fall within a certain set of ranges. We might further limit these scales by appealing to some perspectival tools rather than others: perhaps we should exclude the tools that rely on, say, chemical scales insofar as they are used by palaeobiologists to infer larger-scale phenomena. Regardless, I don't think my arguments rely on specifying a scale or set of scales. The point is just that my position isn't scale-independent.

Another way of putting this is to say that the local epistemic activity of the science of lost worlds licences a local, scale-dependent metaphysics, and I'll restrict myself to that.

1. When Does History Matter?

The claim that "history matters" is common, but its meaning is somewhat obscure. Let's consider a few examples and then zoom in on the most relevant to Professors Ichthy and Ichthyosaurus.

In political science, Charles Tilly argues that "explanatory political science can hardly get anywhere without relying on careful historical analysis" (Tilly 2006, 417). Here, history matters for developing decent explanations of contemporary phenomena. John Tosh argues that "to know that the past can illuminate the contours of the present is to be better equipped to make intelligent decisions about difficult public issues" (Tosh 2019, viii). Here, history matters for informing our future-oriented decision-making. For Gerda Lerner, history matters because it informs and challenges political views, describing her work as taking place in the backdrop of culture wars "being fought about the meaning of history, when the history of minority and marginalized is being challenged for its very existence by those who see history as the property of ruling elites and as a vehicle for the justification of states and other institutions of power" (Lerner 1998, xvi–xvii). Here, uncovering and telling the stories of once-invisible marginalized people can confront and undermine received histories.

Each of these notions, that specific details of history are crucial for making sense of the past and present, that history matters for tackling the future, and that history matters for challenging myth and emphasizing historically silenced perspectives, is worthy of serious attention. Here, I'm interested in something like that first notion. The idea, I take it, is that for some investigations we cannot simply avail ourselves of a temporal snapshot, or the short-term dynamics, of our investigative target. We cannot screen off the past. We also need to incorporate a detailed, long-term perspective on the target's trajectory. We need to ask where our subject came from and what processes maintain it. But this remains ambiguous: how long is "long term"? Why might we be forced to venture into deeper time? And what kinds of investigations do we have in mind?

As a first pass, we might understand history mattering for some target when aspects of history must be incorporated into an *explanation* of that target. But this also doesn't help us much: explanation is notoriously description- and interest-sensitive. If I'm just not interested in those long-term dynamics or the origination of some target, then what is there to force me to consider these elements? Again, comparing Prof. Ichthy with Professor Ichthyosaurus gives some help.

For Prof. Ichthy, the past and present differ significantly. For them, new entities, dynamics, etc., arise over time, and these might underwrite regularities. However, these regularities are fragile. In explaining past denizens, then, we are led into increasing complexity and detail; we must incorporate multiple causal factors, entities, and dynamics. For Professor Ichthyosaur, the differences between past and present are inconsequential: regularities are stable across time, and differences in initial conditions and subsequent paths will nonetheless lead to similar outcomes. As such, much simpler explanations appealing to those laws will do. We're not led to complexity but generality.[1]

In chapter 1's discussion of loss, I mentioned its inverse: novelty and innovation. Over time, we don't simply lose denizens, but new denizens arise.

1 This point could likely be cashed out with reference to my distinction between "simple" and "complex" narratives (Currie 2014, 2019f).

Consider Cuthill and Conway Morris' discussion of rangeomorphs and the fall of the Avalonian: "The unique rangeomorph fronds were fractal, surface area specialists of the Ediacaran. At the Cambrian explosion, the ecological and geochemical conditions to which the rangeomorphs were optimized ceased to exist, and their extraordinary body plan was lost from life's repertoire" (Cuthill and Conway Morris 2014, 13125).

We don't simply see the loss of rangeomorphs, but of the "ecological and geochemical conditions" that underwrote their survival. As we saw in chapter 5, those conditions were replaced by those of the later Ediacaran, which included better-oxygenated oceans and processes of bioturbation generated by the evolution of burrowing metazoans. By the Cambrian, the menagerie included critters with specialized adaptations for movement, sensing, and consumption: scavenging, predation, predator avoidance, and mate location. And these underwrite the diurnal vertical migration from the ocean's surface to its depths.

That's an example of history mattering: a complex sequence of events, a trajectory, takes us from one world order to another. To explain how we got from there to here, we must trace that sequence.

Philosophers often make sense of a lawless, or at least fragile, history by appealing to notions of *contingency* (e.g., Desjardins 2011; Millstein 2011; Currie 2012; Powell 2020).[2] It seems plausible that the loss of the Ediacaran worlds and the arrival of the Mesozoic worlds were not inevitable. It could have been different. But "could," like explanation, is a notoriously tricky idea, highly sensitive to context (Lewis 1976). Do we mean the events were indeterminate? Do we mean they weren't determined by some set of causal factors? Do we just mean we can imagine, or that it is possible, by the laws of physics, that something else could have occurred?

John Beatty divides claims about contingency in biology into two related notions (Beatty 1995, 2006). First, *contingent upon*. This is the idea that the occurrence of some outcome relies upon some earlier event. If Butterfield is right, the evolution of complex metazoans first required the evolution of simpler metazoans, as their structuring of the oxygen environment paved the way to increased complexity. If the biochemical explanation is right, a condition on the evolution of complex metazoans was an increase in global

2 I'm here interested in contingency as a claim about history, particularly macroevolution, as opposed to the contingency of *science* (e.g., Kidd 2011; Soler, Trizio, and Pickering 2016).

oxygen. The views have differing ideas about what the target, the evolution of metazoans, is contingent upon: what enabling conditions were required.

Second, *contingent per se*. This is the idea that some event wasn't guaranteed to occur. A strong version of this idea denies determinism (Beatty 2016). A weaker, and I think more useful, interpretation specifies a set of conditions, and says that those conditions are not sufficient to generate the outcome (Powell and Mariscal 2015). For instance, Alison McConwell and I have argued that palaeobiological arguments denying natural selection's special role in explaining life's shape appeal to contingency per se in this sense. When Stephen Jay Gould and others argue from the contingency of, say, the Cambrian explosion to the autonomy of macroevolution from microevolution, they mean that the dynamics and theories associated with microevolution are insufficient—cannot be "scaled up"—to generate the dynamics of macroevolution (McConwell and Currie 2017). As such, macroevolution is contingent per se regarding microevolutionary processes. Similarly, if Butterfield is right, appeal to biochemical conditions alone is insufficient to generate the metazoan radiations that occurred after the Avalonian. Referring to those conditions, the radiation is contingent per se.

This contingency talk amounts to discussion of the historical conditions required for the emergence of some outcome. The outcome is contingent *upon* a set of conditions, it is contingent *per se* given some (insufficient) conditions. To see why this might matter for our discussion, consider concepts that contrast with contingency, like inevitability and determination. It might be that, say, once oxygen levels rose sufficiently, the evolution of complex metazoans was set: holding fixed some unlikely destabilizing events (say, some kind of extraterrestrial impact), that condition was sufficient for metazoan complexity. Or it could be that no matter what kind of path life took, it would converge on something like complex metazoans (Powell 2020). In these cases, history seems to matter less than cases where the evolution of metazoan complexity requires a set of complex, themselves contingent, preconditions. So, you might think that history matters vis-à-vis some outcome when the conditions of its occurrence are sufficiently contingent. We might say that outcomes that are highly *path dependent*, that is, whose occurrence relies on particular, themselves contingent, sequences of events, are particularly prone to exhibit "historicity" (Ereshefsky 2014).

So, contingency gets us close to what we need. However, as thus far analysed, "contingency" is too focused on outcomes. We don't simply care about contingent events and their fragility, but also about robustness and stability. The well-mixed oxygenated oceans of Earth have been remarkably stable, partly maintained by diurnal migrations for something like 400 million years. But that stability relies on maintenance from ongoing patterns of metazoan migration, and these only arose due to a complex series of events. This is another sense in which history might matter: by explaining stability in terms of potentially fragile maintaining causes.

In earlier work (Currie 2019, 31), I tried to capture this kind of notion with what I'll call *peculiarity*.[3] Here's the definition I'll use here:

> Some target is peculiar to the extent that its actual properties and modal profile are sensitive to the properties of the processes that generate or maintain it.[4] The more processes the target is sensitive to, and how fine-grained the relationship between the target's properties and profile and those generating and maintaining processes, the more peculiar it is.

So, to unpack. We can understand a "modal profile" as what a target can do, what behaviours are open to it, and what dispositions it has. This includes the conditions under which the target's behaviour remains stable and the conditions under which it maintains its identity, that is, remains being the thing that it is. Both a target's modal profile and its actual properties have dependency relations to a set of processes that generate the target (or its properties) and a set that maintains those properties and capacities.

To grab a philosophical mainstay, compare the modal profile of a statue with that of the clay it is made from. The statue has particular generation conditions, say, being shaped by the artist, and particular conditions under which it ceases to exist, say, being crushed and broken.[5] By contrast, the clay's modal profile differs: it needn't be shaped by the artist and can survive being crushed. Different targets, then, are differently peculiar depending on what their modal profiles are sensitive to. Further, the actual properties (as well as

---

3 Recently, Robert Northcott has discussed the properties of sciences under conditions of *fragility* in ways that have significant overlaps with the themes discussed in this chapter. For Northcott, "fragility" differs from "peculiarity" due to the latter's modal character, and the former being about the predictability of token relations. I don't use "fragility" in Northcott's specialized sense in this chapter (Northcott 2025). Rob Wilson's work on the "fragile sciences" takes fragility in a similarly broad way to mine (Wilson 2004, 2005).

4 In 2019, I defined *peculiarity* purely in terms of modal properties. I think this was a mistake: we care about the actual distribution of properties as well.

5 Elizabeth Barnes has a useful discussion of this old chestnut (Barnes 2023).

the modal profile) of the statue might be differently sensitive to generating and maintaining processes. Compare statues produced by a mould with those carved by hand. The actual properties of the moulded statues will presumably be more regular in appearance, while the hand-carved ones will be more varied. This is because the properties of the hand-carved statues will depend on features of the artisans, so their peculiarity is partly determined by those features, whereas the others' properties will depend on the mould. Whether we want to say that one set is more peculiar than the others will turn on factors such as the idiosyncrasies of the craftspeople, the moulds, and so on.

Peculiarity differs from other accounts of contingency (although I take it to be part of the same family). An outcome might be *contingent upon* some previous event or process, or *contingent per se* relative to some process or set of processes, but this is only part of whether or not we should consider the outcome highly peculiar. This will depend on the number of maintaining or generating processes involved and how sensitive the outcome is to these. Peculiarity captures a kind of second-order contingency. Outcomes can be highly robust and still be peculiar: it is simply that their robustness is highly dependent on a set of generating or maintaining processes.

Compare the biochemical explanation of metazoan complexity with Butterfield's. We can understand an explanation's "peculiarity" by the extent to which it appeals to its target's peculiarity. In the biochemical explanation, metazoan complexity is sensitive to the Earth being sufficiently oxygenated, a process that they argue occurred more or less as a result of the changes that occurred during the great oxygenation event. By contrast, Butterfield argues that metazoan complexity is a solution to insufficient oceanic oxygen. In his narrative, the event was sensitive to the evolution of (1) various oxygen-structuring critters (rangeomorphs, sponges, etc.), (2) phytoplankton, (3) burrowing, and (4) key morphological innovations in metazoa. Both explanations are peculiar, but we can understand Butterfield's explanation as being *more peculiar* insofar as it appeals to a wider number of denizens at varying scales, which are sensitive in more fine-grained ways to the properties of those denizens.

This notion of "peculiarity" is a good candidate for understanding when history matters. In circumstances of high peculiarity, explanations are likely to have to appeal to a wider range of particular historical factors, whereas in low peculiarity contexts, significantly less detailed explanations, appealing to broader generalizations, will likely be satisfactory. If Professor Ichthyosaurus is right, the scales and entities of interest to Earth scientists and palaeontologists are *highly unpeculiar.* If Prof. Ichthy is right, the scales and entities of interest to Earth scientists and palaeontologists are *highly peculiar.*

What is the relationship between peculiarity and loss? Peculiarity is a partly modal relational property concerning persistence and origination, while loss concerns the actual existence of a denizen (characterized as a kind or an individual). If the past is highly peculiar, we should expect it to contain highly complex, local dynamics due to the variety of generating and maintaining processes. In such a world, loss might be rampant, due to the fragility of those conditions. Not necessarily, however: the world might have a lot of contingent origination, but much persistence. That world would be peculiar but not contain much loss. Conversely, a world full of loss is likely highly peculiar. So, if we think the world peculiar, we should potentially expect loss, but it is not guaranteed.

I understand peculiarity as a graded and comparative notion. As with loss, claims about peculiarity rely on a particular characterization of the target at hand. For instance, understood as the historical kind *macroevolutionary radiation*, the Cambrian explosion might be relatively unpeculiar: arising due to the normal dynamics of that kind of event. Understood in terms of the specific adaptations and lineages that arose during the Cambrian, the radiation might be highly peculiar, taking the shape it does in virtue of a set of particular local features. But, again, as with loss, this doesn't make peculiarity a mere feature we project onto the world. With a fixed characterization, I take it, there is a fact of the matter about peculiarity. Peculiarity is also comparative: at least for all of the denizens that palaeobiologists and their ilk are concerned with, everything will be peculiar at least to some extent. But some are more peculiar: they have modal profiles and actual properties that are more sensitive to various factors than others.

So, when I say "peculiar world" going forward, I mean *highly* peculiar.

There could be other reasons for history to matter or not. For instance, we might take the *shit happens* view of history: events in the past do not follow any regular patterns, it's just one thing after the other. Or we might think that history unfolds toward inevitable, perhaps predetermined outcomes, as seen in "progressivist" views of evolution. Fortunately, for my purposes, these can be safely dismissed. We clearly have some grip on regularities in history, and the thought that history progresses "toward" anything requires teleological commitments that I'm very happy not to incur. A quite different approach to denying history's importance appeals to epistemic pessimism: we just don't have sufficient evidence to tell a story about peculiarity at geological scales.

On Secord's reading of Lyell, this is just what we get: Lyell "claimed that *any* kind of global narrative would prove impossible to reconstruct, because too much of the record had been lost [italics in original]" (Lyell 1990, 151). Secord's Lyell refuses to commit to a progressivist or steady-state model of life because he thinks the fossil record is too incomplete to support either. No doubt there are cases where pessimism is the right attitude to take about historical knowledge, but no doubt as well, there are cases where optimism is a better bet (Currie 2018a, chapter 12).

So, history matters to the extent that some target is peculiar. Professor Ichthyosaurus takes much of interest to be relatively unpeculiar; Prof. Ichthy takes much of interest to be comparatively peculiar. What makes this a *metaphysical* disagreement? "Metaphysics" is an infamously ambiguous term, but we can make some progress with contrasts. Metaphysical claims are not epistemic but ontic: that is, they don't concern what we know, but what the world is like (or could be like). Metaphysical claims concern the general rather than the particular: that is, they don't make claims about some instances, but systematically. Metaphysical claims concern fundamentality, not superficial appearances: they tell us about the structures underlying the world as it appears to us. This doesn't necessarily make big-picture metaphysical views and empirical science independent, so long as one commits to there being, as John Dupre has put it, "a dynamic interdependence between such general views and the specificities of scientific opinion" (Dupre 2023, 2).

Thus, in claiming that the past at the scales of interest to palaeontologists and Earth scientists is peculiar, I'm saying that the past itself is structured in a peculiar manner. Our fictional ichthyosaurs have a metaphysical disagreement: the two have differing views on the structure of the past at palaeobiological scales. How might we make some headway in distinguishing them? To answer that, we need to consider approaches to the metaphysics of science.

2. Metaphysics and Science

Metaphysics and the philosophy of science have a vexed history, far too dull to rehearse here. Needless to say, how features of science might inform metaphysical questions is an open and much-disputed question (Chakravartty 2010; Hawley 2016). Given my arguments in this book thus far, I'm interested in exploring how the strategies palaeobiologists employ in the face of loss—strategic perspectivalism—might provide inroads into the metaphysical nature of the deep past. Specifically, I'm interested in whether they can underwrite an argument for the peculiarity of the palaeobiological past. To get there, I think it is helpful to situate my approach, which shifts from practice to metaphysics, with others in the metaphysics of science. As such, I'll draw a set of distinctions regarding the relationship between metaphysics and science, zooming in on what I'm interested in. In the next section, I'll make that argument, but also zoom out once more and suggest that a peculiar history can be defended from multiple methodological standpoints.

We might call the most common approach in analytic metaphysics *minimal naturalism*. At base, our metaphysical views should be consistent with the results of our best science. If your view of the metaphysics of time, for instance, requires an absolute present, then this must be reconciled with the lack of absolute simultaneity in relativity theory. This approach is often coupled with the idea that the methods of metaphysics are in some important respect continuous with the methods of science. A popular approach, for instance, argues that both appeal to various theoretical virtues in forming inferences to the best explanation (Swoyer 1999; Paul 2012).

Philosophers of science have denied both the legitimacy of appeals to mere consistency with science and the idea that science and metaphysics are methodologically continuous (Ladyman and Ross 2007; Novick 2017). What distinguishes a *metaphysician of science* from a bog-standard metaphysical naturalist is, as Kerry McKenzie puts it, that the metaphysician of science "does not spin her view out of air but rather develops it in close consultation with the best science of her time" (McKenzie 2025). That is, the views of a metaphysician of science are in some sense inferred from interpretations of scientific results or scientific practices. But which results and which practices?

McKenzie argues that the relationship between science and metaphysics is best understood in terms of falsification. That is, metaphysical ideas are something like theses about the (general, fundamental) nature of the world, and these theses might clash with scientific discoveries. I take it that on this approach, some minimalist commitment to scientific realism is required: we take scientific theories to provide at least an approximately correct view of the world. Thus, a sustained interpretation of such theories is required. These theories, the idea runs, disallow some metaphysical views while accommodating others. It may be that accounts of time that posit an absolute present simply do not align with the best theories of physics, yet those physical theories may still leave open—underdetermine—other mutually exclusive metaphysical options (Ladyman 2012). As such, McKenzie argues, the metaphysician of science should be less in the business of arguing for or against various theoretical virtues—that is, constructing inferences to the best explanation from features of theory to the world—and instead aim at the falsification (or near-falsification) of metaphysical hypotheses. They should seek inconsistencies between our best scientific theories and the world.

But scientific theories are not the only potential sources for informing a metaphysics of science. Some philosophers defend approaches that emphasize scientific "practices," which might include theorizing but are more inclusive (Bausman, Baxter, and Lean 2023; Kaiser and Suarez 2025). As Marie Kaiser and Javier Suarez put it recently:

> Metaphysicians of scientific practice typically consider the results or products of scientific investigation (e.g., scientific theories or explanations), as well as the scientific activities that lead to these results, that is, the different epistemic practices (e.g., measurement practices, reasoning practices, explanatory practices, individuation practices) that characterize a scientific field. Hence, this method builds metaphysical claims on empirical information from scientific practice (results, products) and empirical information about scientific practice (activities, practices). (Kaiser and Suarez 2025, 309–310)

Notice the distinction drawn between scientific *products* and scientific *processes*. The former are outputs of scientific work: theories, evidential arguments, datasets, and so on. This includes Butterfield's narrative trajectory from the "boring billion" to the metazoan-filled, well-oxygenated Mesozoic world is a product, Hoekzema et al.'s model of *Dickinsonia* development via deltoidal addition, and Mitchell et al.'s claim that Avalonian ecosystems are spatially neutral. The latter involves what scientists do to generate those products. As we've seen, historical scientists construct and deploy various models and experimental apparatuses (they artifice), use phylogenies and other narrative devices to link past denizens to each other and the present (they graft), and link together various instances of the phenomena they're interested in (they anchor). More generally, they iteratively apply strategic perspectivalism. These are the practices I've focused on in this book, but of course, there are many more. Consider, for instance, the discovery, extraction, preparation, storage, archiving, and analysis of fossils.

I think that focusing on scientific products leads to much continuity between the methods of the metaphysics of science discussed by McKenzie on the one hand and Kaiser and Suarez on the other. Both require a thorough understanding of the product and asking about its likely truth. At base, we are inferring from how scientists represent the world to how the world is. Butterfield's explanation is highly peculiar, so at least a little inductive weight is given to the idea that the world he's in the business of explaining is a peculiar one.

There are at least three challenges to metaphysical views that proceed from analyses of scientific products. First, why think that our "best" science aligns with metaphysical truths? "Best" need not accord with "true," and we shouldn't accept the best of a bad lot. Second, what scientific theories in fact say or mean is a matter of philosophical interpretation—and it isn't obvious that we can identify a "correct" interpretation (Kinzel 2015; Currie 2025). Second, scientific claims are not on their face general, structural claims about the fundamental nature of reality, so we must generalize and systematize them to generate metaphysical claims, even if we've nailed the interpretation. Third, plausibly scientific theories underdetermine metaphysical theories (McKenzie 2025). We can't, then, take scientific theories to simply mirror metaphysics.

There is nonetheless space for this approach to underwrite peculiarism about the past. Research into the deep past has revealed much that is peculiar: often shifting from simple to complex narratives, suggesting highly sensitive and interdependent dynamics (Currie 2014; Dresow 2021). We've seen plenty of examples throughout the cases in the first half of this book.

However, in the context of my approach in this monograph, I am not well placed to make an argument from scientific products. Astute readers will have noticed that I haven't really committed to any particular scientific claim. I'm not in the business of arguing that, say, Avalonian ecosystems were neutral, that pizza disks are not a true taxon, that *Oryctodromeus* burrowed, or even that dinosaurs felt pain. My arguments have not turned on the truth or otherwise of token propositions about the deep past, but concern the strategies scientists adopt when confronted with loss or its possibility. As such, I'm much better placed to speak to practices than products.

How would understanding scientific practices be a route into metaphysics? The approach must be less direct than a product-driven one. After all, at least sometimes, scientific products are supposed to be about the world, and the inductive metaphysician takes those results and says, "Well, they're true, and thus they partly reveal the metaphysical structure of the world." Practices aren't like this. They do not represent the world in the same way.

However, scientific practices are not independent of the world. Far from it: they are rather part of, and situated within, the world. And they are directed toward discovering features of that world. As such, practices might serve as an indirect way into metaphysics. As Ken Waters has put it: "The best strategy underlying scientific metaphysics, I suggest, is based on the idea that the world provides constraints on scientific inquiry, and that philosophers can inform metaphysics by investigating these constraints" (Waters 2017, 83).

This relies on the idea that not just any scientific approach will succeed for some investigative target. Sometimes, at least the world will resist our attempts to make empirical or theoretical inroads. In the best case, we hope, the practices that scientists adopt are *adapted to* the system they are interested in understanding. As I'll expand below, and following Will Bausman (2023), something analogous to biological adaptation might be applicable. That is, much as we might infer from the platypus having webbed feet that it evolved in aquatic environments, perhaps we can infer from scientists adopting particular practices that there is a world with particular properties they seek to understand.

Ian Hacking's argument for entity realism is potentially an example of this kind of approach (Hacking 1983; Miller 2016). Hacking argued that some unobservable entities posited in physics shouldn't be understood as theoretical postulates—entities inferred from the results of experimental interventions—but instead understood as entities that are *used* to perform experimental interventions. The classic example involves incorporating electrons into an experimental apparatus built to detect properties of other non-observable particles. For Hacking, the existence of those entities, or at least entities with the properties made use of in the experiments, is in some sense presupposed by the experimental practice. Thus, the practice of these experimental scientists is revelatory of the world they examine, in virtue of how the world is incorporated into and used in those practices. We'll consider how exactly that inference might look in the next section, but here Hacking is useful as an example of a potential inference from practice to metaphysics.

Much work along this vein has focused on scientific concepts. Waters, for instance, argues that the flexible concepts of "gene" developed in experimental biology are made sense of in light of a biological world that lacks a general structure. I suspect that much concept-focused work in the philosophy of biology could be understood in something like these terms. If scientific concepts are developed in dialogue with the systems they are intended to make sense of, that is, if scientific concepts are shaped by what we could call "ontic interaction," then the concepts scientists employ will reflect that interaction to some extent. For instance, biologists make use of many differing concepts of "species," some appealing to phylogeny, others to interbreeding, yet others to genetic relatedness, and so on. We might think that this conceptual pluralism is a reaction to the complexity of the world biologists encounter. Something similar could be said of work focusing on classification in science: the plurality of classifications maps onto a complex world (Dupre 1993; Ereshefsky and Reydon 2015; Kendig 2015).

However, in this book, I have not had much to say about concepts in historical science. Although a very loose notion of kinds—token-historical, type-historical, and ahistorical—is useful for understanding characterization, my analysis has not been focused on concepts and their deployment. Rather, my focus is on the strategies these scientists adopt in tackling loss. So, we want to focus on strategies or methods. We might argue that methods, too, are shaped by ontic interaction.

Kirsten Walsh and I have developed an account of what we call *ontic-driven explanations of scientific method* (Currie and Walsh 2018). At least sometimes, when we try to make sense of why scientists adopt the methods they do, we point to particular features of the system they are trying to understand. We explain the notion as such: "An explanation of scientific method is ontic-driven when the explanatory target and the relevant contrasts diverge in virtue of properties of the systems investigated" (Currie and Walsh 2018, 122).

Explanations are often contrastive; we want to know why one thing happened rather than another. In the cases Walsh and I are interested in, we want to know why some group of scientists adopted one method instead of another. The explanation is *ontic-driven* when it appeals to features of the system the scientists are trying to understand: they adopted that method *because* they're trying to understand a system with particular properties that the method is attuned to.

So, ontic-driven explanations target scientific methods. We might ask, say, why scientific community *A* adopts lab-based experimental methods while community *B* uses observational methods. An answer might appeal to many not mutually exclusive factors. These include various social and economic factors: *A* might have the means to develop, house, and maintain expensive experimental apparatuses, while *B* might not. They could also be due to differing epistemic traditions: *A* might value intervention, while *B* values examining their target "in the wild." It could be due to the aims and purposes of the science: *A* might be primarily interested in developing technological applications, while *B* might be interested in uncovering natural patterns. At least sometimes, an important additional factor is *the kinds of systems A and B are interested in.* *B* might be interested in complex, large-scale systems. Community *B* might be targeting, say, meteorology, planetary dynamics, or ecosystems. Meanwhile, community *A* might be interested in more localized mechanistic systems, perhaps developmental systems, or chemical reactions. As such, we might agree that *A*'s experimental approach is due to the kinds of investigations that the system affords, while *B*'s more observational work is adapted to what their target will allow. When we say *A* and *B* differ in their method because of differences in the systems they target, we are providing an ontic-driven explanation. Ontic-driven explanations can also be brought to bear on the success of various methods. A scientific community might have more success with one method rather than another because the former is better suited to their explanatory target.

When we provide an ontic-driven explanation of some scientific method, it seems to me, we also provide prima facie reason to accept the ontology cited in the explanation. If we don't think what we're saying about the world is true, then presumably our explanation of why those scientists use those methods fails. This is highly defeasible. We might just be wrong about the method. It might be that the explanation is wrong, that the method is adopted primarily for some non-ontic (or a different ontic) reason. It could be due more to social or epistemic factors, say, or perhaps the ontology isn't playing the explanatory role we thought. It might also be that the method is inappropriate: that scientists are making some kind of mistake in adopting it. Katherine Valde, for instance, has recently argued that some metaphysical commitments in cancer research can lead to egregious methodological choices, instead suggesting that agnosticism about underlying metaphysics would lead to better methodologies (Valde 2024). But a sufficiently well-supported, ontic-driven explanation, I take it, provides some route into the ontology the method is adapted to.

At this point, the shape of the argument I'm interested in exploring (perhaps even making!) is emerging. Some of the strategies historical scientists adopt, particularly in light of loss or putative loss, are adapted to a peculiar world. And this gives us some reason to think the world is, in fact, peculiar.

3. The Peculiarity of History

I've thus far provided an account of "history mattering" in terms of peculiarity, and zeroed in on a set of pathways from scientific practice to metaphysics: namely, that an ontic-driven explanation of method underwrites a metaphysics committed to that ontology. It's now time to see what headway we can make in applying that approach to peculiarity. Here's a sketch of how that argument might go:

1. Historical scientists are *strategic perspectivalists.*

2. Ontic-driven explanation: strategic perspectivalism is best understood as a response to the peculiarity of the entities and scales historical scientists are interested in.

3. Bridging principle: ontic-driven explanations underwrite metaphysical claims.

4. Therefore, the scales and entities of interest to historical scientists are peculiar.

The first premise was defended through chapters 3, 4, and 5, at least pertaining to the science of lost worlds. In response to loss or potential loss, historical scientists adopt perspectives that facilitate artificing, grafting, and anchoring, perspectives that are enabled by various tools that often partially overlap, thus allowing for iterative progression. My task here, then, is to consider premises 2 and 3. Is there an ontic-driven explanation to be had of strategic perspectivalism and, indeed, is there some principle that might take us from that explanation to the past in fact being peculiar? I suspect so.

Let's remind ourselves of strategic perspectivalism's essential features. "Strategic perspectivalism" is not the same kind of view as "philosophical perspectivalism"—it isn't about the nature of knowledge itself—rather, it is an account of the strategies scientists adopt in tackling loss. This strategy involves (1) characterization: using a perspectival tool to pick out a set of denizens and their properties (and potentially a set of standards and expectations associated with that tool and characterization). This characterization forms the basis of some combination of (2) artificing, grafting, or anchoring. That is, the characterization affords various investigative routes into understanding those denizens. Finally, (3) recontextualization: results from investigations are brought into dialogue with features of the past world targeted (and other perspectives). I've further emphasized the *bespoke* nature of the investigations: in artificing, for instance, the details of the model are often specified to highly particular features of the past world or its traces. I've also emphasized the *pluripotent* nature of the investigations: tools, results, and interpretations often play multiple epistemic roles, informing us about the target, aiding in the construction of models of regularities, and iterating between various perspectival tools.

To what kind of world would this method be best applicable? A highly peculiar one. To see this, consider what an investigation of *unpeculiar* targets might be like. An unpeculiar target is one that is insensitive to generative and maintaining processes; that is, it maintains its properties and persists across a wide range of background conditions, processes, and possible disruptions. For such a target, knowledge of the dynamics of its properties themselves, and the relatively few conditions of its persistence, would be sufficient for explanation and prediction. An investigation into such a system would no doubt involve characterization, but relatively few perspectives would be required to make sense of it. Considering investigations of the past, in an unpeculiar world, a narrow uniformitarian method would work highly reliably. One would be able to investigate current processes and simply project them backward, scaling up to the deep past's horizons.

But a strict uniformitarian approach doesn't work. As Meghan Page (2021) has emphasized, what appear to be stable regularities often only stretch so far into the past. Some examples include the following: assuming niche differentiation in spatial ecologies when interpreting Avalonian rangeomorph-dominated systems would yield the wrong results; the size limits on today's obligate osmotrophs do not necessarily apply to rangeomorphs; and the relationship between the biosphere and atmospheric carbon on Earth is potentially unique to our planet and has been stable only for the last few hundred million years.

In a highly peculiar world, the existence and persistence of denizens will be sensitive to the properties of various *other* denizens. And those *other* denizens will themselves have various sensitivities. The world would both be highly interdependent and highly fragile. The well-mixed oxygenated oceans of the Mesozoic and Cenozoic have been stable for over 400 million years. But if Butterfield is right, that stability is sensitive to processes of bioturbation, the biological pump, large populations of krill and their allies, and so on. Rangeomorphs appeared to play a crucial role in Ediacaran ecologies for something like 30 million years. But their dominance was fragile: their sessility made them unfit in ecologies sporting more dynamic metazoans. As I discussed in previous work (Currie 2019a), from the dawn of the Jurassic until the "Mid-Cretaceous Terrestrial Revolution" around 100 million years ago, the world's macroflora and fauna could be characterized in terms of sauropod dinosaurs consuming gymnosperms. With the advent of angiosperms, we've seen millions of years of flowers and pollinating insects. These patterns are highly stable (often due to the interdependence between their constituent systems) but, nonetheless, are fragile due to reliance on various stabilizing processes. In other words, they are peculiar.

Peculiar worlds, due to their multiple fragilities and interdependencies, must be understood from multiple scales across multiple entities and processes. Many denizens have crucial similarities and continuities between them, but those similarities are typically only partial. More generally occurring regularities will only be capturable by highly abstract models. In trying to understand a peculiar world, scientific tools must be bespoke to capture the specific details and thus be epistemically relevant. Investigations of peculiar worlds will be open-ended as they navigate the various partial similarities that history has generated, and varying tools will be required to characterize the varying denizens and their interdependency. In other words, a peculiar world is best tackled via strategic perspectivalism. Alisa Bokulich (2021a) has argued that Earth scientists navigate among various concrete and conceptual models to investigate phenomena that reach across scales. Models are used to make sense of both hierarchical relations (identifying "the functionally relevant scales given the application of interest, as well as the constraints coming from the higher levels of the hierarchy" (Bokulich 2021a,14179) and threshold effects "identifying the relevant dynamic thresholds in the system where there are qualitatively discontinuous changes in the system's behavior") (Bokulich 2021a, 14179). That is, different tools are used to examine both diachronic dependencies, such as that between the evolution of multicellularity and a well-oxygenated ocean (if Butterfield is right) or fossils and their living progenitors; and synchronic dependencies, such as between an organism's traits and its capacities. In the parlance of this book, the variety and complexity of generating and maintaining processes lead to scale-relative complexity, which requires multiple perspectives to make sense of. The word's peculiarity—or at least its complexity—demands strategic perspectival investigation.

Consider once more Butterfield's arguments concerning the emergence of complex metazoans. Both his narrative and its supporting evidence require switching between a variety of different scales and denizens. We must consider: models and data concerning global biochemical interchange, the optimality or otherwise of various morphological adaptations (often in their biochemical context), the structure of benthic ecologies, niche construction as applied to oxygen environments, and the evolvability of metazoan innovations. We bounce between the global, the ecological, the physiological, and the evolutionary. This switching between and combining of perspectives is, I suggest, necessitated by the past's peculiarity. If the past were *less* peculiar, then we wouldn't need to pick out so many differing sensitivities to explain phenomena.

So, we have at least a putative, ontic-driven explanation of strategic perspectivalism. Historical scientists are strategic perspectivalists due, in a significant way, to the peculiarity of the past worlds they aim to understand. So, how might we link this with the claim that the past world is, in fact, peculiar: what bridging principles between practice and metaphysics are suitable here?

Boaz Miller (2016) has dissected Hacking's argument for entity realism, providing a set of possible interpretations that I'll canvass before drawing on William Bausman's (2024) idea that practice might be adapted to the world via a process analogous to natural selection.

Inference to the best explanation is the most discussed bridge between science and metaphysics. Following Peter Lipton's turn of phrase, an inference to the best explanation takes us from an explanation's loveliness to its likeliness (Lipton 2000). Beyond an explanation's empirical adequacy and its capturing the phenomenon we're interested in, we might also think that certain other properties—elegance, simplicity, and so on—could lead us to think it more likely to be true. Like some philosophers of science, I'm sceptical that inference to the best explanation, as traditional metaphysicians appeal to it, is continuous with the epistemology of science (Novick 2017). Regardless, I think there's a disanalogy between what I'm doing and what philosophers who attempt to argue from scientific results are doing. I'm arguing from a philosophical explanation of scientific practice to a claim about the world: I don't take myself to be making anything like a scientific argument. Rather, I'm making a philosophical argument based on a philosophical explanation of scientific practice. This doesn't commit me to saying that philosophical and scientific arguments aren't in some sense continuous. However, I don't think that appeal to inference to the best explanation (IBE), insofar as it occurs in science, gives philosophical appeals to IBE any special license.

Shifting from IBE specifically, what would we need to establish to connect our ontic-driven explanation to a peculiar ontology? Ideally, such an argument would establish the truth of a conditional along the lines of the following:

> If the past were not peculiar, historical scientists would not be strategic perspectivalists.

However, a strict conditional relationship is too much to hope for, as we'll see. But there is reason to commit to something weaker.

When I considered what the science of a non-peculiar past might be like, I suggested that it would be considerably more uniformitarian. Scientists would seek to discover the basic dynamics by which the past unfolds, and these dynamics would not require bespoke models to capture them (see Walsh and Currie's discussion of Newton's astronomical work, 2018). At least sometimes, then, a lack of peculiarity underwrites a different methodological strategy for scientists.

Thus, we might argue that the best explanation of historical scientists adopting strategic perspectives is that the world is peculiar. To apply this approach, we'd both need to give reason to think it is the best explanation, and reason to think it is, as it were, sufficiently best. Bausman's account is, I think, promising here: "Successful scientific practices are adapted to features of the world for purposes as organisms are adapted to features of their environment for functions" (Bausman 2023, 28). If the sciences follow a process analogous to adaptation while investigating their target systems, then we can make sense of the above conditional: over time, ontic interaction has shaped the methods and practices of historical scientists, thereby fitting them to their target system.

While ontic-driven explanations do not exclude other possible reasons for scientists to adopt a method—explanations can cite multiple factors—there might be explanations that exclude or minimize the importance of peculiarity, thus undermining the argument. In the last section's discussion of ontic-driven explanation, I named a set of non-ontic factors, such as social, economic, or ideological factors, that could explain strategic perspectivalism. Here I'll discuss some ontic alternatives.

A set of potentially pressing alternatives is also ontic-driven. It might not be peculiarity but erasure that underwrites strategic perspectivalism. Historical scientists often face "epistemic scarcity" (Currie 2021a). They tackle a gappy record, with often hard-to-manage specimens and data, and often lack the background theory required to "fill" that record's gaps. In response, the explanation goes, they adopt strategic perspectival methods. As peculiarity underwrites loss, this explanation amounts to claiming that most loss is only apparent and that a lack of uniformity is due to the degradation of traces. In chapter 1, we saw Lyell argue something like this: the apparently "progressive" pattern in the fossil record, taking us from fish-dominated to reptile-dominated to mammal-dominated macrofauna, is due to differential preservation (and is moreover a highly selective reading of the record). We lack sufficient evidence to conclude it is a biological signal.

On this alternative ontic-driven explanation, historical scientists are strategically perspectival because of a lack of traces. This amounts, perhaps, to saying that information preservation is peculiar, but the sources of that information—the non-trace denizens in the past—are not, or at least that we should remain agnostic on that score.

I have an answer, I think, to this alternative. Many of the troubling cases of loss or putative loss we've discussed have been in relatively trace-rich contexts. Take, for instance, Mitchell et al.'s examination of rangeomorph ecosystems. The fossil beds of Mistaken Point are remarkable: there is good reason to think that volcaniclastic processes have preserved the spatial positions of the rangeomorphs *in situ*. However, loss and potential loss nonetheless present major epistemic challenges in this trace-rich context, and Mitchell et al. adopt a paradigmatically strategic perspectival strategy in light of this. David Sepkoski (2019) and I (2018a) have argued that the capacity of historical scientists to "re-read" the fossil record, distinguishing between biological and geological signals and thus partly overcoming a lack of traces, is remarkable. No doubt there are many cases where we should be pessimistic about our knowledge of the past because of epistemic scarcity, but this pessimism should only go so far. To be sure, strategic perspectivalism is in part a response to a lack of traces, but if that were the only driver, we would see historical science become significantly more uniformitarian as records became less gappy or better understood. And I don't see evidence of this.

A further set of alternative ontic-driven explanations might appeal to non-peculiar ways the world might be that strategic perspectivalism is a good response to. One alternative is *complexity*.[6] The world might be highly

6 Thanks to Alisa Bokulich for pushing me on this alternative.

complex and heterogeneous, that is, made up of many different components with a wide variety of arrangements and relations, without being peculiar. These complex properties and relations could have few dependencies between generating and maintaining processes; they would be complex, but nonetheless not peculiar. Plausibly, under those conditions, successful scientists would also adopt strategic perspectivalism. I consider this a live and alternative hypothesis, although I suspect that such a world would demand explanation in a way that a peculiar world does not. As I'll argue in the next section, a peculiar past *makes sense of* a complex and heterogeneous present. At the very least, a complex-but-unpeculiar world would require some explanation.

So, although these alternatives undermine a strict conditional relationship between peculiarity and strategic perspectivalism, I do think there are reasonable grounds to take at least some of the practices of historical science to be adapted to a peculiar world.

Miller discusses two alternative readings of Hacking's argument that are worth mentioning. He calls these the "indispensable" and "transcendental" arguments. Both routes more or less argue that a particular ontological posit is required for a scientific practice. In Hacking's case, if there were no entities like electrons, then apparatuses like electron microscopes wouldn't make sense: without committing to the existence of electrons, the practice is rendered incoherent. Similarly, the notion of *electron* (or at least some entities with electron-like-properties) is indispensable to the use of electron microscopes. To apply these approaches to the case at hand, we'd need to establish that the coherence of strategic perspectivalism turns on there being a peculiar past. I think there is potentially *something* to this kind of argument: historical scientists tackle the past with the expectation and understanding of peculiarity. And understanding peculiarity makes sense of why they use the strategies that they do. But I don't think anything as strong as indispensability can be established. As we've seen, there are plenty of other contenders.

Overall, linking peculiarity and strategic perspectivalism is best done by arguing that the method is adapted to a peculiar past. I imagine that scientists who begin investigation taking the world as non-peculiar—and thus adopting a uniformitarian approach—would quickly find their investigation either stalling or gradually becoming increasingly perspectival. Although, as we've seen, historical science is often very concerned with understanding the present and linking it to the past in various ways, attempts to uncover non-bespoke general regularities have typically failed. There are at least two ways one might try to strengthen this argument. First, one might take it as a serious historical claim and go looking for cases across the history of palaeontology to see whether such a pattern presents itself (I suspect claims that historical explanations increase in complexity over time could be co-opted here: see Currie 2014; Dresow 2021). Second, one might take it as a thought experiment aiming to pump and explore our intuitions about the connection between method and the kind of system we're targeting.

Bausman distinguishes between inferring from practice, the metaphysics of a particular domain, and inferring across domains. Waters' argument that there is no general structure, for instance, first involves establishing that the practices of geneticists are not adapted to a generally structured world, and then requires universalizing that claim. I am interested in the former but not the latter inference. That is, I am interested in the kinds of scales investigated by palaeobiologists and Earth scientists (already a pretty rich domain!). There might be ways of expanding this—maybe peculiarity in one domain entails peculiarity in others—but I think such arguments require care. Fragility and sensitivity in one domain need not guarantee the same at "lower levels" of description (maybe those atoms behave highly regularly), but also not at "higher" levels (order can arise out of chaos; see Strevens 2009).

Of course, this doesn't mean that I am not making various kinds of inductions. I've hardly conducted a survey of practices across the Earth sciences to show that strategic perspectivalism is the norm. Although I'm being metaphysically fairly cautious, I'm being rather reckless in my claims about methodology.

None of my arguments would force a really determined detractor to agree with me on pain of irrationality. But I don't think metaphysical arguments need to be "knock-down" in that sense. Nor need we rely on a single argument for a conclusion. Nancy Cartwright has decried the use of "long, skinny" deductive arguments on the basis of their fragility—lose one premise, and the whole shebang is up in logical smoke (Cartwright 2015). Instead, she recommends arguments that are perhaps less logically impeccable, but are significantly more robust. I see the argument here as following the latter approach. (1) That historical scientists have discovered peculiarity in the past over and over again, (2) that an ontic driven explanation of their method appeals to peculiarity, and (3) that their method is plausibly adapted to a peculiar world, are all imperfect arguments pointing toward the same metaphysical conclusion: at the scales that interest palaeobiologists and Earth scientists at the very least, the past is peculiar. That is, history matters.

4. From Peculiar to Dappled

> There is a weighty pause whilst Prof. Ichthy awkwardly shuffles on their lower flippers, mentally corralling their own paragraphs. Scratching their lower jaw, they respond to the *Allosaurus*.
>
> "I don't think the world is pure chaos: history is highly structured in various ways. Broadfoot, we have good reason to believe, lived in shallow fresh water. We think they used their echolocation to detect prey, the best bet being small crustaceans. They existed, then, in an environment that they were adapted to, and were linked with other organisms through various trophic interactions. I don't know if there was anything we might call a 'harmony' in that ecosystem, but there was at least some kind of dynamic stability. And, I would say, that stability itself had a history: we can make sense of why the mammals arose to such prominence and perhaps make sense of why they fell.
>
> "History, then, is neither chaotic nor ordered, but a bit of both. And I like to think that understanding ourselves in light of that history—how the order and, yes, the chaos arises from complex historical processes—is a way of making sense of our lives. I guess that's likely as philosophical as I can get, I'm afraid."
>
> Prof. Ichthy and the *Allosaurus* both half-shrug as if to signal a conversational cease-fire.

Everyone agrees that the representational power of contemporary science provides us with, at best, a patchwork. Insofar as science provides an empirical grip on reality, it does so impartially, imperfectly, requiring various fudges and kludges to scale the gaps between perspectives. Science is disunified in that sense. Descriptive claims concerning science's disunity have been traditionally drawn on to undermine unificationist and reductionist conceptions of science. No doubt, seeking underwriting mechanisms for various processes—understanding the "macro" through the "micro"—is often a successful scientific strategy (Machamer, Darden, and Craver 2000), but it doesn't often lead to more general unification, and nor does it often lead to applications that can be applied "off the shelf," as it were, to domains beyond those that they were developed to make sense of.

A common way of making sense of disunity appeals to the limits of human knowledge. A philosophical perspectivalist, for instance, might infer from the claim that all knowledge is from a perspective—that is, requiring both a characterization and a standpoint—to the idea that all knowledge is partial. Further, we might point to the limits of representation generally to make sense of science's disunity (Mitchell 2002). Our theoretical understanding is necessarily partial, so no wonder the picture of the world provided by science is a patchwork.

One needn't lean on representationalist accounts of scientific knowledge nor on theoretical understanding to provide an epistemic explanation of science's disunity. We might also appeal to the varying aims of science and how these aims shape scientific research into differing forms. Alan Love's notion of a *problem agenda* (Love 2008), for instance, conceptualizes scientific research as driven by particular questions or "problems" (say, understanding rangeomorph feeding). Tackling these questions sets an agenda, which requires marshalling multidisciplinary evidence and structuring that evidence into various roles (traces that inform us about morphology, geometric models, studies of osmotrophy, etc.). As differing questions—differing aims—require differing complexes of disciplines and evidence, we should expect science to take the form of a complex web of disunities and integration (see also Brigandt 2013).

Finally, in chapters 6 and 7, I defended a form of *artifactualism* about the imagination of historical scientists and the modal knowledge they provide. An artifactualist explains the justification of modal knowledge by appeal to the dispositions of the artifacts scientists construct. We might generalize this idea, claiming that scientific knowledge is *fundamentally constrained by its medium*, that is, the physical thing that bears or instantiates the dispositional properties that constitute scientific knowledge. Different media instantiate different modal properties, and this necessitates a diversity of media. As these media exemplify slightly different modal spaces (see Currie 2020), knowledge is always partial due to its medium-boundedness. I have some sympathy for this idea, particularly if coupled with a know-how account of understanding. If artifactualism about scientific knowledge generally could be defended (and doing so is well beyond the scope of this book! Maybe the next one . . .), it seems to follow that this knowledge is of necessity a patchwork.

In the metaphysical spirit of this chapter, we might go beyond these epistemic explanations of science's disunity to construct an ontic-driven explanation. It isn't only constraints on knowledge, nor the nature of scientific knowledge, that underwrite disunity. Additionally, science is disunified because *the world is disunified.* John Dupre (1993), Nancy Cartwright (1999), and (as we've seen) Ken Waters (2017) have argued along these lines. Cartwright, focused on economics and physics, argues that the theories of those sciences, despite the best efforts of the scientists working on them, fail in their goals of generality. And this failure is explained by the world being "dappled": there is one material world, which is to say whatever there is, is spatio-temporally connected; however, the various relations between those things cannot be systematically described.

> The sciences are each tied, both in application and confirmation, to the same material world; their language is the shared language of space-time events. But beyond that there is no system, no fixed relations among them . . . Their boundaries are flexible: they can be expanded or contracted; they can even come to cover some of the same territory. But they undoubtedly have boundaries. There is no universal cover of law. (Cartwright 1999, 6)

Dupre focuses primarily on how biologists go about carving up the world: the categories they employ are often not intertranslatable, cross-cut one another, and serve different purposes. For Dupre, there is no basis for metaphysically privileging one way of carving over another. When we find order, "There is no reason to suppose that whatever uniformities apply to a particular ordered domain will apply to every such pocket of order" (Dupre 1993, 4). As such, we should commit to what he calls *promiscuous realism*: the world is constituted by a wide variety of complex, overlapping kinds, all of which deserve to be taken metaphysically seriously.

Certainly, Cartwright and, to an extent, Dupre, provide product-first metaphysical arguments, and as such face some of the same challenges I described above for that route from science to metaphysics. Waters, as we've seen, argues along similar lines to mine: from the practices of scientists. What unifies these views, however, is their *synchronic* nature. They paint the world as a complex patchwork, but have little to say about *why* the world might be such a patchwork. Let's quickly point to some synchronic arguments for disunity, before turning to a diachronic account.

Taking the metaphysical view of the causal closure of the physical as our target, we can discern several synchronic arguments for the world's disunity. Causal closure is the idea that the causal powers of whatever the smallest bits in the world are—at least in principle, the things studied by current or future fundamental physics—exhaust the causal powers of everything else. Arguments for metaphysical disunity often involve denying causal closure in some way (Dupre 2023). One argument is *from context*: the behaviour of many low-level entities turns not just on their internal properties, but often on highly context-sensitive factors within their environment. These contextual factors are crucial for understanding how messy the world is (Dupre 1993; Havstad 2018; Baxter 2023). Gene expression doesn't just rely on that particular gene, but on the influence from other genes, on the biochemical environment they are in, and on interactions from various other factors in the developmental sequence (Griffiths and Gray 1994). Another argument is from explanatory autonomy: if some explanations at a higher level capture explanatorily ineliminable information that is not discernible from the lower level (Jackson and Pettit 1992; Sterelny 1996; Grantham 1999)—and we read explanation in ontic terms—then either causation doesn't exhaust ontology, or there are "higher-level" causes. Further, there are arguments from reductive failure: there is a long list of entities that were once taken as putative places where the world might bottom out, or that might be the underlying determinant of some system, which, on investigation, have turned out to be significantly messier. The relationship between Mendelian and molecular genetics is perhaps the classic instance of these phenomena (Kitcher 1984).

These synchronic arguments can be supplemented with a diachronic argument, which we might call *the argument from peculiarity*. If we think the past is peculiar, then we should expect the present (and future!) to be dappled and disunified. Let's see why.

A peculiar world is one in which new kinds arise as new conditions for existence emerge. The post-Avalonian evolution of metazoans blessed with movement and mouths represented not only a set of new individual taxa but also enabled new ecological dynamics: predation and scavenging, for instance. These are often highly stable, but also highly peculiar. This should lead to a differential structuring of different parts of the world at different times. Globally, the Earth has housed fairly stable relationships between atmosphere and biosphere for 400 million years. But there are more local variations within that global story, and earlier there were different global stabilities. Peculiarity, then, provides diachronic grounding for a dappled world: the world is disunified *because* the past is peculiar.[7] I'll illustrate this with a simple ecological case.

7 In section 4 of chapter 5, I mentioned Lange's distinction between temporary and eternal but time-dependent laws. One might insist that the right set of eternal but time-dependent

Aotearoa (New Zealand), prior to the arrival of our species and their accompanying troop of placental mammals, boasted a fauna populated by birds and insects. Islands like Aotearoa often have weird animals, and one way of making sense of these weirdos is by appeal to familiar niches being occupied by surprising taxa due to founder effects. As such, Aotearoa is full of birds filling niches typically associated with mammals. One of the most iconic is the flightless ratite moa, which ranged in size from around a meter to the enormous 3.6-meter-tall *Dinoris*. They were highly populous (Gemmell, Schwartz, and Robertson 2004) and took on many ecological roles, from sheep-like grazing to deer-like browsing (Wood et al. 2020). So, various moa species acted something like deer or sheep, fulfilling grazing and browsing roles. Meanwhile, their equally iconic relatives, the kiwis, are birds attempting to act like nocturnal burrowing insectivores, such as badgers. It is sometimes assumed that the introduction of mammals into Aotearoa involved "ecological replacement," that is, mammals took up the various roles that birds once filled. At an extremely coarse-grained level of analysis, I suppose something like this is right: deer and moa are both browsers after all. This perspective runs into both philosophical and empirical trouble, however.

We might have philosophical worries about this notion of "niche." As Dennis Walsh puts it: "A (nonmetaphorical) niche is a space, or a recess, into which something (say, a statue) might fit. Like a real niche, an evolutionary niche is a set of properties of an environment, to which an organismal form may fit" (Walsh 2015, 169). However, Walsh points out, this seems to assume that niches somehow exist independently from their occupants, as if Aotearoa was "waiting" for kiwis to evolve. Richard Lewontin ties this independent and concrete conception of niches to adaptationist thinking: "To make the metaphor of adaptation work, environments or ecological niches must exist before the organisms that fill them. Otherwise, environments couldn't cause organisms to fill those niches. The history of life is then the history of coming into being of new forms that fit more closely into these pre-existing niches" (Lewontin 2001, 63).

I think Walsh and Lewontin's philosophical skepticism has some merits, although I doubt it applies across all niche-talk among scientists. In the spirit of strategic perspectivalism, we might think that conceiving of niches in coarse-grained ways might gain epistemic purchase in some contexts.

laws, coupled with sufficiently complex initial conditions, would render a non-dappled, law-like world.

Regardless, there are empirical problems with applying ecological replacement to Aotearoa, and this has everything to do with the peculiarity of trophic interactions.

The transition from a bird-rich to a mammal-rich fauna drastically changed some of Aotearoa's trophic dynamics. Here's one example, taken from Atkinson and Greenwood (1989). In this study, they test replacement by seeing if it can make sense of the changes that occurred in Aotearoa when mammals were introduced. "Discussions of the effects of introduced browsing mammals on New Zealand plants and vegetation have sometimes included the suggestion that they are not fundamentally different from those formerly exerted by moas. This suggestion is treated here as a null hypothesis and tested by some comparisons between moas and mammals" (Atkinson and Greenwood 1989, 67).

A fair few of Aotearoa's plants exhibit a form of cryptic mimicry where healthy parts appear to be unhealthy or dead. For instance, the leafless pōhuehue (*Muehlenbeckia ephedoides*) has lower shoots that take on the appearance of dead twigs. This is presumably an adaptation, a surprising kind of mimicry that makes the plant seem less palatable to grazing animals. This makes sense for a bird-rich biota, as birds primarily hunt by eyesight (with kiwis being a notable, mammal-like exception). However, looking dead is very little help if you're trying to avoid being eaten by mammals like sheep, who rely not on their eyes but primarily on their noses. So, the leafless pōhuehue is adapted to the bird world, and sadly maladapted to the mammal world. Atkinson and Greenwood provide many examples of this phenomenon: although mammals and birds both act as grazers and browsers, they do so in quite different ways, and these differences matter.

> It is unlikely that moas bit and stripped bark from stems, in the way that deer and goats bark Pseudopanax spp. and some podocarps. It is also hard to imagine them ploughing the soil and destroying roots in the way that pigs do even though they may have scratched and torn at the ground with their feet. Moas were presumably not capable of climbing trees, so there would be no parallel with the arboreal feeding of brush-tailed possums. (Atkinson and Greenwood 1989, 89)

A picture emerges: Aotearoa's flora and fauna co-evolved; thus, truths about bird-rich Aotearoa do not carry over into mammal-rich Aotearoa. A peculiar process—co-evolution—leads to a disunified world. The leafless pōhuehue is adapted towards a bird-filled ecology in which visual mimicry is a good strategy. Much less so in a mammalian ecology. So, the regularity *visual mimicry is a good way of avoiding grazing and browsing animals* held true in Aotearoa prior to the arrival of humans and other mammals, but no longer held true afterward. 1500 years ago, it was true in Aotearoa but false in, say, North America.

One might complain that more coarse-grained conditional regularities nonetheless held true, say, "If grazers rely primarily on vision, then playing dead is a good strategy." However, this conditional itself captures the results of peculiar historical processes. Peculiarity doesn't necessarily undermine fairly coarse-grained conditional properties or relations. But note that many of the properties above, e.g., "vision," "browsing," and "playing dead," are themselves the outcomes of fairly peculiar processes, and how they play out—what acts as vision, how one browses, what it takes to play dead—are often highly sensitive to evolutionary histories. Due to the co-evolution of biota, such regularities only gain purchase under specific local circumstances. That is, the peculiarity of the past leads to a dappled picture.

I want to close with two discussions. The first concerns the limits of experimentation, and the second concerns the relationship between what I've established here and a more general metaphysical view: processualism.

Much has been made of the importance of experimentation to science (rightly, I think: Weber 2004; Currie and Levy 2019); however, philosophers committed to a dappled or disunified world sometimes point to the limitations of experimental systems, Cartwright most paradigmatically: "Physics in its various branches works in pockets, primarily inside walls: the walls of a laboratory or the casing of a common battery or deep in a large thermos, walls within which the conditions can be arranged just so, to fit the well confirmed and well-established models of the theory, that is, the models that have proved to be dependable and can be relied on to stay that way" (Cartwright 1999, 2).

Similarly, in biology, Jim Griesemer has emphasized the relationship between experimental systems and mathematical models, explaining how the design of experimental systems involves the application of scientific agency toward that system behaving in ways related to mathematical models (Griesemer 1988; Griesemer and Wade 1988). The expected experimental behaviour is not necessarily analogous to behaviour in the natural target: "Expectations about effects stemming from knowledge of causes in the laboratory system are applied to natural systems through the use of the same data model. To justify using the same data model, the experimental design structure of a laboratory model must be recognized in the natural system" (Griesemer and Wade 1988, 84).

Experiments give the results that they do because, in part, they have been designed in a particular way to "capture" the regularities at hand—they provide "walls," as Cartwright puts it—so, if they are to apply to natural systems, the differences between the structure of the lab and the world must be understood.

This take on experiments can be understood using peculiarity: established experimental regularities are highly peculiar to human-manufactured conditions. As such, understanding the limits and otherwise of an experimental system is extremely difficult from *within the system*. This is why experimental approaches need to run in parallel with non-experimental, more "natural" observations. It is also why history is so important in peculiar worlds, one reason why history matters. It is, in some sense, "easy" to establish a regularity experimentally (not *practically* easy!): we set up the conditions and control much of the subsequent behaviour with much tweaking. No wonder, after all this labour, that we get something relatively well-behaved. But in a peculiar world, those experimentally established regularities will only extrapolate inelegantly and to a highly restricted range of targets. Both the peculiarity of the world and the peculiarity of experiments themselves restrict the epistemic power of experiment.

Let's finish with processualism.

I've argued that the world at scales investigated by the Earth and historical sciences is peculiar; that is, the world's actual and modal features are highly sensitive to features of a large number of generating and maintaining processes. I've further argued that this provides a diachronic argument for a disunified present. This is closely aligned with an old philosophical view that's been reinvigorated in the last ten years or so: *processualism* (Nicholson and Dupre 2018; Dupre 2025). In this final discussion, I'll argue that committing to "peculiarism" aligns well with processualism—they are fellow travellers, if you will—but nonetheless one may adopt my view without thereby being a processualist. And indeed, the scale-relativity that underwrites peculiarism as I've induced it from strategic perspectivalism might motivate refusing to swallow processualism whole hog.

I understand *processualism* as a metaphysical view about the fundamental constituents of reality, according to which the world is built of processes.

What, then, is a *process*? The concept is often explained in contrast to *things*, where "things" are taken to be intrinsically stable objects with essences. Take the early-modern view *corpuscularianism*, according to which the world is constituted by units ("corpuscles") that have particular essential properties, such as size and shape. Accordingly, observable behaviours and effects are due to the interactions of corpuscles. This is a thing-like view. By contrast, processualists deny there are stable entities such as corpuscles (or atoms, or whatever); rather, there are, well, dynamic, shifting, not-things. One way of capturing the notion of processes, and one that well-distinguishes processes from things, is to define them as "entities for the identity of which change is essential" (Meincke 2023, 234). Things are what they are in virtue of stable properties, and processes are what they are because of how they change.

There is a lot for someone fascinated by lost worlds to like about processualism. In particular, the processualist assumption of change and flux highlights the importance of stability as worthy of explanation. One needn't be a processualist to explain surprising stability, but it is nonetheless a useful perspective correcting a bias toward change as explananda. Further, analyzing the science of lost worlds highlights the importance of the historical individuals and long-term trajectories that are both so crucial for grafting strategies, and a processualist metaphysics naturally cashes out individuals not as things but as temporarily stable homeostatic processes or trajectories. Further, historical kinds are what they are in virtue of the processes by which they form, not their essential properties: processualism thereby highlights these.

However, despite the appeal to process, if you agree with the metaphysics of this chapter, you're not thereby committed to processualism, at least not by the definition of peculiarity. This is because admitting that an individual is sensitive to generating and maintaining conditions does not thereby commit you to thinking an individual is thus *metaphysically subservient* to those processes. Causal dependence is not in itself sufficient for metaphysical dependence. It seems we are logically and metaphysically free to describe, say, various taxa in terms of stability—as things—or in terms of various processes, while happily admitting their peculiarity.

Further—and this is the step that pulls me back from the processualist brink—once we have the idea of a "thing ontology" and a "process ontology" on the table, it is natural to think that whether we should use "thing-language" or "process-language" turns a fair bit on the scale at which we work. When Cuthill and Conway Morris model the development of rangeomorphs, it seems natural to describe their target as temporarily stable processes, whose temporary stability is made sense of in light of developmental (and other) processes. When Mitchell et al. model rangeomorph ecology, it seems natural to think of rangeomorphs as a population of things. After all, in that context, the rangeomorphs are treated as objects with set properties, that is, a bunch of individuals in a population.

To establish processualism, we need an argument to say that the levels of description, temporal duration, or scales by which things are best described as processes should take metaphysical priority, and I'm unsure why this should be the case. It might be that at the widest possible scale, we find processes, not things. This is an empirical conjecture, and I'm not sure how we might establish it, but let's go along with it for the argument's sake. Why would processes dominating the widest scale lead us to say processes are metaphysically prior? We'd need an argument to say that the widest possible scale is the most important, a claim I find symmetrical to, and problematic for the same reasons as, claiming that the "lowest" level of description, the base of the mereological heap, should have ontological priority. Another approach might be that the monism and unity of processualism itself is reason to prefer it. Even if you're someone who sees these as theoretical virtues, I'm not sure why we should think a monistic or unified world is more likely due to those virtues. And further, such an argument seems to favour a thing-ontology as much as a process-ontology.

Further, recall the commitment to strategic perspectivalism that underwrote my argument for peculiarity. I suggested that strategic perspectivalism is best explained as an adaptation to a peculiar world. And strategic perspectivalism shifts between thing-ontology and process-ontology (and probably a bunch of others as well!), so adopting a monist ontology in its name seems, well, against the spirit.

This is not to say that processualists and those who commit to peculiarity do not significantly agree; indeed, processualists should expect a lot of peculiarity. Recently, John Dupre has argued that processualism is another route to providing a diachronic argument for disunity: "Process ontology provides the diachronic dimension that unifies the disordered world" (Dupre 2023, 13). In this instance, I think it is fair to think of the argument of this chapter as an *application* of processualism to the scales of interest to palaeontologists and Earth scientists. But a commitment to peculiarity doesn't require one to commit to processualism. Thus, I'll keep to agnosticism about things beyond the scales of the Earth and historical sciences, and stick to a claim about peculiarity rather than processualism.

So, I take myself to have provided an argument to favour Prof. Ichthy's view of the world over Professor Ichthyosaurus'. I've argued we should understand their differences in terms of how peculiar they take the past to be. Insofar as the science of lost worlds has uncovered much loss, and loss is anticipated by a peculiar world, we've reason to think the world peculiar. Insofar as historical investigation uncovers peculiarity, we've reason to think the world is peculiar. Insofar as strategic perspectivalism is adapted to a peculiar world, we've reason to think the world is peculiar. It doesn't follow from this that all things in the past are highly peculiar, nor that they are peculiar to the same extent. But, if we want to commit to a metaphysical view about the structure of the past at the scales of interest to palaeobiology and related sciences, we have a host of reasons to pick peculiarity.

9

Unprecedented Worlds

A large pterosaur[1] sporting a teardrop fedora impatiently waves a notepad and pen clutched in the claws at the end of their enormous wings.

"Prof. Ichthy . . ." they begin, speaking at a rapid pace, "You've said that knowing about the past is important because it can help us think about the future. Do you really think that? Can you back that up? And can you say something concrete? Can you go on record for *Eons'* readers, explaining how all this great progress about knowing how things were might tell us about how things will go? Would you be willing to take any bets on the future? Will you, Professor, put your money where your mouth is?"

"The thing is . . ." the pterosaur's speaking pace slows, and she sheepishly shrugs, "I find my readers are worried about the future; they worry about it a whole lot. And I don't just mean small-time, personal worries for their fortunes and families and such, but big-time worries about where we'll be in ten years, or fifteen years, or a hundred years. People feel like the social world, the political world, the technological world, and even the natural world are all spinning out of control."

The pterosaur puts pen to notepad expectantly as they finish their question: "In light of all of this anxiety, is there any advice you can give for facing that future? What has all your learning about the past given you in terms of lessons for thinking about the future?"

The professor grimaces, takes a pause.

"Well, you're calling me out a bit here. We academics are often encouraged to frame what we're doing in as earth-shattering a way as possible. And this can sometimes lead to us overdrawing

1 A member of the *Dearc* genus, to be taxonomically exact if metaphysically dubious.

> the bow on topics that exceed our expertise. I'm afraid I am no expert on the future. Advising your readers in specific, concrete ways is somewhat above my pay grade. But I'll try nonetheless to corral some thoughts. I think I have three things to say."

Alert readers will have noticed that my definitions of "past world" and "denizen" are only restricted to the past by conceptual stipulation. While past worlds are, by definition, subsets of the actual past, nothing is stopping us from considering denizens occupying subsets of the current world or denizens occupying subsets of the future. This is a feature, not a bug, as similar epistemic challenges arise for "lost" denizens in the present and in the future as for those in the past. For instance, a major challenge in understanding denizens of the contemporary deep sea comes from that environment's oddities—a lack of sunlight, extremely high pressures—which lead many of the properties of those denizens to be "lost" relative to the environments we're used to or can easily examine. When biologists are trying to understand denizens of the deep, such as the colossal squid, they adopt strategies that rely on the examination of proxies, such as smaller and less pressure-adapted squid (Levy and Currie 2015). I think it is reasonable to consider such practices as examples of grafting toward understanding a denizen of the "lost world" of today's deep oceans.

What about the future? What can we say about future worlds, particularly *unprecedented* worlds, in light of my discussion of past lost worlds?

Concerns about rapid, global, previously unseen changes occurring now or in the nearish future, and these escalating, have become so prevalent that it is difficult to talk about them without relying on a dystopian stock of near-clichés involving tipping points and catastrophic collapse. The increasing pace of technological and social change, coupled with our increasing impact on the environment around us, makes it increasingly possible, even plausible, that we are calling forth an unprecedented future. That is, we can perhaps expect a future where previous certainties can no longer be taken for granted. A future that obeys different rules from the present. Further, even putting aside an anthropogenic perspective, if something like chapter 8's arguments hold water, then there's no particular reason to think the often-shifting, peculiar past shouldn't continue to surprise us into the future.

Let's define an *unprecedented future*, then, symmetrically with lost worlds.

> As a past world is composed of past denizens that do not exist in the present, an unprecedented future is composed of future denizens that do not exist in the present.[2]

This conceptual symmetry suggests an epistemic symmetry between lost and unprecedented worlds. Worlds are lost when their denizens are—that is, when some denizen, under some characterization, existed then but does not exist now. So, lost and unprecedented worlds only differ in their temporal orientation. This suggests that, perhaps, the strategies historical scientists adopt in trying to uncover lost worlds might apply to unprecedented futures. In this chapter, I'll suggest that this epistemic symmetry only partially holds.

Metaphysically speaking, you might be committed to a determined, fixed future, and thus think that—in principle at least—sufficient knowledge from the present could track the future; you might be committed to an open, probabilistic future, and thus think that—again in principle—there is simply no way of knowing some future outcomes. In keeping with the methodology of the last chapter, I don't think the analysis of scientific strategies I've provided can ground commitments either way on this score. Further, I'm not convinced that these metaphysical views generate such epistemic conclusions, at least at the scales that concern us here. I'll briefly return to metaphysical asymmetries at the chapter's close, but, for my purposes, I'm primarily interested in analysis not only at an abstract epistemological level, but in terms of the strategies, methods, and practices available to us for uncovering the future. As such, the partial asymmetry I'll try to establish concerns epistemic practice, not metaphysical questions about determinism or otherwise.

I'll start by discussing various ways in which the knowledge we've generated about the past might be helpful for thinking about the future. Don't expect any big surprises, although my emphasis on the modal nature of historical scientific knowledge perhaps lets me push the boat out for the importance of past knowledge a little further than others have. I'll then ask whether my answers concerning knowledge of loss carry over to unprecedented futures. Do the epistemic strategies that enable historical scientists to uncover lost worlds apply to unprecedented worlds as well? I'll provide an example of how careful characterization can open research into improbable but catastrophic

2 As with lost worlds, unprecedented futures are profligate, due to a lack of restriction on what kinds of entities, processes, and dynamics can be "denizens." I don't see this as a problem. As with the past, we'll be interested in *significantly* unprecedented futures, where a lack of denizens makes epistemic trouble for our knowledge of the future.

global risks in ways analogous to the approaches of historical science. So, I'll suggest that very similar strategies apply to unprecedented worlds as to lost worlds: both are amenable to strategic perspectivalism.

However, I'll then turn to an epistemic asymmetry between investigations of lost and unprecedented worlds. I've argued that the success of the science of lost worlds doesn't just turn on clever characterization, but on *re-contextualization*, that is, the results of investigations being situated within an actual history as revealed by, for instance, the fossil record. Further, I've argued that traces play a central role in the productively iterative investigations characteristic of the science of lost worlds. It is unclear whether there are "records" of the future. Without such "records," the power of iteration and recontextualization is prima facie undermined. So—perhaps surprisingly—at least under some circumstances and about some questions, we should be more pessimistic about our knowledge of the unprecedented future than the lost past.

I'm on record as not thinking there's anything that in principle distinguishes historical from non-historical research. The epistemic differences within those categories are greater than the differences between them: there's nothing epistemically distinctive about historical research per se (Currie 2019a). My claim about our knowledge of unprecedented futures requires a slight revision to this way of thinking: I think access to records does mean that a scientist of the past differs from a scientist of the future in a particular way. Specifically, the existence of records allows scientists of the past to map out the past's surprising peculiarities and to explore the possibilities closely aligned with those. These are not available, or at least not to the same extent, for the future, and—unless other approaches and epistemic tricks are discovered—something like traces are required for rich knowledge of unprecedented futures. As we'll see, I consider this conclusion revisable, based as it is on our current methods and approaches to uncovering the future.

Before diving in, it is worth noting a rather obvious further disanalogy between our knowledge of the past and our knowledge of the future. We can change the future, but we cannot change the past (except in rather metaphysically inconsequential ways: Currie and Swaim 2021). I'll incorporate this further disanalogy into the discussion as we go.

1. Past-Derived Knowledge and the Future

> "The first, and I think quite obvious thing to point out is that the past can be a guide to the future. And not just because the past provides us relevant data or analogues, but because the past gives us knowledge of what is possible."

This chapter primarily concerns whether the knowledge-generating strategies applicable to lost worlds can be applied to unprecedented worlds, so this is not the place to develop a particularly long-winded account of how the past might epistemically inform us about the future, but it is nonetheless important to sketch these ways. I'll draw on a geological example to provide some illustrative, slightly caricatured examples.

In 1982, Christopher Scotese (at the time, a graduate student at the University of Chicago) speculated about the future formation of a supercontinent he called *Pangea Proxima* (Scotese 2018).[3] The name indicates the new supercontinent's similarity to Pangea, the supercontinent that broke apart at the close of the Jurassic period. This new Pangea would be formed by the closing of the Atlantic and Indian oceans, thus unifying the Old and New Worlds (that is, the Americas along with Europe, Africa, and Asia). This is postulated to occur about 250 million years from now (Romano and Cifelli 2015).

Scotese's speculation about continental arrangement 250 million years in the future was based on the *supercontinent cycle*, the idea that there is a regular cycle of supercontinental assembly and break-up with something like a 500-million-year turnover (Nance, Murphy, and Santosh 2014; Mitchell et al. 2021). The last few billion years have seen three supercontinents, from Columbia to Rodinia to the most recent, Pangea. Continental movement is largely governed by subduction processes, where one plate slides under another to be consumed by the mantle. When the mechanisms of plate tectonics began is disputed, with estimates ranging from when indisputably subducted rocks appear during the Neoproterozoic just under a billion years ago to the Mesoarchean around 3 billion years ago (Stern 2018; Palin and Santosh 2021). So far as I can tell, earlier dates are mostly favoured (although there remains plenty of uncertainty, Harrison 2024). Prior to subduction taking off, it is thought that there were patterns of continental formation for "supercratons," but as these plates were segregated, the pattern seen in supercontinental

3 Originally *Pangea Ultima*, but he changed the name later.

assembly wasn't detected. Scotese more or less took current continental positions and extrapolated forward based on the supercontinent cycle.

With our case on the table, we can ask: how might what we know about the past, especially lost worlds, help us understand the future? Items in the following list overlap in various ways, but I think it useful to articulate them separately.

First, we can do so by identifying regularities and establishing their modal profile by testing them against a wide range of conditions. The supercontinent cycle is a long-term pattern, but is nonetheless one that has a history—it began at some point—and emerged due to a range of fragile conditions, such as the formation of a melted mantle and subduction processes kicking off. Our understanding of this regularity and how it might play out in the future depends on our studying long temporal scales. Understanding a process' modal profile by testing it against many conditions enables us to understand how that process might work in the future (for discussion of the philosophical point, see Jeffares 2008; Currie 2018a, chapter 7; Page 2021).

Second, this can be done by tracing local dynamics in a historical individual that may be projected into the future. In addition to general regularities and mechanics, historical individuals sometimes themselves unfold in predictable patterns. Scotese took the current position of the continents as a starting point, say, North America's current position and trajectory, and extrapolated into the future. Insofar as historical individuals follow peculiar trajectories, knowing their past and present patterns and projecting forward is a route into the future.

Third, this can be done by providing partial analogues to particular possible future scenarios or conditions. Scotese's name for the postulated continent hints at the importance of Pangea as a model for imagining the kinds of conditions the formation of *Pangea Proxima* would herald. For instance, more exposed land might increase surface weathering, thus storing more carbon in geological processes, which seems to have occurred for Pangea. As such, Pangea acts as an analogue for the hypothetical future continent. It may be, then, that at least some features of the Jurassic would return if we returned to Jurassic continental conditions (for discussion of the philosophical point, see Wylie 1985; Currie 2018a, chapter 8; Wilson 2023; Watkins 2024c).

Fourth, both the second and third strategies—extrapolating local trends into the future and identifying partial analogues—can aid in discovering the future by suggesting new things to look for, new ways of conceptualizing the entities populating the present, and how they might lead us into the future.

For instance, understanding that the continents shift on tectonic plates underwrites characterizing continents as (if you want) "tectonic individuals." Another set of examples are studies of mass extinctions—uncovering these in the deep past opens the door to asking whether we are currently undergoing one, for instance.

Fifth, providing information for calibrating measurements and models can help us understand the future. We might be interested in rates of continental dispersal at various scales. For instance, Condie et al. (2015) examine long-term trends across the Earth's history to argue that rates of supercontinental assembly and dispersal diverge from the velocity of typical plate movement. To do this, they needed to calibrate measurements of subduction and then compare them to large-scale trends. The development and calibration of such measurements require a long-term historical perspective, often involving identifying and developing proxies for the intended measurement (for discussion of the philosophical point, see Bokulich 2020; Bocchi et al. 2022; Wilson and Boudinot 2022; Watkins 2024b).

Sixth, providing a range of conditions relevant to global or local situations can help us understand the future. By comparing various global supercontinental states, the transitions between them, and various rates of dispersal, arrangement, and an understanding of the mechanisms driving them, we build a rich set of resources for imagining future possible Earths and comparing our system to others. Earth is the only planet we know of with active plate tectonics (although it has been postulated for Europa; see Gaidos and Nimmo 2000), and this potentially has major implications for the kinds of peculiar dynamics on our planet—particularly for making sense of the particular stabilities and features seen between the lithosphere, atmosphere, and biosphere that are (so far as we know) unique to our planet. Further, these rich resources can be used to imagine our planet in various states, actual or otherwise.

Seventh, studies of the past can provide a guide and inspiration for imagining possible futures. In chapter 7, I argued that scientific imagination is constrained, enabled, and enacted by the artifacts scientists construct. These artifacts range from particular specimens to data collection, tabulation, and analysis to the models and other devices constructed to represent and examine data, hypotheses, and specimens. There is a wide variety of models and simulations constructed and being constructed to understand the continental future of our planet (e.g., Yoshida 2016; Broussolle 2022), all of which draw on information from past continental positions, as well as tectonic mechanisms informed by historical data. More ephemerally, understanding both the range of ways the Earth's lithosphere has been, and objective modal knowledge concerning the ways it could have been, underwrites our capacity to imagine how it could be.

Overall, I take this list to demonstrate that past-derived knowledge, both about actual past worlds and possible worlds, has critical and kaleidoscopic upshots for thinking about the future. If you're called upon to defend the epistemic importance of understanding the past for knowing the future, you have bountiful things to say. Some of this knowledge, but not all, concerns significantly lost worlds, so let's turn to ask whether the strategies used to uncover lost worlds might apply to unprecedented worlds.

2. Strategic Perspectivalism and Unprecedented Worlds

> "The second thing to say is that many of the ways that we scientists try to find out about the past, the strategies we adopt, if you want, are also applicable to thinking about the future. Many of the ways I've figured out surprising things about Broadfoot and their kin might be turned to thinking about what the future is like."

Throughout this book, I've argued that historical scientists, through devious use of strategic perspectivalism, gain rich knowledge of lost worlds. Can the same strategy be applied to unprecedented worlds? In this section, I'll draw on some examples from existential and catastrophic risk to say:

> Yes.
>
> However, in the next section, I'll pivot to a:
>
> Yes, but . . .

Let's begin with some common-sense ideas about *prediction*. A prediction takes a set of information, some concerning initial conditions, and some concerning the laws, regularities, and rules governing how those initial conditions might unfurl. These are brought together to show what state of affairs or outcomes we should expect to follow. In simple contexts, the combination of regularities and initial conditions guarantees the predicted outcome, or near enough: epistemic possibility and actuality converge. In trickier contexts, the relationship between prediction, regularities, and initial conditions becomes more complex: ranging from being less guaranteed, that is, providing a set of possible outcomes with associated probabilities, to difficult cases involving ribbons of possible outcomes with uncertain relationships, possibilities, and probabilities. In complex cases, even knowing which variables might be relevant becomes extremely hard to determine.

Our capacity to intervene is a double-edged sword in these contexts. Controlling initial conditions and intervening as events unfurl can enable us to follow desired paths to bring about the outcomes we desire—as is paradigmatically the case in experimental contexts. But this relies on us both understanding the effects of our actions and having some handle on what our future actions will be like. Often, the things we care about in the future cannot or should not be controlled like an experiment. And ongoing intervention often requires ongoing coordination, political will, and economic and resource capacity. These can be fragile indeed.[4]

All of this, then, requires deep modal knowledge not only of the relationship between initial conditions and the various regularities governing them, but also of our own agency in those conditions and regularities. While (unless our technological prowess becomes impressive indeed) it is unlikely we will have the ability to make much difference to the tectonic fates of continents, there are plenty of global processes that we have plenty of capacity to interfere with.

One area where consideration of unprecedented worlds is prominent concerns *existential* and *catastrophic risk*. These risks often come in familiar lists, ranging from events we have limited control over, such as asteroid impacts, megavolcanoes, and so on, to events that we can take pretty direct blame for, such as nuclear catastrophes, malicious (or indifferent) artificial superintelligence, runaway synthetic biology, the downstream effects of extreme climate change, and the degradation of the planet's biosphere and atmosphere more generally (Bostrom 2002; Bostrom and Cirkovic 2011; Rhodes 2024). Existential risk has been characterized in a variety of ways (Torres 2023): from relatively straightforward human extinction (e.g., Tonn and Stiefel 2013), to including civilizational collapse (Beard, Rowe, and Fox 2020), to the more esoteric, such as closing off the possibility of our descendants' flourishing in an intergalactic post-human utopia.

4 This, I take it, is what lies behind the difference between a "prediction" and a "projection."

The events and processes associated with existential and catastrophic risk are often considered to be low probability but high impact (e.g., Bussiere and Fratzscher 2008), although these days the "low probability" assumption might be questionable for at least some of the risks. The chances that the effects of climate change are so bad that they fundamentally destroy our species' capacity to survive on this planet, or that a megavolcanic eruption will occur over the next few centuries, are low. But the occurrence of such events would be so bad that we should nonetheless put significant effort into understanding, avoiding, or mitigating them.

I tend to think "existential risk" is particularly interesting because the risk pertains to the *existence of something*, as opposed to, say, risk as understood in terms of financial costs, opportunity costs, or other measures (Currie, under review). I take a personal existential risk by riding a bike rather than driving a car; languages or even whole cultures often face existential risks; our actions have led many species to face existential risk; our putting various features of the biosphere and atmosphere under existential risk puts our own species at existential risk. Despite its wide applicability, the concept is almost always applied to thinking about risks to the extinction of our own species. This is partly because existential risk is often considered in terms of the risk to our species failing to achieve its potential and the loss of future generations.

There are a fair few avenues for criticizing the perspectives often adopted by existential and catastrophic risk researchers (see, for instance, Cremer and Kemp 2021; Schuster and Woods 2022; Sundaram, Maas, and Beard 2024; Corsico 2024; McLaughlin 2025; Currie, under review). First, there are many outcomes from processes like climate change, the global adoption of large-scale computing, and environmental degradation that fall short of global catastrophe (and certainly of species-level extinction), that nonetheless involve an unacceptable amount of suffering, death, and other significant badness. An overemphasis on species-level extinction, especially when backed up by a simplistic moral calculus, can lead to recommendations ranging from mildly embarrassing to horrific. Second, the vast majority of that much more likely suffering will be the burden of more vulnerable people. To some extent, worrying about existential risk at the level of our species is the domain of the privileged, and that shows. Third, the philosophical underpinnings of existential risk work often draw on fairly fringe ideas from the perspective of academic philosophy, what Cremer and Kemp characterize as "techno-utopian" (Cremer and Kemp 2021). Fourth, the focus on catastrophes as caused by calamitous events has at least two deleterious and, I think, related upshots:

a focus on the long term rather than the medium term, and a tendency to conceive of these impacts as one-off "bolts from the blue." Let's focus on the latter point.

Much work on existential risk in the first fifteen years of this century focused on particular, hard-to-predict events and their avoidance. This is most stark in the discussion of the possibility of runaway effects of the development of artificial intelligence. Generally, we might think of these as existential risks caused by a single, precipitating event. One advantage of this perspective is the capacity to carefully uncover various causal features of those events (I make a point like this regarding historical science in Currie 2019e). There is, however, a major disadvantage, as Liu et al. point out: "A high profile 'one-hit-KO' existential risk such as global nuclear war or a pandemic may constitute only one avenue towards that 'adverse outcome,' and concentrating predominantly upon (ways to intervene in) its origin and direct pathway, risks overshadowing other potential paths or disaster interaction effects that functionally converge towards that same disastrous outcome" (Liu et al. 2018, 7).

An existential calamity might not come about due to a particular powerful trigger but due to a slow decline or a perfect storm (Kuhlemann 2018). To take an infamous palaeontological example, the collision occurring around the K-Pg event may have been a "one-hit" event, which more or less alone triggered the subsequent mass extinction, or it could have acted as the "last straw" to an already weakened biota, or it could have been the impact's interaction with other events, such as continental break-up and a period of increased volcanism, that together led to the extinction. There are, then, multiple kinds of routes to catastrophic outcomes. These slower, more complex, and less sexy catastrophes raise further epistemic issues (Baum 2020). Thus, in addition to narrow perspectives on particular risks, we need perspectives that can productively think about risks in more general ways that are both empirically and politically informative.

Many existential risks to our own species are unprecedented, at least in terms of scale and target. Some are due to unprecedented features of our own species: the extent to which we (intentionally or otherwise) engineer global systems, global interconnectedness, the speed of information exchange, and so on. So, the targets of these investigations are highly peculiar and, to varying levels of characterization, unprecedented—they concern denizens not currently present in our world in an epistemically deleterious way.[5] As such, trying to understand them has at least some analogies to how we might understand lost worlds.[6] According to strategic perspectivalism, when faced with a challenge from a lost, or putatively lost, denizen, scientists will adopt—or construct—perspectival tools that productively characterize their target. We see something similar, I think, in some studies of existential and catastrophic risk. An example is Shahar Avin and colleagues' classification of global catastrophic risks.

Similarly to Liu et al., Avin et al. characterize earlier work as "identifying individual GCR [global catastrophic risk] scenarios, and in compiling lists of the scenarios of greatest concern," but they don't yet see "a comprehensive, interdisciplinary view of severe global catastrophic risks" (Avin et al. 2018, 20). This is the aim of their classification. Instead of focusing on particular sources of threat—big rocks hitting the earth, rampant artificial intelligence, etc.—they instead draw our attention to the kinds of systems that could break down and how we might fail to stop these.

Avin et al.'s framework has three dimensions.

First, they provide an analysis of *critical systems* that operate within particular safety boundaries. Here, inspiration is drawn from the literature on "planetary boundaries" (Rockstrom et al. 2009; Steffen et al. 2015). The thought behind that literature is that human life depends on various systems remaining sufficiently stable: on a global level, the biosphere and atmosphere are dependent on various processes, such as the oxygen cycle we discussed at the beginning of chapter 5. This perspective takes human life as dependent on various, often-interdependent, fragile regularities. It emphasizes the peculiarity of global systems. Avin et al. extend the notion of safe planetary boundaries to include various intertwined hierarchical levels. These include, for example, ecological levels such as trophic webs relying on primary producers; social levels like food production and dissemination; and individual levels including cardiovascular, immunological, and other systems. All of these processes might be disrupted by potential threats.

5 Lalitha Sundaram and I (2024) have argued that a science of existential and catastrophic risk shouldn't trade in quantified probabilities and prediction narrowly understood, but instead walk in the ambiguities of possible scenarios, the relationships between these, heuristics guiding actions, and warning signs heralding bad outcomes. Even this less ambitious picture of our knowledge of unprecedented worlds requires modal understanding.

6 In Currie (2019e), I compare the epistemic situation of palaeontology to existential risk science.

Second, Avin et al. point out, to reach global scales, system failures require mechanisms of spread. Some large-scale events, such as significant meteor impacts or megavolcanoes, have fairly immediate spread mechanisms. Others, such as pandemics, might rely on human-built dissemination structures such as global travel or cyberattacks through information networks. Avin et al. identify three sets of such mechanisms: "natural" processes on a global scale, such as air or water-based dispersal, anthropogenetic networks such as communication systems and via various forms of replication, such as biological, digital, or cultural.

Finally, third, Avin et al. include possible failures of prevention and mitigation. The approach here is to identify "potentially fragile areas in the global risk prevention and mitigation system" (Avin et al. 2018, 23). As in their analysis of critical systems, this takes the form of an intertwined hierarchy. Individual failures, such as failures in risk perception or cognitive bias, stand alongside interpersonal failures in, say, trust, reputation, or conflict resolution. On higher levels, institutional failures include inappropriate incentive structures, adaptability, and decision-making processes, while extra-institutional factors might include governance, diversity, and enforcement.

A major benefit of the framework is that it allows some threats to be bundled together: "Scenarios with significantly different primary causes could manifest their GCR [Global Catastrophic Risk] potential through a similar mechanism" (Avin et al. 2018, 24). This both simplifies and helps guide research and resource efforts toward critical systems that might be the target of multiple disruptions. Avin et al. also highlight how such a framing makes space for interdisciplinary collaboration and the identification of relevant expertise for mitigating and tackling risks. Their final application is of particular import to us here: "The system is a tool to highlight highly uncertain or neglected corners of the GCR possibility space, and guide research efforts towards these corners, in the hope of discovering unknown unknowns. The combinatorial nature of the classification systems provides a natural way of progressing from well-known systems and mechanisms to a vast and as-yet largely unexplored space of possible GCRs" (Avin et al. 2018, 25).

The framework captures a complex modal space, intended to convey the various ways in which critical systems might fail, and those failures might spread, and how we might fail to stop them. Some spaces within that framework are familiar and well-studied; others are not. If it works well, the framework would help identify these unknowns and reveal some unknown unknowns, guiding research into those areas.

We can think of Avin et al.'s classification as a perspectival tool: it *characterizes* catastrophic risk scenarios in a way intended to unify various discipline-bound theories and data, such that it helps identify and focus research and policy efforts. In its exploratory guise, we can understand it in terms of the artifactualist account of imagination defended in chapter 7. The framework is a mode, consisting of three complex dimensions, that may be instantiated across various media. I've used a casual linguistic medium here, but Avin et al. also present it using various diagrams and tables, which enable comparisons as well as tracing possible routes from system failure to spread mechanism to mitigation failure, and considering other possible GCR sources. It thus enables exploration of the space of GCRs, scaffolding our otherwise unsystematic or isolated work on particular threats. It is, in other words, a tool for the imagination.

There are further similarities between Avin et al.'s framework and many of the artifacts we've seen in the historical sciences. Mitchell et al. make use of a model of ecological space differentiation, which brings out the complex relationships among population size, niche stability, and other factors to map out possible conditions under which we would expect neutrally structured ecosystems. Both Mitchell et al. and Avin et al.'s possibility spaces should not be understood as targeting mere epistemic possibility. We're not just interested in what will happen, but what might. Further information might shape or complexify the model, but it won't shrink toward the actual. Further, Avin et al. explicitly emphasize the potentially iterative relationship between their framework and the various studies that could be carried out, for instance, that it could "both inform, and be informed by, different 'foresight' tools" (Avin et al. 2018, 25). These "foresight" tools are themselves ways of imagining and monitoring future possibilities, typically involving various expert opinion-gathering and amalgamating activities, as well as various kinds of simulations (Conway 2006; Kemp 2024).

So, in tackling the unprecedented future, we see researchers strategically adopting perspectives that enable fruitful, iterative studies and constrain and enable epistemically relevant imagining.

You might worry this conclusion is unsurprising, as strategic perspectivalism is so general a strategy that it is no wonder that it applies across cases. If we define the strategy in highly abstract terms, specifically, as the practice of characterizing a target and constructing a tool to understand it, then this is right. However, I've understood strategic perspectivalism more narrowly, construed as a way of uncovering lost worlds. As such, I've emphasized the role of artificing, grafting, and anchoring, and the iterative studies they generate, toward understanding denizens unfamiliar to our present. It is in this more narrow sense that Avin et al. exemplify many of the features of strategic perspectivalism. By identifying critical systems, they are able to integrate—or graft—current systems with future possible states. Mechanisms of spread can be understood via various artificing strategies, particularly simulations. The framework can work as a structure supporting various iterative studies.

In developing their perspectival tool, then, Avin et al.'s approach makes use of many of the strategies discussed in the first half of the book. Artificing, especially of global-level models, but also of various scenarios used in policy planning, enables us to create partial simulacra of unprecedented worlds, and, as we've seen, extending trends from previous records into the future enables us to graft unprecedented worlds, and various anchors can be found throughout history. In short, faced with potentially unprecedented denizens, we seek characterizations that link them in various ways with what we do know.

Our potential knowledge of unprecedented worlds, then, is great. And I think the analysis I've provided in this book can make some sense of that knowledge. So that's my "Yes." However, it's time for the "But."

3. Records of the Future?

> "The third, and I'm afraid, much less happy point, involves thinking about the crucial importance of things like this," Ichthy gestures to the skull in the glass. "I've said a few times that I couldn't have imagined Broadfoot without this specimen, that nothing about today other than this specimen speaks to something like this animal existing. It is fossils, and records built of objects like them, that help us find surprises in the past. So, if the future is supposed to be as surprising as the past, how are we supposed to detect those surprises? That is, what could play the role of a fossil for our investigation of the future? I don't think there is such a thing. And if there isn't," Ichthy sighs, "then I suppose I cannot provide as optimistic a story about our ability to understand the future as we can the past. Unless. Unless there could be something like a record of the future . . .?"

I've argued that many of the tools of strategic perspectivalism are available for scientists of the unprecedented, so many of the lessons of the first half of this book carry over to future-oriented projects. However, I have a hunch—one

I can't quite kick—that there is nonetheless an asymmetry concerning our knowledge of the past and the future. This difference is grounded in the importance of records. There are different ways of establishing the asymmetry, and I'll sketch a few in this section. Even if the negative point doesn't hold water—and for our sake, I hope it doesn't—the discussion nonetheless serves to capture what is so important about traces for understanding lost worlds.

Recall the (hopefully) commonsensical account of prediction I introduced earlier: predictions work by taking some initial conditions and, in combination with regularities governing what unfolds given those conditions, estimating some later set of conditions. Trace-based reasoning, one might think, is simply a temporally-reversed example of this. One takes some current conditions, the existence of a fossil, say, and based on regularities pertaining to the formation of that kind of thing, estimates some set of earlier conditions. A "trace" is some current observation that, in combination with a justified theory capturing the dependency between the observation and its causal or informational descendants, licenses an inference to the past (Currie 2018a, chapter 3). So, a "future trace" would simply be a current observation that, in combination with the relevant theory, licenses an inference into the future. At a certain level of description, that's correct, and it suggests a symmetrical treatment: just as fossils and other traces allow us to build records of the past, we can treat the current state of the world as "traces" or "records" of the future. Let's call that principle *symmetry*: epistemically speaking, knowledge of both the past and the future is gained by making use of contemporary observations, and extrapolating either forward or backward from those observations using regularities about how those phenomena transform over time.

I don't think this symmetry works *at the level of epistemic practice*. First, traces can tell us *what is surprising* about the *actual* past, while extrapolating from present observations can, at best, provide some *possible* surprises in the future. Second, traces enable *recontextualization* with an actual past, while extrapolation from current observations at best enables probabilistic integration of possible futures. Third, traces are perspectival tools that enable iterative, constrained imaginings tied to an actual past; observations of current phenomena don't seem to enable that kind of investigation.

Let's start by considering a few features of trace-based reasoning.

In this book's first chapter, I suggested that an ironic downside of over-emphasizing traces in the philosophy of the historical sciences, that is, focusing on information decay at the expense of loss, is that the value of traces themselves is not sufficiently captured. Why do I think that? Consider (again!) some of the cases of loss throughout the book: rangeomorph spatial neutrality, morphology, and feeding strategies; the unique Avalonian conditions—bacterial mats, heterogeneous oxygenation—that led to ivesheadiomorph fossils; the sequence of changes that took us from the Avalonian to the more familiar Phanerozoic. These are all circumstances that depart from the usual running of things in our current world. They are surprising in a particular kind of way. If we took the world around us as representative of the past, we wouldn't expect to find those denizens. Plausibly, our knowledge of these surprising denizens is grounded in traces. As Prof. Ichthy emphasized throughout their lecture, were it not for the fossil of Broadfoot, they wouldn't have imagined that creature. Without fossils, I don't think we would have imagined many dinosaurid forms, and certainly wouldn't have thought those imaginings to track an actual past.

So, traces matter because they tell us what is surprising about the past. This is what makes them so crucial for driving the otherwise impoverished imaginations of scientists.

I've further emphasized the importance of *recontextualization*. Strategic perspectivalists often characterize their targets in ways amenable to artificing, anchoring, and various forms of data analysis. This typically involves isolating and abstracting various properties from traces; for instance, Mitchell et al. isolated the spatial and taxa-specific properties of Avalonian rangeomorphs. While this isolation is powerful, it's also necessary to resituate those properties within the past world. Consideration of features of rangeomorph environments, such as resource availability, and further information about their ecological situation, such as their small populations, was required to align the various anchors and models Mitchell et al. used to the actual past world, and thus generate their explanation of ecological neutrality in the Avalonian. So, recontextualization brings the fruits of various more-isolated perspectives into contact with an actual history and provides further opportunities for integration and iteration.

So, traces matter because they provide a link, often multiple links, to the actual past, which enables recontextualization.

Compare now the sources of uncertainty we might have of the past and of the future. In trace-based reasoning, we must be sure that we've characterized the trace correctly (say, that it is in fact a fossil) and that the various theories required to interpret it are well-founded, thus allowing us to infer that the fossil formed at a particular time, is of a particular taxon, and so on. When these are relatively secure, we can say with relative confidence that *this taxon* existed in *this location* at *this time*—we can treat it as a fixed point.

Switching to the future, from some current observation, and our understanding of how things unfold, we are often able to lay out possible outcomes. Sometimes this is highly constrained: if a trajectory continues, the outcome is certain—a species will go extinct so long as its population continues to decline. Some constraints on the future are fairly powerful: for instance, wooden structures won't last as long as metallic structures. Our knowledge of the future is based on (1) the current states, (2) extrapolating from historical trends, and (3) constraints from the regularities we've identified. In trying to understand future continent formations, we rely on the mechanisms of tectonics we've managed to generate (and these not breaking down), past patterns in tectonics and current continent positions (the initial conditions), and what we're able to glean from various artificing and anchoring procedures.

On the face of it, for prediction, there doesn't seem to be epistemic objects that concretely play the specific roles I've emphasized for traces. That is, how does analysis of the present reveal surprising—i.e., unprecedented—things about the future? And further, how does analysis of the present provide iterative links with an actual future?

Our knowledge of the future is strongest when targets unfurl in regular, patterned, and convergent ways, and when the records we have of the past can be extended into the future. For plate tectonics, we have an enormous amount of historical data, a good understanding of the mechanisms, and so on. This allows us to model the future using "conservative extensions" of existing theoretical and empirical knowledge. But if the world is often peculiar, those regularities will be highly fragile, and what new regularities might arise will be extremely hard to predict. Many of the various scenarios of existential and catastrophic risk are highly peculiar. Food security, for instance, depends on production methods, the economics, legal structures, and material conditions of trade, local and global weather patterns, soil conditions, and so on. All of these are deeply interdependent and in constant flux. Predicting the future of food security to any degree of accuracy requires getting a handle on all of these highly peculiar systems. In contrast, our knowledge of the past—deeply fallible though it might be—sometimes includes records that can ground knowledge of what we are often unable to detect in the future, that is, the surprising and the peculiar. Even in the case of tectonics, we are left less with a concrete hypothesis of what will happen in 250 million years, and more with a range of options, often highly conditional on what comes before—and that's without worrying about unprecedented surprises.

Further, the artifacts that ground and direct historical science are not exhausted by extrapolating from existing trends, the models of artificing, or the various analogues of anchoring. They also include intimate, productively iterative analysis of specimens and traces. As I said earlier, records are, from one perspective, current objects construed as data relevant to past inferences. However, in chapter 7, I argued that they are also objects that we have constructed to be revelatory of times and possibilities in the past. Mitchell et al.'s exploration of neutral evolution couldn't have occurred without careful data gathering and recording, and then using this to reconstruct a past world grounded in the Mistaken Point fossils. These frozen(ish) time-slices of lost denizens partially ensured the relevance of their exploration of the possible to an actual past. For future records, we might examine contemporary objects and imagine them as revelatory of the future. Some metals degrade more slowly than others, and as such, some of our metallic constructions are likely to survive into the future, and we can imagine how those might constrain and affect the peculiar futures we might move into. But these are typically unsurprising futures. While current objects can be iteratively examined to understand how the future might be, the specimens that typically enliven historical research stand in often obstinate testament to various past events and, as such, strongly link investigation to the actual past.

These arguments rely on an approximate distinction between two kinds of evidence from the present concerning other times. Sometimes, we might take the state of the world as *representative* of other times. This is roughly uniformitarian: take the denizens of today and anticipate them as continuing to exist in future times. Sometimes, we might take particular components of the current world as processually connected with components of the world at other times. This is what we see in trace-based reasoning and grafting more generally. Considering unprecedented and lost worlds, the former approach cannot be taken as a steady guide. The latter is, or so it seems, more available for the past than for the future. If that is right, our knowledge of past worlds is more powerful than our knowledge of unprecedented worlds. The symmetry breaks in that sense. Our imaginations lack the material constraints required to make them relevant to the actual future or possibilities relevant to it.

We further might distinguish between two kinds of *surprise*, implied above. In scientific contexts, we might standardly think of surprise as relative to our knowledge—something is surprising if it bucks our empirical and theoretical expectations. Triceratops, at least concerning our most well-established knowledge of them, are no longer surprising in this sense. However, there is another sense of surprise—the sense of being *unfamiliar*—in which triceratops are still surprising. This sense of surprise takes the present and compares it to other times, possibilities, or circumstances. Traces enable knowledge of things that are surprising in this latter sense. Without traces, our capacity to identify such future surprises is reduced.[7]

A single trace, then—when background theory is sufficient—nails down a part of a past world and provides a pivot point for developing modal knowledge. A pattern of traces, linked together, provides a trajectory—a story—through the actual past. These patterns provide powerful access to aspects of the past that are no longer represented in the present—that is, to lost worlds. While something like this might sometimes be available for the future, when our theoretical constraints are sufficient to extrapolate to highly constrained paths beyond the present, or when future paths converge in cases of low peculiarity. But these are significantly more tied to our current world and offer significantly fewer opportunities to explore possibility in a way grounded in, relevant to, and guided by future actuality.

So, in some limited ways, the metaphor of "records" might be applied to our knowledge of the future in some circumstances, mostly when we're extending a set of historical patterns from one location into that location's future, or when we adopt an extremely abstract perspective. But what concerns us often goes beyond a location or takes the location into futures where the intertwining possibilities stretch epistemic possibility beyond what we can handle; and what concerns us is often the particular, the finer-grained, rather than what can be delivered by abstract modelling of possibility or extrapolated from current patterns.

Depending on how central we take traces to be in the strategy of uncovering lost worlds, this might lead us to conclude that, in fact, strategic perspectivalism is *not* a strategy adoptable for much future-oriented work. This turns on whether we think trace-based reasoning is essential to strategic perspectivalism. I'm not sure what exactly hangs on that discussion in the current context, but I am open to being convinced of this point.

7 There's likely a lot more to be said about "surprise" as discussed here, especially regarding its phenomenological and aesthetic senses. See Currie (2018b) for epistemic surprise, and French and Murphy (2023) for surprise in science more generally.

One objection to the asymmetry I've identified comes from *epistemic function*. You might argue that while at least sometimes for past knowledge we care about the actual past—we want something like a retrodiction—for the future all we need are projections, that is, what would happen under certain policies or interventions. Because we only want projections concerning the future, the distinction I've drawn is irrelevant. In response, I'm not convinced the difference between a retrodiction or a prediction on the one hand, and projections on the other, is so stark. A projection is a *conditional prediction*, and although this means the prediction coming true depends on that conditional occurring, it isn't obvious to me why this modal character makes such a profound epistemic difference. Even if it does, and that exploring possibilities is all we need for (some) future-oriented science, this doesn't undermine the idea that there is knowledge we can have of the past that we cannot have of the future.

In his discussion of betting on science's progress, Derek Turner distinguishes between *methods neutral* bets and *current methods* bets (Turner 2016). The former bets that some knowledge won't be gained regardless of whatever progress we might make on methods. The latter bets that, given our current technological methodological capacities, some knowledge is unlikely to be gained. This provides a few ways of construing my arguments. On a *current-methods* reading, I'm arguing that, given the current epistemic resources available to us, absent either different ways of uncovering the future or some way of developing records of the future, our past knowledge will outstrip our future knowledge. On a *methods-neutral* reading, I'm further claiming that we *won't* develop further ways of understanding the future or records of the future. I take myself to be making something like the weaker bet, and why is revealing. To see this, let's consider what might underwrite the epistemic asymmetry.

What accounting might we give of the epistemic asymmetry? One might underwrite this asymmetry via a metaphysical claim: if the past is truly closed and the future truly open, then as a matter of fact what has happened is a series of fixed points and processes and what will happen is in some sense indeterminate (Lewis 1979); if, as Carol Cleland has argued (2002), the laws of physics have temporal asymmetries embedded in them, and these laws in some regard govern everything else we might care about, then epistemic consequences might follow. I don't think those sorts of claims are appropriate here. I'm sceptical that we have good grounds to commit to a metaphysically open future, am sceptical that the derivations of physical theory must apply so uniformly, and (especially) doubt that such temporal asymmetries make a profound difference to the knowledge of historical scientists like palaeontologists and archaeologists.

However, the *peculiarity* of the past (and the future!) is a metaphysical claim I'm happy to entertain. In a peculiar world, whatever regularities we encounter depend on complex sets of generating and maintaining processes. This can ground local asymmetries in various ways. Peculiar patterns could be *path-dependent* in an *accumulative* way. That is, the previous set of states might strongly constrain the direction of future states (this, I think, is captured by Bill Wimsatt's notion of "generative entrenchment" (2007)). In that context, access to previous states and some handle on how these constrain the future could lead to something like a trace of the future. Because a current state is necessary but insufficient for later states, with the right understanding of those constraints, we can study that current state as bearing a mark of the future. However, peculiar patterns can also be wildly disrupted or follow chaotic dynamics. Here, previous states do not constrain later states in any strong way. Under these circumstances, present states are nothing like traces of the future, as there isn't a clear inferential path between that putative trace and the future. As we saw in chapter 1, philosophers sometimes discuss trace-based reasoning in terms of whether various historical processes are information-preserving or destroying. These two kinds of peculiarity provide us with a future-oriented framing of this distinction. If peculiarity is accumulative, then processes carrying signals of the future often preserve information; if peculiarity is disruptive, then those processes will often be information- destroying.

On this kind of framing, whether peculiarity metaphysically grounds local temporal asymmetries turns on a further question concerning the kind of peculiarity we face. There's a lot more to say on this score, which I'll leave for now. For our purposes here, suffice to say, appeal to peculiarity alone won't ground a sufficiently systematic asymmetry.

Instead, we could explain an imperfect and revisable temporal asymmetry by appeal to our epistemic situation. That is, rather than pointing to the structure of the world we study, we point to features of ourselves as knowers. There are weaker and stronger versions of this thought. On weaker versions, we point to the current scientific toolkit we happen to possess. We have powerful theories geared toward understanding the processes by which traces form. But we lack powerful theories, at least often, geared toward understanding how current states might unfurl into future states—at least surprising future states. On stronger versions, we might say that regardless of the general structure of the world, the knowledge that we generate depends in part on our temporal nature. Regardless of the metaphysics of time's arrow, our *experience* and *agency* operate in one narrow temporal direction. We are thus epistemically bound by our subjective temporal straightjacket.

I'm not sure if ultimately much sense can be made of either of these suggestions, but regardless, it strikes me that both may be overcome in principle. Weaker views require new approaches and techniques to thinking about the future. Stronger views require us to find ways to epistemically transcend the limitations of our temporal agency. This isn't as obscure as it sounds: we overcome our spatial limitations through various detection and visualization techniques that bring things too big or too small to human levels; we do the same with durations that are too long or too short (see Rheinberger's discussions in 1997, 2010, 2023).

Just as I'm not confident in our ability to get a good grip on a peculiar future generally, I also doubt we're in a good position to predict the path of peculiar science.[8] Thus, I'm left with a *current methods* bet. Crucially, this transforms an apparently pessimistic argument into a challenge.

The pterosaur scribbles on her notepad, disappointment creeping across her face. "So," she says, "the future is just unknowable?"

Prof. Ichthy waves their flippers conciliatorily, "No, no, not at all! It is really hard to know about the future, but I have great faith in our ingenuity, if perhaps not as much in our wisdom. As I said, I'm no expert on the future, so cleverer and more informed heads than my own might be able to provide a more reassuring answer. It's worth noting that some things we can be pretty sure of, but some other things we can't." Ichthy switches from talking directly to the reporter to the audience more generally.

"I'm sometimes asked to make predictions about the future of science. What new discoveries might we make? What questions will remain unanswered? Here, I think we have good reason to be optimistic that we'll discover a lot of things. The history of science, the ingenuity of scientists, and so forth, all provide good reason for optimism. I think this is true even of my subject, the deep past. Although fossils provide only partial information, our ongoing capacity to rejig our perspectives via new sources of evidence, new finds, and new technologies is likely to continue

8 That is, science itself is contingent (Kidd 2011; Rheinberger 2023, chapter 8).

ILLUSTRATION 9.1 Robo-Ichthy

into the future. If someone were to say that we won't know much about the past, or that some particular mystery will never be resolved, they're placing a pretty risky bet." Ichthy nods emphatically, their posture indicating a debate long-ago played out from their perspective.

"So, I think our relationship with the future—and scientific ingenuity—makes it an incredibly risky bet to say that some particular question will go unanswered. However, trying to tell what exactly will be discovered—how future science will play out—is a different question altogether. Another question I'm asked is about how we scientists will do our science in the future. Against the backdrop of technological advancement, what new ways of doing science will emerge? To be a little topical: as technology advances, will scientists like me even be required? We can imagine scientific futures dominated by automation, where fancy machines whir up answers that now rely on scientific thinking, or great machines do our fieldwork rather than our eyes and flippers. I admit to finding such futures deeply unattractive, but I also think it is very difficult to make predictions about them. Partly this is due to our influence over the future—do we want a future with automated science?—but that is not the only source of difficulty. I think we know so much about the past in part because we have records of it. So, can we develop records of the future? Or can we show that we don't need them?"

Still unsatisfied, the pterosaur responds, "Okay, well, I'm sure *Eons*' readers will want to know, what should we do? If there's such uncertainty, not just in the future generally but in science's future, what policies or paths should we follow?"

Ichthy again grimaces, "We're once again far outside of my wheelhouse. Decision-making under uncertainty, the details of policy, and certainly the ethics and justice of the whole question bring us well outside of what I could say anything sensible about. I don't really see myself as the kind of academic who speaks confidently outside of my expertise. I'm sorry to let you down."

A further silence stretches across the audience. One attendee whispers to another, "It'd be a shame to finish on such a bummer . . ."

10

Loss and Wonder

Catching the awkward lull after Ichthy's exchange with the reporter, the Chair moves to wrap up the lecture. Before they can, the graduate student clutching the toy-like Broadfoot model interjects.

"What will you do now?" Many attendees turn to the young theropod in the mildly accusing way of an audience ready to go home, before switching their attention to Ichthy as they take up the question with a last wind of gusto.

"Now?" they ask. "Now I wish to return to the field, and return to the lab, and return to my conversations, and return to my work. I have so many questions still about Broadfoot. Did they burrow? Were they really oviparous? What role did their venom play? I hope that some of those questions will be answered by returning to the site where I found it and undertaking new excavations. And indeed I hope we'll find new questions to ask! With new excavations, we might find further evidence about the organisms that share Broadfoot's environment, as well as information about that environment itself. Broadfoot was presumably part of some ecosystem, and I want to know more about that ecosystem's nuts and bolts. In addition to these new planned excavations—which I admit aren't yet funded—"Ichthy shoots a meaningful look at some of the university administrators in the room—"there's still plenty of material excavated from the site that produced Broadfoot that we've not looked at. Imagine what surprises could be waiting in storage? And further, what new tests, theories, models, and ideas might myself and my colleagues"—another gesture to the cluster of graduate students—"come up with to build upon what we already know?"

"Much scientific work is careful, systematic—monotonous, even—but . . ." Prof. Ichthy's characteristic grin returns, "those moments when you find something new, when you are confronted by something from the past that makes you see everything—even yourself—as if for the first time, that's when I love it."

ILLUSTRATION 10.1 Prof. Ichthy in the Broadfoot Den

It's traditional to finish a book with a summation of the main arguments, and that's my task in section 1 of this chapter. In section 2, I'll expand on the value of studying the deep past, taking the discussion in one final direction. An emotional one. You might think nostalgia or grief are the most obvious emotions for a book about loss, and no doubt there's a lot to be said about these. But I'm not in the mood for the often regressive indulgence of nostalgia nor the necessary, heart-rending work of grief (for the latter, see Jorgensen 2019). Instead, I want to talk about wonder. To underwrite both our recap and discussion, I'll turn to another famous case in the history of the Earth sciences: the Reverend William Buckland's 1821 description and analysis of a Pleistocene hyena den in Yorkshire.

1. The Lost World of Kirkdale

Buckland is a usual suspect in any history of 19th-century English geology. According to Pemberton, he "possessed one of the most innovative minds ever to grace the science of geology, but he was also one of its most eccentric characters" (Pemberton 2010, 261). It is tempting to go on and on about Buckland's eccentricities (if you have an entire book of doggerel verse published in your honour you know you've led an interesting life), but I'll just indulge myself with Lyell's description of Buckland's presentation of his paper to the Royal Society on February 21, 1822 (deep breath) *Account of an assemblage of Fossil Teeth and Bones of Elephants, Rhinoceros, Hippotamus, Bear, Tiger, and Hyaena, and sixteen other animals; discovered in a cave at Kirkdale, Yorkshire, in the year 1821: with a comparative view of five similar caverns in various parts of England, and others in the Continent. By the Rev. William Buckland, F. R. S. F. L. S. Vice President of the Geological Society of London, and Professor of Mineralogy and Geology in the University of Oxford, andc. andc. andc.* "Buckland in his usual style, enlarged on the marvel with such a strange mixture of the humorous and the serious that we could none of us discern how far he believed himself what he said" (Lyell 1822, quoted in Pemberton 2010, 246).

Whatever the apparent levity of Buckland's delivery to the society, the published text itself (as well as the book he published the following year) is serious business.

Buckland describes a large collection of disarticulated bones, partly encased by stalagmites, under thick layers of mud. The bones were from a startling menagerie, from small rodents to ducks to deer to elephants, rhinoceroses, hippopotamuses, and, in great quantity, large hyenas. At the time, the find had a handy explanation: the (or a) great diluvian flood swept the collection of tropical animals northward.[1] Instead, Buckland argued that these animals were part of the normal ecology of *pre-diluvian* (or "antediluvian" in the language of the time) England, and that the menagerie was explained by, as Alan Bewell has put it, "The catholic tastes and scavenging habits of hyenas [which] made them terrific natural history collectors: any animal would do for dinner" (Bewell 2014, 384). That is, Buckland claimed that antediluvian England was populated by species more typical of the tropics, some of which—like

1 The notion of a cooling Earth was also popular at the time, although Buckland doesn't discuss it in the Royal Society publication.

the large hyena that populated the cave—were now extinct. That is, Buckland described a lost world (these days we date it to around 130 kya).

Buckland's work was remarkable for several reasons. It was one of the first detailed reconstructions of a palaeontological site. To achieve this, he combined the comparative anatomy of Cuvier with the *in situ* analysis of geological features, becoming distinctive of geological science. It also represented, for those in Britain, a remarkably well-preserved set of perspectival tools underwriting an evocative realization of a lost world. Antediluvian England was not the same as postdiluvian England.

Jim Robins' image of Prof. Ichthy above is inspired by a caricature of Conybeare's, depicting Buckland and his lost world. It represents Buckland, crawling through the cave to be confronted not with disarticulated bones, but the hyenas themselves. As Rudwick describes it: "The picture is no simple reconstruction. Into the cave crawls Buckland himself, candle in hand, illuminating this scene from deep time with the light of science, penetrating the epistemic barrier between the human world and the prehuman, looking perhaps as surprised to see the hyenas as they are to see him" (Rudwick 1992, 40).

Rudwick highlights the caricature as the first visual representation of a lost world in palaeontology. Although textual descriptions existed much earlier, he argues that it wasn't until sufficient knowledge and material evidence had accumulated that the depth of and confidence for pictorial representation of lost worlds were possible. "Once the reality of extinction was generally accepted, these fossils provided in principle the raw material for reconstructed scenes that would differ significantly from any scene drawn from the world of the present day" (Rudwick 1992, 28).

As Rudwick chronicles, once this knowledge had sufficiently accumulated, representations of lost worlds took off, forming a lineage leading to the traditions of palaeoart that diverted us in chapter 7. Additionally, these worlds became increasingly diverse: it wasn't just that there was a world before the flood, but many antediluvian worlds (Rupke 1983). In addition to fossils themselves, palaeoartistic images are the major source for public imaginaries of pre-human times. This reiterates the historical discussion of the first chapter: early geology and palaeontology were motivated by and celebrated for revealing lost worlds.

But what makes a world *lost*? What made the past revealed by the Kirkdale cave a *lost* past? I've defined "loss" in largely ontological terms: something from the past is lost just in case it doesn't exist now. Hyenas, elephants, and hippopotamuses were known to a greater or lesser extent to scientists of the

FIGURE 10.1 Conybeare's Image of Buckland in the Hyena Cave. Wikimedia Commons

time, so in what sense did they no longer exist? *Loss*, recall, is sensitive to description. Although hyenas are not lost, they are certainly lost to England's wilderness (as is that wilderness itself). Buckland reported that "the fossil hyaena was nearly one third larger than the largest of the modern species" (Buckland 1822, 189). Thus, this species of hyena was hitherto unknown and long extinct. Further, the ecology and environment these organisms indicated spoke of an England sporting a much hotter climate and a wildly different fauna. As with all lost worlds we've examined, the Kirkdale cave combines the familiar with the unfamiliar.

In reconstructing the cave's formation and ecology, Buckland faced two kinds of challenges. One stems from *erasure*: the information provided by traces was incomplete. For instance, the bones were not preserved in beautifully articulated completeness, but in scattered pieces, "strewed all over like a dog kennel" (Buckland 1822, 183), as Buckland (rather leadingly) described them. The second relates to *loss*: England's caves no longer house hyenas, elephants, hippopotamuses, or rhinoceroses. With loss, we apparently lack the resources required to develop the understanding needed to infer from traces to the past, or to expand on that knowledge; with erasure, we lack the traces to make those inferences, even if we had the requisite understanding. While these challenges are deeply intertwined, this book has focused on loss. So, given that Buckland and his contemporaries were faced with an apparently unknown world, how did they empirically ground their imagining and reconstructing it? I've argued that to understand this, we must understand how scientists confront both the possibility and actuality of loss and erasure—and we see just this in Buckland's approach. So, let's start with the combination of three strategies: grafting, anchoring, and artificing.

Grafting involves linking traces and the denizens they reveal with their particular histories. This includes trace-based reasoning. Using geological theories about cave formation and deposition, brought into deep engagement with the traces, Buckland argued that the cave had been untouched since its discovery, discounted diluvian explanations of its formation, and generally told its history. The close fit between the hyena teeth and marks on the bones provides reason to think the remains were consumed by hyenas. The particular position of the bones, the stalagmites they were clustered among, and the mud and gravel atop them, underwrote his reconstruction. He linked both the cave's particular story with the shared geological history being uncovered across Northern Europe and America.

> The accumulation of these bones, then, appears to have been a long process, going on during a succession of years, whilst all the animals in question were natives of this country. The general dispersion of similar bones through the diluvian gravel of high latitudes, over a great part of the northern hemisphere, shows that the period in which they inhabited these regions, was that immediately preceding the formation of this gravel. (Buckland 1822, 201)

So, Buckland told a local story about the cave, but also integrated it into a broader narrative, grafting all the way.

A distinctively biological instance of grafting includes comparative thinking. Buckland's examination of the anatomy and habits of hyenas involved comparison with currently living examples. In addition to the size comparison we saw above, he also drew on behaviour. For instance, you might worry about the presence of macrofauna in the Kirkdale cave. While the numerous deer are highly plausible hyena prey, are we supposed to believe that these hyena, large as they may be, were in the habit of taking down rhinoceroses or elephants? Not at all, Buckland reassures us:

> It is foreign to the habits of the hyaena to prey on the larger pachydermata, their young perhaps excepted. No other solution of the difficulty presents itself to me, than that the remains in question are those of individuals that died a natural death; for though an hyaena would neither have had the strength to kill a living elephant or rhinoceros, or to drag home the entire carcase of a dead one, yet he could carry away, piecemeal, or acting conjointly with others, fragments of the most bulky animals that died in the course of nature. (Buckland 1822, 198)

Here, the possible behaviour of extinct hyenas is understood by considering the behaviour of extant hyenas. Buckland is working prior to the emergence of modern notions of evolutionary ancestry (although transformationist ideas were about), so forgive some wince-worthy anachronism: through the lens of contemporary biology, we'd read him as situating his extinct hyenas with their closest relatives via their shared evolutionary history.

Anchoring involves examining other cases with specific similarities to the lost target to build a model of the relevant properties. Buckland canvases descriptions of other caves across Europe, noting similarities and drawing on differences to build his case.

> The history of the diluvian gravel of the Continent, and of the animal remains contained in it, appears altogether identical with that of our own; and with respect to the bones that occur in caverns, the chief difference seems to be, that on the Continent some of the caves have their mouths open, and have been inhabited in the post-diluvial period by animals of now existing species [italics in original]. (Buckland 1822, 216)

The mixture of pre- and post-diluvian fauna in many caves, but not in Buckland's, is explained by the Kirkdale cave being undisturbed since its antediluvian burial. Further, differences in the markings and constitution of the bones are explained by the differences between the continent's bear-dominated caves and England's hyenas: "Such variation is consistent with the different habits of bears and hyaenas, arising from the different structure of their teeth and general organization; from which it follows, that bears prefer vegetable food to that of animals, and, when driven to the latter, prefer sucking the blood to eating the flesh, whilst hyaena are beyond all other beasts addicted to gnawing bones" (Buckland 1822, 221).

Thus, by navigating between various analogues (anchors), Buckland is able to both provide inductive strength to his hypothesis, while also deepening and developing it, along the way finding opportunities for testing by noting and explaining away differences. In a sense, he builds a comparative model of what we should expect from hyena-dominated and bear-dominated caves.

Artificing tackles loss by constructing simulacra of lost properties and examining them. Buckland reports that his hypothesis was "rendered almost certain by the discovery I made, of many small balls of the solid calcareous excrement of an animal that had fed on bones" (Buckland 1822, 186). Buckland was able to study a living hyena, "Billy," when a travelling menagerie made its way past Oxford. He fed Billy ox bones, subsequently investigating both the remnant bones themselves and the droppings. "It was at first sight recognized by the keeper of the Menagerie at Exeter Change, as resembling, both in form and appearance, the feces of the spotted or Cape Hyaena, which he stated to be greedy of bones, beyond all other beasts under his care" (Buckland 1822, 187).

So, faced with a putative trace fossil, Buckland brought about something like the conditions under which such a thing would form and examined it. That is, he engaged in an early example of experimental taphonomy. Indeed, according to his son, Buckland's examinations of Billy played a large role in underwriting his analysis.

> Billy cracked the marrow-bones of oxen, and refused those bones which contained no marrow, exactly as did his ancestors ages before him in the wilds of Yorkshire, as yet untrodden by the foot of man. So wonderfully alike were these bones in their fracture, that, judging from this point alone, it was impossible to say which bone had been cracked by Billy, and which by the aboriginal hyaena of Kirkdale. Again, Billy polished with his feet and hide the sides and floor of his den of wood, as his ancestors did the sides and floor of their den of stalactite in the Yorkshire hills; and as the ancient beasts deposited "album graecum" in abundance after a dinner of bones, so did Billy deposit pounds of the same substance. (F. Buckland 1873, 52)

Here, artificing and grafting blur: Buckland both examines Billy as approximately the same taxon as his extinct cousins and sets up circumstances whereby similar properties are instantiated. As we've seen throughout this book, the three strategies are often deeply intertwined, and it is in their combination that their epistemic prowess takes full flight.

In adopting these strategies, Buckland deftly demonstrates what I've called *strategic perspectivalism.* He characterized his target in an impressive variety of ways, thus isolating properties, which enabled him to connect his lost target to existing empirical information. He reconstructed the geology of a *pre-diluvial cave*, examined teeth both as those of *hyenas* and of *bone-crushers*, considered them as *partially preserved vertebrates*, as *hyenas*, as *scavengers*, and so on. These various isolated properties were then knitted together—recontextualized—to provide a coherent narrative about the cave, its inhabitants, and their collective history.

Further, these processes involved frequent *iteration* between isolated properties, ideas about them, and the various comparators. For instance, in his comparison between the Kirkdale cave and others in Britain, he didn't simply learn about his target, but also the other caves, "the origins of which though not before satisfactorily made out, becomes evident as a corollary from the proofs afforded by the cave at Kirkdale"(Buckland 1822, 208). That is, Buckland's Kirkdale interpretation makes sense of various other caves across Britain, not as well-preserved and untouched as Kirkdale, but nonetheless carrying the (now) telltale signs of the antediluvian lost world. The model, then, is built by navigating between various interpretations of the various caves.

It's striking that, although Buckland worked before many of the theoretical edifices of modern geology and palaeontology had been erected, using what are now outdated ideas about the Earth's history, he nonetheless adopted similar strategies, as we've seen modern palaeontologists use to uncover lost worlds. Although he wouldn't have seen his comparisons between extinct and extant hyenas in terms of evolutionary homologies and would have seen the cave as part of a diluvial history, these strategies enabled him to get a lot right. It is plausible that we see here the employment of a *style of reasoning* (Hacking 1992) that has been more or less stable across two hundred years of scientific development.[2]

---

2 It is worth noting that artificing is significantly more limited in Buckland's approach compared to the modern cases that have dominated this book. David Sepkoski (2019) has emphasized the role of modelling and more formal approaches to data analysis through the Palaeobiological Revolution. It may be that artificing only really came into its own once those theoretical and technical capacities were developed and adopted into palaeobiology.

This style of reasoning, and its ongoing success, provides reason to think that the Earth sciences reveal a highly *peculiar* past. Buckland's analysis relied on a world boasting fragile regularities and flux. Patterns are explained in terms of the various generating and maintaining processes that account for them and their exceptions. The Kirkdale cave's being undisturbed contrasts it with many others and explains its special epistemic status; the pre- and post-diluvial worlds were deeply different—played by different rules and populated by different denizens—but were not so different that Buckland was left epistemically rudderless. Further, the knowledge he generated was deeply *modal*: Buckland's analysis didn't simply tell us what happened to the cave, but gave insight into how the cave might have been different. Had it been disturbed, say, we wouldn't have seen the same bone composition or distribution; had it been the domicile of bears rather than hyenas, we would have seen differing patterns in the bone. And this modal knowledge is not simply epistemic, but objective: when successful, it tells us how, in fact, things would have been under different conditions.

Commentators on Buckland's work and its reception have emphasized imagination. As Ralph O'Connor says of Conybeare's depiction: "Buckland's work enabled people to imagine a past neither seen by man nor revealed by Scripture—an act of visualization whose strangeness is conveyed by the need to represent the present-day geologist 'watching' the events depicted" (O'Connor 2015, 69).

I've argued that we should understand scientific imagination in terms of *artifactualism*. That is, Buckland and his contemporaries' ability to imagine and depict lost worlds was embedded in and carried by deep engagement with both the traces of the past and the various linguistic, theoretical, and formal models that they employed to understand them. Without the revelation of the Kirkdale cave and sites like it, it is highly unlikely that Buckland and company would have imagined such an antediluvian world—their imaginations were impoverished—and, had they happened to, there would be no reason to take such imagining particularly seriously—their imaginations were too arbitrary. However, with those traces, their analysis, and the deployment of various other perspectival tools, their capacity to so imagine becomes relevant and revealing of rich pasts.

Artifactualism is one way of pulling focus from the individual scientist and embedding them in the material culture of their work, as well as drawing attention to the collective efforts involved. Reflecting on Conybeare's caricature, again, consider Manias' discussion:

> Fossil excavations were the work of many people. While the famous caricature by William Conybeare depicted Buckland's investigation of the Kirkdale "hyena den" as the personal journey of a gentleman scientist into a former world, other images illustrate more collective work . . . Geological surveying, infrastructural projects, and mining were the most extensive acts of digging in the nineteenth century, and a frequent source of fossils. While miners and quarry workers were often occluded in scholarly texts, they were often the people most responsible for excavating and finding fossils. (Manias 2023, 31)

That our narratives of scientific discovery frequently make invisible many of the workers crucial to it is a familiar theme in the sociology and history of science (Shapin 1989; Barnett 2020; Wylie 2021). An artifactualist account of imagination highlights the discovery, invention, preparation, construction, and maintenance of the artifacts that constitute scientific work. Presumably to his credit, Buckland's account of his discovery names not only broad-shouldered giants like Cuvier, but also figures like W. Salmond, Esq., who "has been engaged with much zeal and activity in measuring and exploring new branches of the cave, and making large collections of the teeth and bones, from which I understand he also intends to enrich our public cabinets in the metropolis" (Buckland 1822, 181–182), although the "workmen" who found the cave remain nameless.

So, in Buckland's two-century-old interpretation, we can sketch many of the contours I've developed throughout this book. He adopted perspectives strategically—grafting, anchoring, and artificing—before knitting it all together into a narrative concerning a lost world. In doing so, he used the artifacts of his work—the fossils, the cave itself, analyses of other caves, living hyenas, and reports about the hyenas—to facilitate his imagination and an exploration of possibility deeply rooted in an actual past. This lost world was peculiar, an ecosystem dependent on England's past environment and the contingencies of history. Hyenas lived, died, and gnawed bones in the cave for generations, but this stability was fragile. In the Kirkdale cave's uncovering and its reception, we can also draw out my final theme: the relationship between loss and wonder.

2. Lost Worlds and Seeing Anew

> Historicity is negated by the assumption that the world is "already there," whereby its "thereness" is taken for granted as the background of action in the present. To see the world as if for the first time is to notice that which is there, is made, has arrived, or is extraordinary. Wonder is about learning to see the world as something that does not have to be, and as something that came to be, over time, and with work. (Ahmed 2014, 179)

Buckland's analysis of the Kirkdale cave underwrote a new imagining of England, one with a dramatically different past. Unsurprisingly for the era, this was often approached with a heavy dash of the romantic (Sommer 2003). The heroic wildness of this imagined antediluvian age and the romance of its loss is captured in Philip Duncan's poem, *The Last Hyena* (one of those published in honour of Buckland):

> High on a rock, which o'er the ragin flood
>
> Reared its bleak crag, the last Hyaena stood;
>
> Beneath his paws a kindred skull was seen,
>
> And he, with commons short, looked grim and lean;
>
> Potent his jaw to crack his bony rapine,
>
> Potent his stomach as the Pot of Papin;
>
> O'er this last bone of many a murdered brother
>
> He growled, for he in vain had sought another.
>
> Full oft, like Captain Franklin, did he prey,
>
> On bones rejected in a former day;
>
> But now the whelming surge had buried all,
>
> In caves below, of beast both great and small,

But e'er it rose to mix him with the rest,
Thus did he growl aloud his last request:
"My skull to William Buckland I bequeath." –
He moaned, and the ocean's wave he sank beneath;
Southward the flood from Yorkshire chances to travel,
And rolled the monster deep in Rugby gravel.
(Duncan 1869, quoted in Pemberton 2010, 255)

In a style reminiscent of romantic evocations of the loss of what we'd now call Indigenous cultures in the wake of colonialism, coupled with shades of the "heroic" sacrifices sometimes made by those colonists themselves, Duncan describes England's final hyena, starved of its usual diet and driven to cannibalism, being overtaken by the great deluge whilst giving its mortal remains to the future. Duncan's poem links the heroic past with its remains. The last hyena provides its skull to Buckland and is buried in Yorkshire, underneath familiar gravel. Understanding the past, and how it might have been different, can lead us to see the present and how it could be different, anew. We gain Turner's "stereo-temporal" vision. Where quarrymen now labour, once the hyena prowled. The "light of enlightenment," represented by Buckland's torch in Conybeare's picture, doesn't just fall on the hyena, but on the cave, and on Buckland himself.

In chapter 9, I sketched a set of ways in which knowledge of the past might be epistemically valuable. I want to add a final piece to this by highlighting a further element: *wonder*. In doing so, I'll largely draw on Helen de Cruz's recent *Wonderstruck*, which, generally speaking, provides a broad account of wonder, awe, and related emotions, as well as their importance in the history of thought and human flourishing (de Cruz 2024).

De Cruz defines wonder as an epistemic emotion involving both confronting the *unknown* and a subsequent desire for *accommodation*, that is, making sense of the phenomenon in some way. As they put it, "*Wonder* is the emotion that arises from a glimpse at the unknown terrain which lies just beyond the fringes of our current understanding [italics in original]" (de Cruz 2024, 4). Wonder is *epistemic* insofar as it involves some kind of desire to know, or at the very least the recognition of mystery. Wonder is *self-transcendent* insofar as it involves opening ourselves to possibilities beyond what we usually consider, further helping "us move away from a focus on ourselves and our concerns" (de Cruz 2024, 6). Wonder underwrites and motivates a drive for discovery and pursuit by making our ignorance salient and guiding investigation. Wonder, then, is closely related to imagination in the sense I targeted in chapter 7: it involves the generation of new ideas, approaches, and concepts, as opposed to mental imagery.

Both de Cruz and Derek Turner have argued for a close connection between knowledge and various kinds of emotional and aesthetic affect. Turner, in *Paleoaesthetics* (2019), argues that knowing the history of some object of appreciation can enhance, deepen, or otherwise "make better" our judgements and appreciation of the object's aesthetic qualities. On his view, say, seeing the Kirkdale cave as once inhabited by hyena better places us to appreciate what is aesthetically wonderful about it (see also Currie 2023c). The layers of mud, strewn bones, and stalagmites are the result of a particular history. And we can learn to see that history in it. Attending to those details helps us see it not as a mess, but the remains of a past world; or—perhaps—a mess, but a fascinating, beautiful mess (or, depending on our view of hyenas, perhaps a bit terrifying). De Cruz points out that epistemic emotions, such as awe and wonder, play important roles in motivating and shaping science: "Awe and wonder are not only required for the day-to-day practitioner of science; they are crucial to help reorient scientists' thinking . . . They provide constant emotional motivation for scientists to continue their work, and instill openness to scientific ideas in the public" (de Cruz 2024,152). Rather than focus on motivation, let's consider how the investigation of past worlds can generate "newness."

Wonder is closely associated with *newness*, or what de Cruz calls "firstness." Some philosophers, most prominently Descartes, took wonder to require the novelty of the first encounter. On this view, wonder is fundamentally transient. We are wonderstruck by a Britain populated by hyenas, but this fades as we gain new understanding. By contrast, de Cruz doesn't tie wonder to the *literal* newness of first discovery. They instead emphasize how wonder can emerge from seeing the familiar *as if* new. "My concept of firstness is inspired by and closely allied to Descartes's view of the function of wonder, but slightly expanded so as to allow for the idea that we wonder not only at what we apprehend for the first time, but also at what we apprehend when we take the attitude that it is happening for the first time" (de Cruz 2024, 15).

"Firstness," then, might not simply involve surprising encounters with the unknown, but surprising encounters of the known seen anew. Further, the connection with knowledge is more complex than the Cartesian might suppose—that firstness disappears as knowledge increases—because knowledge is *scaffolded* (Currie 2015, 2018a). Famously, learning something often opens new vistas of enticing ignorance. Sometimes we wonder at how what we have just learned demonstrates to us how much more we have to learn. There doesn't seem to be a simple inverse relationship between knowledge and wonder: rich background knowledge can further deepen our wonder. Further, seeing something "as if" for the first time doesn't erase our knowledge of its history, but rather reframes it. As Sara Ahmed puts it:

> Does such an impulse to wonder require an erasure of history, by forgetting that one has seen the world before, or even that the world precedes the impulse to wonder? It could be assumed that the "as if" functions as a radical form of subjectivism, in which the subject forgets all that has taken place before a given moment of contemplation. But I would suggest that wonder allows us to see the surfaces of the world as made, and as such wonder opens up rather than suspends historicity. (Ahmed 2014, 179)

So, we can interpret "seeing anew" not as necessarily aligned with our first experience of something, but in terms of the object of wonder being *recast* in light of its history, in particular, the contingency (or I would say peculiarity) which that history reveals.

As we saw in the first chapter, the rhetoric of early geology was tied to its capacity to uncover lost worlds. This often included an appeal to the wonder these discoveries generated. As O'Connor puts it, "From the 1830s onwards, the story revealed by geology was promoted not only for upstaging ancient history—the 'facts' around which epic poems were spun—but also for equalling poetry in the wonder it evoked. With an increasing repertoire of bizarre monsters to call up before the mind's eye . . ." (O'Connor 2008, 214). In his aptly named popular lecture series, *The Wonders of Geology Or, A Familiar Exposition of Geological Phenomena,* published in 1839, Gideon Mantell repeatedly emphasized the number and scale of diverse discoveries the Earth sciences have provided us: "From this general view of the physical records of the mutations which the surface of our globe has undergone, we learn how numerous and important are the phenomena comprised within the sphere of geological inquiry, and how vain is the attempt to offer more than an epitome of its wonders in the brief space allotted to a popular course of lectures" (Mantell 1839, 181).

The past isn't just different, but diversely so. Mantell further emphasizes how revolutionary and sudden our realization of these lost pasts has been.

> Should an inhabitant of the new country express his wonder how brittle shells, delicate leaves, and bones of animals had become imbedded in the solid rock, and if when we stated the manner in which those changes had been effected, he should not only refuse his assent, but insist that the shells, leaves, and bones, were merely accidental forms of the stone, should we not feel astonished at his ignorance and prejudice? Yet not a century since, and such an opinion almost universally prevailed, and is even still entertained by many! (Mantell 1839, 320)

Such wonder-inducing changes to our own perception of the past continue apace. For instance, consider the last twenty years' discovery of the bushiness of our recent species' history. One hundred thousand years ago, *Homo sapiens* shared the Earth not only with Neanderthals (with whom, it turns out, we both bred and culturally mixed), but at least five other species of closely related members of the *Homo* genus (Galway-Witham, Cole, and Stringer 2019). In the astonishingly recent past, we shared the planet with diverse human-like lineages. That in itself boggles my imagination, but so too does the recentness of this discovery. Like Prof. Ichthy, I also wonder what revelations concerning our understanding of the past await us.

The connection between wonder and imagination—and our sometimes discomfort with it—is noted by Rudwick:

> Buckland, and probably others too, were fully aware that the construction of any scene from deep time involved an act of imaginative epistemic penetration from the known present into the unknown past. Certainly that was what Conybeare made explicit in his cartoon of Buckland crawling into his own antediluvial cave. That cartoon's magical or fairy-tale quality highlighted the problematic character of the act of penetration, as well as the sense of wonder that its achievement evoked. (Rudwick 1992, 57)

O'Connor and Rudwick emphasize the importance of Conybeare placing Buckland in the image, as a way of representing both the strangeness of his discovery and the risky inferences involved. The fantastical caricature of Buckland encountering the hyena makes the epistemic risk of the image, as well as its wondrous quality, explicit. The first half of this book sketches and I hope defends the effectiveness of the strategies scientists adopt when penetrating from the known present to the lost past. In chapter 7's discussion of palaeoart, I highlighted the challenge between getting the science as right as possible while also representing extinct organisms as real. Conybeare's image is an ingenious way of doing both.

In light of the connection with wonder, I'd add the stereo-temporal nature of Conybeare's image. The cave isn't simply depicted as it *is*, a muddy, bone-strewn mess into which Buckland crawls, but also as it *was*, hyenas and all. It also represents the apparent surprise—the shock—of the two coming together. Buckland's discovery, then, isn't only represented as changing our view of the past, but changing our view of the present as well. Thinking about the cave in light of its past helps us see it anew. It infuses the apparently familiar with the "firstness" de Cruz associates with wonder. In Ahmed's terms, we see the world not as given, but made.

Aiming for true pictures about the past isn't the only way to generate firstness, but—again reflecting on the arguments from chapter 7—there is reason to see it as an excellent one. If our unaided, present-oriented imaginations are impoverished, then we are badly placed to explore possibility. By aiming to get the past right, we generate rich perspectival tools that enable more diverse and wider imaginings of how the world could have been and, thus, how it might be now or might be in the future. Further, such tools provide ways of rendering fathomable the great scales of deep time, stretching and expanding our imaginative capacities.

So, discovering that the world was once different, often dramatically so, enables us to see the present anew. I've emphasized how discovering lost pasts requires that we "characterize" denizens to make sense of them in familiar terms, and how strategic perspectivalism enables us to move beyond the familiar to gain knowledge of lost worlds. Our wonder at strange past denizens motivates accommodating loss in terms of what we do know. But I've also emphasized the iterative nature of these investigations. In characterizing the past, we often re-characterize the present, and this provides bountiful experiences of firstness. A cave in Kirkdale was once haunted by hyenas; the Earth's now-separated continents have several times clustered together; the apparently stable relationship between global oxygen and sea life is the result of the behaviour of microorganisms dead for millions of years; the familiar complex metazoa of our world were preceded by beautiful yet sessile, comparatively humble organisms attached to the seafloor. Further, our world as it is was in no way inevitable, but is as it is due to a bewildering array of past denizens arising, interacting, and falling away.

I know of no better source for seeing the present as if for the first time than the investigation of lost worlds. It is often discomforting in its revelations of contingency and fragility (as well as the often risky epistemic bets that Rudwick highlights), often fills us with awe at the depths of time and contingency revealed—as well as our connection to those depths—and leads us to further wonder about what could have been in the past and what it might further tell us about our present and future.

For me, there is no better source of wonder than our hard-won knowledge of the deep past.

> The applause drawing a final curtain across Prof. Ichthy's lecture ranges from enthusiastic to somewhat desultory (it has been a *long* lecture, after all), and Ichthy leaves the stage glad—grateful—of the attention they've received. They also carry the hope that some of the audience was perhaps sufficiently moved, sufficiently enticed, to themselves wonder about the deep past, as well as about the ingenious, creative—imaginative—work required to make sense of it. Ichthy scoops up the replica skull and shuffles away from the podium, joined by a gaggle of graduate students, as well as several family members, as they head into an adjourning room where presumably less formal discussion, congratulations, and most likely inebriation will take place.
>
> The lecture, then, draws to a close with Ichthy looking both into the future and into the past: the future and its potential discoveries, the past with its surprising secrets, and Ichthy in the middle—the present—seeing each anew.

Bibliography

Ahmed, Sara. 2014. *The Cultural Politics of Emotion*. Routledge.

Ambrosio, Chiara. 2014. "Objectivity and Visual Practices in Science and Art." In *New Directions in the Philosophy of Science*, edited by Maria Carla Galavotti, Dennis Dieks, and Marcel Weber, 353–368. Springer International Publishing.

Ambrosio, Chiara, and Julia Sánchez-Dorado, eds. 2024. *Abstraction in Science and Art: Philosophical Perspectives*. Routledge.

Ansón, Marco, Manuel Hernández Fernández, and Pedro A. Saura Ramos. 2015. "Paleoart: Term and Conditions (A Survey among Paleontologists)." In *Current Trends in Paleontology and Evolution*, edited by Laura Domingo et al. Cercedilla, Spain: XIII Encuentro en Jóvenes Investigadores en Paleontología (XIII EJIP).

Ashby, Jack. 2021. "The Political Platypus and Colonial Koala—Decolonising the Way We Talk About Australian Mammals." *Journal of Natural Science Collections* 9: 35–45.

———. 2022. *Platypus Matters: The Extraordinary Story of Australian Mammals*. University of Chicago

Atkinson, Ian Athol Edward, and Rupert M. Greenwood. 1989. "Relationships Between Moas and Plants." *New Zealand Journal of Ecology* 12: 67–96.

Avin, Shahar, Wintle, Bonnie C., Julias Weitzdörfer, Sean S. Ó hÉigeartaigh, William J. Sutherland, and Martin J. Rees, 2018. "Classifying Global Catastrophic Risks." *Futures* 102: 20–26.

Baker, Victor Reginald. 1998. "Catastrophism and Uniformitarianism: Logical Roots and Current Relevance in Geology." *Geological Society, Special Publications*, 143 (1): 171–182.

Bakker, Robert T. 1975. "Dinosaur Renaissance." *Scientific American*, 232 (4): 58–79.

Baleka, Sina, Luciano Varela, P. Sebastián Tambusso, Johanna L. A. Paijmans, Dimila Mothé, Thomas W. Stafford Jr., Richard A. Fariña, and Michael Hofreiter. 2022. "Revisiting Proboscidean Phylogeny and Evolution Through Total Evidence and Palaeogenetic Analyses Including *Notiomastodon* Ancient DNA." *iScience* 25 (1): 103559.

Bamforth, Emily L., and Guy M. Narbonne. 2009. "New Ediacaran Rangeomorphs from Mistaken Point." *Journal of Paleontology* 83 (6): 897–913.

Barnes, Elizabeth. 2023. *Health Problems: Philosophical Puzzles about the Nature of Health*. Oxford University Press.

Barnett, Lydia. 2020. "Showing and Hiding: The Flickering Visibility of Earth Workers in the Archives of Earth Science." *History of Science* 58 (3): 245–274.

Bartha, Paul. 2022. "Analogy and Analogical Reasoning." In *The Stanford Encyclopedia of Philosophy*, Summer 2022 Edition, edited by Edward N. Zalta. https://plato.stanford.edu/archives/sum2022/entries/reasoning-analogy/.

Batterman, Robert W., and Colin Rice. 2014. "Minimal Model Explanations." *Philosophy of Science* 81 (3): 349–376.

Baum, David A., and Susan Offner. 2008. "Phylogenics and Tree-Thinking." *The American Biology Teacher* 70 (4): 222–229.

Baum, Seth D. 2020. "Quantifying the Probability of Existential Catastrophe: A Reply to Beard et al." *Futures* 123: 102608.

Bausman, William C. 2023. "How to Infer Metaphysics from Scientific Practice as a Biologist Might." In *From Biological Practice to Scientific Metaphysics*, edited by William C. Bausman, Jacob K. Baxter, and Oliver M. Lean. University of Minnesota Press.

Bausman, William C., Janella K. Baxter, and O. Matthew Lean. 2023. "Introduction: Toward a Scientific Metaphysics Based on Biological Practice." In *From Biological Practice to Scientific Metaphysics*, edited by William C. Bausman, Janella K. Baxter, and Oliver M. Lean. University of Minnesota Press.

Baxter, Janella. 2023. "Just How Messy Is The World?" In *From Biological Practice to Scientific Metaphysics*, edited by William C. Bausman, Jacob K. Baxter, and Oliver M. Lean. University of Minnesota Press.

Beard, S. J., Thomas Rowe, and James Fox. 2020. "An Analysis and Evaluation of Methods Currently Used to Quantify the Likelihood of Existential Hazards." In *An Anthology of Global Risk*, edited by S.J. Beard and Tom Hobson. Open Book Publishers.

Beatty, John. 1995. "The Evolutionary Contingency Thesis." In *Concepts, Theories, and Rationality in the Biological Sciences: The Second Pittsburgh–Konstanz Colloquium in the Philosophy of Science*, edited by Gereon Wolters and James G. Lennox, 45–64. UVK Universitätsverlag Konstanz.

———. 2006. "Replaying Life's Tape." *The Journal of Philosophy* 103 (7): 336–362.

———. 2016. "What Are Narratives Good For?" *Studies in History and Philosophy of Science Part C: Studies in History and Philosophy of Biological and Biomedical Sciences* 58: 33–40.

Bewell, Alan. 2014. "Hyena Trouble." *Studies in Romanticism*, 53 (3): 969–397.

Bishop, Peter J., Karl T. Bates, Vivian R. Allen, Donald M. Henderson, Marcela Randau, and John R. Hutchinson. 2020. "Relationships of Mass Properties and Body Proportions to Locomotor Habit in Terrestrial Archosauria." *Paleobiology*, 46(4): 550–568.

Bobrovskiy, Ilya, Janet M. Hope, Andrey Ivantsov, Benjamin J. Nettersheim, Christian Hallmann, and Jochen J. Brock. 2018. "Ancient Steroids Establish the Ediacaran Fossil *Dickinsonia* as One of the Earliest Animals." *Science* 361 (6408): 1246–1249.

Bocchi, Federica. 2022. "Biodiversity vs. Paleodiversity Measurements: The Incommensurability Problem." *European Journal for Philosophy of Science* 12 (4): 64.

Bocchi, Federica, Alisa Bokulich, Leticia Castillo Brache, Gloria Grand-Pierre, and Aja Watkins. 2022. "Are We in a Sixth Mass Extinction? The Challenges of Answering and Value of Asking." *Synthese* 200: 303.

Boden, Margaret A. 2004. *The Creative Mind: Myths and Mechanisms*. Routledge.

Bokulich, Alisa. 2014. "How the Tiger Bush Got Its Stripes: 'How Possibly' vs. 'How Actually' Model Explanations." *The Monist* 97 (3): 321–338.

———. 2020. "Calibration, Coherence, and Consilience in Radiometric Measures of Geologic Time." *Philosophy of Science* 87 (3): 425–456.

———. 2021a. "Taming the Tyranny of Scales: Models and Scale in the Geosciences." *Synthese* 199 (5): 14167–14199.

———. 2021b. "Using Models to Correct Data: Paleodiversity and the Fossil Record." *Synthese* 198 (Suppl 24): 5919–5940.

Bokulich, Alisa, and Naomi Oreskes. 2017. "Models in Geosciences." In *Springer Handbook of Model-Based Science*, edited by Lorenzo Magnani and Tommaso Bertolotti, 891–911. Springer.

Bonnin, Thomas. 2019. "Evidential Reasoning in Historical Sciences: Applying Toulmin Schemes to the Case of Archezoa." *Biology & Philosophy* 34 (2): 30.

Booth, Austin, and W. Ford Doolittle. 2015. "Eukaryogenesis, how Special Really?" *Proceedings of the National Academy of Sciences* 112 (33): 10278–10285.

Borg, George. 2025. "Measurement, Decomposition and Level-Switching in Historical Science: Geochronology and the Ontology of Scientific Methods." *Studies in History and Philosophy of Science* 109: 123–131.

Bostrom, Nick. 2002. "Existential Risks: Analyzing Human Extinction Scenarios and Related Hazards." *Journal of Evolution and Technology* 9 (1).

Bostrom, Nick, and Milan M. Cirkovic, eds. 2011. *Global Catastrophic Risks*. Oxford University Press.

Boudinot, F. Garett, and Joseph Wilson. 2020. "Does a Proxy Measure Up? A Framework to Assess and Convey Proxy Reliability." *Climate of the Past* 16 (5): 1807–1820.

Boyd, Nora Mills. 2018. "Evidence Enriched." *Philosophy of Science* 85 (3): 403–421.

Boynton, Helen, and Trevor D. Ford. 1979. "Pseudovendia Charnwoodensis—A New Precambrian Arthropod from Charnwood Forest, Leicestershire." *Mercian Geologist* 7: 175–177.

Brandon, Robert N. 1990. *Adaptation and Environment.* Princeton University Press.

Brandstetter, Thomas. 2011. "Time Machines: Model Experiments in Geology." *Centaurus* 53 (2): 135–145.

Brasier, Martin D., and Jonathan B. Antcliffe. 2008. *Dickinsonia* from Ediacara: A New Look at Morphology and Body Construction. *Palaeogeography, Palaeoclimatology, Palaeoecology* 270 (3–4): 311–323.

Brigandt, Ingo. 2013. "Explanation in Biology: Reduction, Pluralism, and Explanatory Aims." *Science & Education* 22 (1): 69–91.

Briggs, Derek E. G. 1995. "Experimental Taphonomy". *Palaios* 10 (6): 539–550. https://doi.org/10.2307/3515093.

———. 2015. "The Cambrian Explosion." *Current Biology* 25 (19): R864–R868.

Briggs, Derek E. G., and Richard A. Fortey. 2005. "Wonderful Strife: Systematics, Stem Groups, and the Phylogenetic Signal of the Cambrian Radiation." *Paleobiology* 31 (S2): 94–112.

Brocks, Jochen J. 2018. "The Transition from a Cyanobacterial to Algal World and the Emergence of Animals." *Emerging Topics in Life Sciences* 2 (2): 181–190.

Broussolle, Arnaud. 2022. "Mathematical Modelling Reveals Potential Acceleration of the Supercontinent Cycle." *Scientific Reports* 12 (1): 17391.

Brown, James Robert. 1991. *The Laboratory of the Mind: Thought Experiments in the Natural Sciences.* Routledge.

Browning, Heather. 2018. "Won't Somebody Please Think of the Mammoths? De-Extinction and Animal Welfare." *Journal of Agricultural and Environmental Ethics* 31 (6): 785–803.

Browning, Heather, and Jonathan Birch. 2022. "Animal Sentience." *Philosophy Compass*, 17 (5): e12822.

Brysse, Keynyn. 2008. "From Weird Wonders to Stem Lineages: The Second Reclassification of the Burgess Shale Fauna." *Studies in History and Philosophy of Science Part C: Studies in History and Philosophy of Biological and Biomedical Sciences* 39(3): 298–313.

Buckland, Francis T. 1873. *Curiosities of Natural History.* Richard Bentley and Son.

Buckland, William. 1822. "Account of an Assemblage of Fossil Teeth and Bones of Elephant, Rhinoceros, Hippopotamus, Bear, Tiger, and Hyæna, and Sixteen Other Animals, Discovered in a Cave at Kirkdale, Yorkshire, in the Year 1821." *Philosophical Transactions of the Royal Society of London* 112: 171–236.

Buskell, Andrew, and Adrian Currie. 2021. "Uniqueness in the Life Sciences: How Did the Elephant Get Its Trunk?" *Biology & Philosophy* 36: 1–24.

Bussiere, Matthieu, and Marcel Fratzscher. 2008. "Low Probability, High Impact: Policy Making and Extreme Events". *Journal of Policy Modeling* 30 (1): 111–121.

Butterfield, Nicholas J. 2009. "Oxygen, Animals and Oceanic Ventilation: An Alternative View." *Geobiology* 7 (1): 1–7.

———. 2017. "Oxygen, Animals and Aquatic Bioturbation: An Updated Account." *Geobiology* 16 (1): 3–16.

Cain, Victoria. 2010. "'The Direct Medium of the Vision': Visual education, Virtual Witnessing and the Prehistoric Past at the American Museum of Natural History, 1890–1923." *Journal of Visual Culture* 9 (3): 284–303.

———. 2011. "The Art of Authority: Exhibits, Exhibit-Makers, and the Contest for Scientific Status in the American Museum of Natural History, 1920–1940." *Science in Context* 24 (2): 215–238.

Calcott, Brett, and Kim Sterelny, eds. 2011. *The Major Transitions in Evolution Revisited.* MIT Press.

Caporael, Linnda R., James R. Griesemer, and William C. Wimsatt, eds. 2014. *Developing Scaffolds in Evolution, Culture, and Cognition.* Vol. 17. MIT Press.

Carroll, John W. 2020. "Laws of Nature." In *The Stanford Encyclopedia of Philosophy*, Winter 2020 Edition, edited by Edward N. Zalta. https://plato.stanford.edu/archives/win2020/entries/laws-of-nature/.

Cartwright, Nancy. 1999. *The Dappled World: A Study of the Boundaries of Science.* Cambridge University Press.

———. 2015. "Philosophy of Social Technology: Get on Board." Paper presented at the American Philosophical Association Pacific Division 89th Annual Meeting, Vancouver, Canada, April 1–4.

Chakravartty, Anjan. 2010. "Metaphysics Between the Sciences and Philosophies of Science." In *New Waves in Philosophy of Science*, edited by P. D. Magnus and Jacob Busch, 59–77. Palgrave Macmillan.

Chang, Hasok. 2004. *Inventing Temperature: Measurement and Scientific Progress.* Oxford University Press.

Chapman, Robert, and Alison Wylie. 2018. *Evidential Reasoning in Archaeology.* Bloomsbury Publishing.

Chemero, Anthony. 2011. *Radical Embodied Cognitive Science.* MIT Press.

Clark, Andy, and David Chalmers. 1998. "The Extended Mind." *Analysis* 58 (1): 7–19.

Clark, Eric, Andrew D. Peel, and M. Akram. 2019. "Arthropod Segmentation." *Development* 146 (18): dev170480.

Cleland, Carol E. 2002. "Methodological and Epistemic Differences Between Historical Science and Experimental Science." *Philosophy of Science* 69 (3): 474–496.

———. 2011. "Prediction and Explanation in Historical Natural Science." *The British Journal for the Philosophy of Science* 62 (3): 551–582.

———. 2013. "Common Cause Explanation and the Search for a Smoking Gun." *Geological Society of America Special Papers* 502: 1–9.

———. 2019. *The Quest for a Universal Theory of Life: Searching for Life as We Don't Know It.* Cambridge University Press.

Cole, Devon B., Daniel B. Mills, Douglas H. Erwin, Erik A. Sperling, Susannah M. Porter, Christopher T. Reinhard, and Noah J. Planavsky. 2020. "On the Co-Evolution of Surface Oxygen Levels and Animals." *Geobiology* 18 (3): 260–281.

Colombetti, Giovanna, and Joel Krueger. 2015. "Scaffoldings of the Affective Mind." *Philosophical Psychology* 28 (8): 1157–1176.

Condie, Kent C., Sergei A. Pisarevsky, Jun Korenaga, and Simon Gardoll. 2015. "Is the Rate of Supercontinent Assembly Changing with Time?" *Precambrian Research* 259: 278–289.

Conway, John, C. M. Kosemen, Darren Naish, and Scott Hartman. 2013. *All Yesterdays: Unique and Speculative Views of Dinosaurs and Other Prehistoric Animals.* Irregular books.

Conway, Maree. 2006. "An Overview of Foresight Methodologies." *Thinking Futures* 10: 1–10.

Corsico, Paolo. 2024. "Existential Risk and the Justice Turn in Bioethics." *Journal of Medical Ethics* 50 (12).

Crasnow, Sharon. 2021. "Coherence, Objectivity, and Measurement: The Example of Democracy." *Synthese* 199 (1): 1207–1229.

Cremer, Carla Zoe, and Luke Kemp. 2021. "Democratising Risk: In Search of a Methodology to Study Existential Risk." https://doi.org/10.48550/arXiv.2201.11214.

Crook, Ian G. 1975. "The Tuatara." In *Biogeography and Ecology in New Zealand*, edited by G. Kuschel, 331–352. Springer.

Cunningham, John A., Imran A. Rahman, Stephan Lautenschlager, Emily J. Rayfield, and Philip C. J. Donoghue. 2014. "A Virtual World of Paleontology." *Trends in Ecology & Evolution* 29(6): 347–357.

Currie, Adrian. 2012. "Convergence, Contingency & Morphospace." Review *of Convergent Evolution: Limited Forms Most Beautiful*, by G. R. McGhee. *Biology & Philosophy* 27: 583–593.

———. 2013. "Convergence as Evidence." *The British Journal for the Philosophy of Science* 64 (4): 763–786.

———. 2014. “Narratives, Mechanisms and Progress in Historical Science.” *Synthese* 191 (6): 1163–1183.

———. 2015. “Marsupial Lions and Methodological Omnivory: Function, Success and Reconstruction in Paleobiology.” *Biology & Philosophy* 30 (2): 187–209.

———. 2017a. “Hot-Blooded Gluttons.” *The British Journal for the Philosophy of Science* 68 (4): 929–952.

———. 2017b. “From Models-as-Fictions to Models-as-Tools.” *Ergo: An Open Access Journal of Philosophy* 4 (27): 759–781.

———. 2018a. *Rock, Bone, and Ruin: An Optimist's Guide to the Historical Sciences.* MIT Press.

———. 2018b. “The Argument from Surprise.” *Canadian Journal of Philosophy* 48 (5): 639–661.

———. 2019a. *Scientific Knowledge and the Deep Past: History Matters.* Cambridge University Press.

———. 2019b. “Paleobiology and Philosophy.” *Biology & Philosophy* 34 (2): 1–13.

———. 2019c. “Epistemic Optimism, Speculation, and the Historical Sciences.” *Philosophy, Theory, & Practice in Biology* 11 (7). https://doi.org/10.3998/ptpbio.16039257.0011.007

———. 2019d. “Mass Extinctions as Major Transitions.” *Biology & Philosophy* 34 (2): 1–29.

———. 2019e. “Existential Risk, Creativity & Well-Adapted Science.” *Studies in History and Philosophy of Science Part A* 76: 39–48.

———. 2019f. “Simplicity, One-Shot Hypotheses and Paleobiological Explanation.” *History and Philosophy of the Life Sciences* 41 (1): 1–24.

———. 2020. “Bottled Understanding: The Role of Lab Work in Ecology.” *The British Journal for the Philosophy of Science* 71: 905–932.

———. 2021a. “Stepping Forwards by Looking Back: Underdetermination, Epistemic Scarcity and Legacy Data.” *Perspectives on Science* 29 (1): 104–132.

———. 2021b. *Comparative Thinking in Biology.* Cambridge University Press.

———. 2021c. “Science & Speculation.” *Erkenntnis* 88: 597–619.

———. 2022. “Speculation Made Material: Experimental Archaeology and Maker's Knowledge.” *Philosophy of Science* 89 (2): 337–359.

———. 2023a. “Of Records and Ruins: Metaphors about the Deep Past.” *Journal of the Philosophy of History* 17 (1): 154–175.

———. 2023b. “Towards a New Aesthetics of Science: Aesthetic Cultures and the Processes and Objects of Regard.” In *The Aesthetics of Scientific Experiments*, edited by Milena Ivanova and Alice Murphy, 112–134. Routledge.

———. 2023c. “Epistemic Engagement, Aesthetic Value, and Scientific Practice.” *The British Journal for the Philosophy of Science* 74 (2): 313–334.

———. 2024a. "Fossils, Modality & Central Subjects in Palaeobiological Reconstruction." *Philosophy, Theory, and Practice in Biology* 16 (2).

———. 2024b. "Modelling the Deep Past." In *The Routledge Handbook of Philosophy of Scientific Modeling*, edited by Tarja Knuuttila, Natalia Carillo, and Rami Koskinen, 497–510. Routledge.

———. 2025. "Thinking With Case Studies in the Philosophy of Science." In *Methods in the Philosophy of Science: A User's Guide*, edited by Sophie J. Veigl and Adrian Currie. MIT Press.

———. Under review. "Existential Risk: Perceived, Real and Democratized." Unpublished manuscript.

Currie, Adrian, and Anton Killin. 2019. "From Things to Thinking: Cognitive Archaeology." *Mind & Language* 34 (2): 263–279.

Currie, Adrian, and Arnon Levy. 2019. "Why Experiments Matter." *Inquiry* 62 (9–10): 1066–1090.

Currie, Adrian, and Kim Sterelny. 2017. "In Defence of Story-Telling." *Studies in History and Philosophy of Science Part A* 62: 14–21.

Currie, Adrian, and Daniel Swaim. 2021. "Past Facts and the Nature of History." *Journal of the Philosophy of History* 16 (2): 179–206.

Currie, Adrian, and Kirsten Walsh. 2018. "Newton on Islandworld: Ontic-Driven Explanations of Scientific Method." *Perspectives on Science* 26 (1): 119–156.

Currie, Adrian, and Kirsten Walsh. 2019. "Frameworks for Historians and Philosophers." *HOPOS* 9 (1): 1–34.

Cuthill, Jennifer F., and Simon Conway Morris. 2014. "Fractal Branching Organizations of Ediacaran Rangeomorph Fronds Reveal a Lost Proterozoic Body Plan." *Proceedings of the National Academy of Sciences* 111 (36): 13122–13126.

Darwin, Charles. 1964. *On the Origin of Species: A Facsimile of the First Edition.* Cambridge, MA: Harvard University Press. Originally published 1859.

Daston, Lorraine, and Peter L. Galison. 2021. *Objectivity.* Princeton University Press.

De Cruz, Helen. 2024. *Wonderstruck: How Wonder and Awe Shape the Way We Think.* Princeton University Press.

de La Rocha, Christina L., and Uta Passow. 2014. "The Biological Pump." *Treatise on Geochemistry* 6: 83–111.

de Oliveira, Guilherme S. 2022. "Radical Artifactualism." *European Journal for Philosophy of Science* 12 (2): 36.

de Regt, Henk W. 2017. *Understanding Scientific Understanding.* Oxford University Press.

Debus, Allen A., and Diane E. Debus. 2002. *Paleoimagery: The Evolution of Dinosaurs in Art.* McFarland.

Desjardins, Eric. 2011. "Historicity and Experimental Evolution." *Biology & Philosophy* 26: 339–364.

DiFrisco, James, Alan C. Love, and Günter. P. Wagner. 2020. "Character Identity Mechanisms: A Conceptual Model for Comparative-Mechanistic Biology." *Biology & Philosophy* 35 (4): 44.

Dray, W. H. 1957. *Law and Explanation in History*. Oxford University Press.

———.1968. "On Explaining How-Possibly." *The Monist*: 390–407.

Dresow, Max. 2021. "Explaining the Apocalypse: The End-Permian Mass Extinction and the Dynamics of Explanation in Geohistory." *Synthese* 199 (3): 10441–10474.

———. 2023. "Actualism and Uniformitarianism: From Abstract Commitments to Forms of Practice." *Philosophy of Science* 90 (4): 994–1000.

———. 2025. "Ediacaran Enigma: Uncertainty and Underdetermination in Precambrian Paleontology." *History and Philosophy of the Life Sciences* 47 (3): 38.

Droser, Mary L., and James G. Gehling. 2015. "The Advent of Animals: The View from the Ediacaran." *Proceedings of the National Academy of Sciences* 112 (16): 4865–4870.

Dunn, Frances S., and Alexander G. Liu. 2019. "Viewing the Ediacaran Biota as a Failed Experiment Is Unhelpful." *Nature Ecology & Evolution* 3 (4): 512–514.

Dunn, Frances S., Philip R. Wilby, Charlotte G. Kenchington, Dmitriy V. Grazhdankin, Philip C. J. Donoghue, and Alexander G. Liu. 2019. "Anatomy of the Ediacaran Rangeomorph Charnia masoni." *Papers in Palaeontology* 5 (1): 157–176.

Dupré, John. 1993. *The Disorder of Things: Metaphysical Foundations of the Disunity of Science*. Harvard University Press.

———. 2023. "The Disunity of Science and the Unity of the World: Presidential Address, PSA 2022." *Philosophy of Science*: 1–15.

———. 2025. *Everyone Flows: A Process Philosophy of Human Life*. Oxford University Press.

Dupret, Vincent. 2004. "The Pleurosaurs: Anatomy and Phylogeny." *Revue de Paléobiologie*, vol. spéc. 9: 61–80.

El Skaf, Rawad, and Michael T. Stuart. 2025. "Scientific Models and Thought Experiments." In *The Routledge Handbook of Philosophy of Scientific Modeling*, edited by Tarja Knuuttila, Natalia Carillo, and Rami Koskinen, 325–340. Routledge.

Elgin, Catherine Z. 2017. *True Enough*. MIT Press.

———. 2024. "Selective Disregard." In *Abstraction in Science and Art: Philosophical Perspectives*, edited by Chiara Ambrosio and Julia Sánchez-Dorado, 17–29. Routledge.

Elliott, Kevin C. 2012. "Epistemic and Methodological Iteration in Scientific Research." *Studies in History and Philosophy of Science Part A* 43 (2): 376–382.

Ereshefsky, Marc. 2014. "Species, Historicity, and Path Dependency." *Philosophy of Science* 81 (5): 714–726.

Ereshefsky, Marc, and Thomas A. Reydon. 2015. "Scientific Kinds." *Philosophical Studies* 172: 969–986.

Ereshefsky, Marc, and Derek Turner. 2020. "Historicity and Explanation." *Studies in History and Philosophy of Science Part A* 80: 47–55.

Eriksson, Mats E. 2014. "Master of Puppets: Sculpting Ancient Worlds." *Geology Today* 30 (3): 98–104.

Farlow, James O., Peter Dodson, and Anusuya Chinsamy. 1995. "Dinosaur Biology." *Annual Review of Ecology and Systematics* 26 (1): 445–471.

Fenchel, Tom. 2011. "Anaerobic Eukaryotes." In *Anoxia: Evidence for Eukaryote Survival and Paleontological Strategies*, edited by Alexander V. Altenbach, Joan M. Bernhard, and Joseph Seckbach, 3–16. Springer.

Finkelman, Leonard. 2018. "De-Extinction and the Conception of Species." *Biology & Philosophy* 33 (5–6): 32.

Fisher, Charles K., and Pankaj Mehta. 2014. "The Transition between the Niche and Neutral Regimes in Ecology." *Proceedings of the National Academy of Sciences* 111: 13111–13116.

Forber, Patrick. 2010. "Confirmation and Explaining How Possible." *Studies in History and Philosophy of Science Part C: Studies in History and Philosophy of Biological and Biomedical Sciences* 41 (1): 32–40.

Forber, Patrick, and Eric Griffith. 2011. "Historical Reconstruction: Gaining Epistemic Access to the Deep Past." *Philosophy and Theory in Biology* 3: 1–19.

Forgeard, Marie J. C., and James C. Kaufman. 2016. "Who Cares about Imagination, Creativity, and Innovation, and Why? A Review." *Psychology of Aesthetics, Creativity, and the Arts* 10 (3): 250–269.

Fox, Craig W. 2023. "Does the Present Overdetermine the Past?" In *Theory, Evidence, Data: Themes from George E. Smith*, edited by Marius Stan and Christopher Smeenk, 83–94. Springer International Publishing.

French, Steven, and Alice Murphy. 2023. "The Value of Surprise in Science." *Erkenntnis* 88 (4): 1447–1466.

Friend, Stacie. 2020. "The Fictional Character of Scientific Models." In *The Scientific Imagination*, edited by Arnon Levy and Peter Godfrey-Smith, 102–127. Oxford University Press.

Frigg, Roman, and James Nguyen. 2017. "Scientific Representation Is Representation-As." In *Philosophy of Science in Practice: Nancy Cartwright and the Nature of Scientific Reasoning*, edited by Hsiang-Ke Chao and Julian Reiss, 149–179. Springer International Publishing.

Frigg, Roman, and James Nguyen. 2020. *Modelling Nature: An Opinionated Introduction to Scientific Representation*. Springer International Publishing.

Gaidos, Eric J., and Francis Nimmo. 2000. "Tectonics and Water on Europa." *Nature* 405 (6787): 637–640.

Galway-Witham, Julia, John Cole, and Chris Stringer. 2019. "Aspects of Human Physical and Behavioural Evolution during the Last 1 Million Years." *Journal of Quaternary Science* 34 (6): 355–378.

Gelfert, Axel. 2019. "Probing Possibilities: Toy Models, Minimal Models, and Exploratory Models." In *Model-Based Reasoning in Science and Technology: Inferential Models for Logic, Language, Cognition and Computation*, edited by Ángel Nepomuceno-Fernández, Lorenzo Magnani, Francisco J. Salguero-Lamillar, Cristina Barés-Gómez, and Matthieu Fontaine, 3–19. Springer International Publishing.

Gemmell, Neil J., Michael K. Schwartz, Bruce C. Robertson, et al. 2004. "Moa Were Many." *Proceedings of the Royal Society B: Biological Sciences* 271 (suppl. 6): S430–S432.

Gibbons, Matilda, Andrew Crump, Meghan Barrett, Sajedeh Sarlak, Jonathan Birch, and Lars Chittka. 2022. "Can Insects Feel Pain? A Review of the Neural and Behavioural Evidence." *Advances in Insect Physiology* 63: 155–229.

Giere, Ronald N. 2006. *Scientific Perspectivism*. University of Chicago Press.

Godfrey-Smith, Peter. 2006. "The Strategy of Model-Based Science." *Biology & Philosophy* 21: 725–740.

Gold, David A., Bruce Runnegar, James G. Gehling, and David K. Jacobs. 2015. "Ancestral State Reconstruction of Ontogeny Supports a Bilaterian Affinity for *Dickinsonia*." *Evolution & Development* 17 (6): 315–324.

Goodman, Nelson. 1972. *Problems and Projects*. Bobbs–Merrill.

Gould, Stephen Jay. 1966. "Allometry and Size in Ontogeny and Phylogeny." *Biological Reviews* 41: 587–638.

———. 1988. *Time's Arrow, Time's Cycle: Myth and Metaphor in the Discovery of Geological Time*. Harvard University Press.

Grantham, Todd A. 1999. "Explanatory Pluralism in Paleobiology." *Philosophy of Science* 66 (S3): S223–S236.

Green, Richard E., Edward L. Braun, Joel Armstrong, Dent Earl, Ngan Nguyen, Glenn Hickey, et al. 2014. "Three Crocodilian Genomes Reveal Ancestral Patterns of Evolution among Archosaurs." *Science* 346 (6215): 1254449.

Griesemer, James R. 1988. "Causal Explanation in Laboratory Ecology: The Case of Competitive Indeterminacy." In *PSA 1988: Proceedings of the Biennial Meeting of the Philosophy of Science Association*, 337–344.

Griesmer, James R., and Michael J. Wade. 1988. "Laboratory Models, Causal Explanation and Group Selection." *Biology & Philosophy* 3 (1): 67–96.

Griffiths, Paul E. 1996. "The Historical Turn in the Study of Adaptation." *The British Journal for the Philosophy of Science* 47 (4): 511–532.

Griffiths, Paul E., and Russell D. Gray. 1994. "Developmental Systems and Evolutionary Explanation." *The Journal of Philosophy* 91 (6): 277–304.

Gruber, Jacob W. 1991. "Does the Platypus Lay Eggs? The History of an Event in Science." *Archives of Natural History* 18 (1): 51–123.

Gutarra, Susana, and Imran A. Rahman. 2022. "The Locomotion of Extinct Secondarily Aquatic Tetrapods." *Biological Reviews* 97 (1): 67–98.

Haber, Matthew H. 2019. "Species in the Age of Discordance." *Philosophy, Theory, and Practice in Biology* 11 (021).

Hacking, Ian. 1983. *Representing and Intervening: Introductory Topics in the Philosophy of Natural Science*. Cambridge University Press.

———. 1992. "'Style' for Historians and Philosophers." *Studies in History and Philosophy of Science Part A* 23 (1): 1–20.

Harnad, Stevan. 2016. "Animal Sentience: The Other-Minds Problem." *Animal Sentience* 1 (1): 1.

Harrison, T. Mark. 2024. "We Don't Know When Plate Tectonics Began." *Journal of the Geological Society* 181 (4): jgs2023–212.

Havstad, Joyce C. 2018. "Messy Chemical Kinds." *The British Journal for the Philosophy of Science* 69 (3): 657–679.

———. 2019. "Metaphorical Ripples." *Philosophy, Theory, and Practice in Biology* 11.

Hawley, Katherine. 2016. "Science as a Guide to Metaphysics?" *Synthese* 149: 451-470.

Hearn, Les, and Amanda C. de C. Williams. 2019. "Pain in Dinosaurs: What Is the Evidence?" *Philosophical Transactions of the Royal Society B: Biological Sciences* 374 (1785): 20190370.

Hills, Alison, and Alexander Bird. 2019. "Against Creativity." *Philosophy and Phenomenological Research* 99 (3): 694–713.

Hoekzema, Renee S., Martin D. Brasier, Francis S. Dunn, and Alexander G. Liu. 2017. "Quantitative Study of Developmental Biology Confirms *Dickinsonia* as a Metazoan." *Proceedings of the Royal Society B: Biological Sciences* 284 (1862). https://doi.org/10.1098/rspb.2017.1348.

Hull, David L. 1976. "Are Species Really Individuals?" *Systematic Zoology* 25 (2): 174–191.

Ishigaki, Shinobu, and Martin G. Lockley. 2010. "Didactyl, Tridactyl and Tetradactyl Theropod Trackways from the Lower Jurassic of Morocco: Evidence of Limping, Labouring and Other Irregular Gaits." *Historical Biology* 22 (1–3): 100–108.

Ivanova, Milena, and Alice Murphy, eds. 2023. *The Aesthetics of Scientific Experiments*. Taylor & Francis.

Ivantsov, Andrey, and Maria Zakrevskaya. 2022. "*Dickinsonia*: Mobile and Adhered." *Geological Magazine* 159 (7): 1118–1133.

Jablonski, David. 1989. "The Biology of Mass Extinction: A Palaeontological View." *Philosophical Transactions of the Royal Society B: Biological Sciences* 325 (1228): 357–368.

———. 2005. "Mass Extinctions and Macroevolution." *Paleobiology* 31 (S2): 192–210.

Jackson, Frank, and Philip Pettit. 1992. "In Defense of Explanatory Ecumenism." *Economics & Philosophy* 8 (1): 1–21.

Jackson, Stephen T., and Jessica L. Blois. 2015. "Community Ecology in a Changing Environment: Perspectives from the Quaternary." *Proceedings of the National Academy of Sciences* 112: 4915–4921.

Jeffares, Ben. 2008. "Testing Times: Regularities in the Historical Sciences." *Studies in History and Philosophy of Science Part C: Studies in History and Philosophy of Biological and Biomedical Sciences* 39 (4): 469–475.

Johnson, Kristin. 2007. "Natural History as Stamp Collecting: A Brief History." *Archives of Natural History* 34 (2): 244–258.

Jones, Elizabeth D. 2019. "Ancient Genetics to Ancient Genomics: Celebrity and Credibility in Data-Driven Practice." *Biology & Philosophy* 34 (2): 27.

———. 2022. *Ancient DNA: The Making of a Celebrity Science*. Yale University Press.

Jones, Max, and Tom Schoonen. 2024. "Embodied Simulation and Knowledge of Possibilities." *Philosophical Psychology*: 1–25.

Jørgensen, Dolly. 2019. *Recovering Lost Species in the Modern Age: Histories of Longing and Belonging*. MIT Press.

Kaiser, Marie I., and Javier Suárez. 2025. "Metaphysics of Scientific Practice." In *Methods in the Philosophy of Science: A User's Gui*de, edited by Sophie Veigl and Adrian Currie. MIT Press.

Katz, Eric. 2022. "Considering De-Extinction: Zombie Arguments and the Walking (and Flying and Swimming) Dead." *Ethics, Policy & Environment* 25 (2): 81–103.

Kemp, Luke. 2024. "Foreseeing Extreme Technological Risk." In *Managing Extreme Technological Risk*, edited by Katherine Rhodes, 87–108. World Scientific Publishing Company.

Kemp, Luke, Chi Xu, Joanna Depledge, Kristie L. Ebi, Goodwin Gibbins, Timothy A. Kohler, et al. 2022. "Reply to Kelman: The Foundations for Studying Catastrophic Climate Risks." *Proceedings of the National Academy of Sciences* 119 (42): e2214794119.

Kenchington, Charlotte G., and Philip R. Wilby. 2014. "Of Time and Taphonomy: Preservation in the Ediacaran." *Geological Society of America Special Papers*, 20. https://doi.org/10.1017/S1089332600002825.

Kendig, Catherine. 2015. “Editor’s Introduction: Activities of Kinding in Scientific Practice.” In *Natural Kinds and Classification in Scientific Practice*, edited by Catherine Kendig, 1–14. Routledge.

Khalidi, Muhammad Ali. 2021. “Etiological Kinds.” *Philosophy of Science* 88 (1): 1–21.

Khalifa, Kareem, and Sanford C. Goldberg. 2022. “Socio-Functional Foundations in Science: The Case of Measurement.” *Philosophical Issues* 32 (1): 382–397.

Kidd, Ian James. 2011. “The Contingency of Science and the Future of Philosophy.” *Essays in Philosophy* 12 (2): 313–329.

Kinberg, Ori, and Arnon Levy. 2023. “The Epistemic Imagination Revisited.” *Philosophy and Phenomenological Research* 107 (2): 319–336.

Kind, Amy. 2016. “Imagining under Constraints.” In *Knowledge through Imagination*, edited by Amy Kind and Peter Kung, 145–159. Routledge.

———. 2018. “How Imagination Gives Rise to Knowledge.” In *Perceptual Imagination and Perceptual Memory*, edited by Fiona Macpherson and Fabian Dorsch, 145–159. Oxford University Press.

Kinzel, Katherina. 2015. “Narrative and Evidence: How Can Case Studies from the History of Science Support Claims in the Philosophy of Science?” *Studies in History and Philosophy of Science Part A* 49: 48–57.

Kitcher, Philip. 1984. “1953 and All That: A Tale of Two Sciences.” *The Philosophical Review* 93 (3): 335–373.

Klein, Nico, and Torsten M. Scheyer. 2017. “Microanatomy and Life History in Palaeopleurosaurus (Rhynchocephalia: Pleurosauridae) from the Early Jurassic of Germany.” *The Science of Nature* 104: 1–8.

Kleinhans, Maarten G., Chris J. Buskes, and Henk W. de Regt. 2005. “Terra Incognita: Explanation and Reduction in Earth Science.” *International Studies in the Philosophy of Science* 19 (3): 289–317.

Knuuttila, Tarja. 2005. “Models, Representation, and Mediation.” *Philosophy of Science* 72 (5): 1260–1271.

———. 2011. “Modelling and Representing: An Artefactual Approach to Model-Based Representation.” *Studies in History and Philosophy of Science Part A* 42 (2): 262–271.

———. 2021. “Imagination Extended and Embedded: Artifactual versus Fictional Accounts of Models.” *Synthese* 198 (Suppl. 21): 5077–5097.

———. 2024. “The Artifactual Approach to Modeling.” In *The Routledge Handbook of Philosophy of Scientific Modeling*, edited by Tarja Knuuttila, Natalia Carrillo, and Rami Koskinen, 111–125. Taylor & Francis.

Kolesnikov, Anton, Taniel Danelian, Maxime Gommeaux, Andrey V. Maslov, and Dmitriy V. Grazhdankin. 2017. “Arumberiamorph Structure in Modern Microbial Mats: Implications for Ediacaran Palaeobiology.” *Bulletin de la Société Géologique de France* 188 (1–2): 5.

Kosemen, C. M., Darren Naish, and John Conway. 2017. *All Your Yesterdays: Extraordinary Visions of Extinct Life from a New Generation of Palaeoartists.* Irregular Books.

Kosso, Peter. 1993. "Historical Evidence and Epistemic Justification: Thucydides as a Case Study." *History and Theory* 32 (1): 1–13.

———. 2001. *Knowing the Past: Philosophical Issues of History and Archaeology.* Bloomsbury Academic.

Krueger, Joel. 2014. "Affordances and the Musically Extended Mind." *Frontiers in Psychology* 4: 1003.

Kuhlemann, Karin. 2018. "Complexity, Creeping Normalcy and Conceit: Sexy and Unsexy Catastrophic Risks." *Foresight* 21 (1): 35–52.

Kung, Peter. 2016. "Imagination and Modal Epistemology." In *The Routledge Handbook of Philosophy of Imagination*, edited by Amy Kind, 437–450. Routledge.

Ladyman, James. 2012. "Science, Metaphysics and Method." *Philosophical Studies* 160: 31–51.

Ladyman, James, and Don Ross. 2007. *Every Thing Must Go: Metaphysics Naturalized.* Oxford University Press.

Laflamme, Marc, and Guy M. Narbonne. 2008. "Ediacaran Fronds." *Palaeogeography, Palaeoclimatology, Palaeoecology* 258 (3): 162–179.

Laflamme, Marc, Jonathan D. Schiffbauer, Guy M. Narbonne, and Derek E. G. Briggs. 2011. "Microbial Biofilms and the Preservation of the Ediacara Biota." *Lethaia* 44: 203–213.

Laflamme, Marc, James D. Schiffbauer, and Guy M. Narbonne. 2012. "Deep-Water Microbially Induced Sedimentary Structures (MISS) in Deep Time: The Ediacaran Fossil Ivesheadia." *Palaeogeography, Palaeoclimatology, Palaeoecology* 347–348: 1–12.

Laflamme, Marc, Shuhai Xiao, and Michał Kowaleski. 2009. "Osmotrophy in Modular Ediacara Organisms." *Proceedings of the National Academy of Sciences* 106 (34): 14438–14443.

Lampert, Winfried. 1993. "Ultimate Causes of Diel Vertical Migration of Zooplankton: New Evidence for the Predator-Avoidance Hypothesis." In *Diel Vertical Migration of Zooplankton*, edited by Joop Ringelberg, 79–88. Springer.

Lange, Marc. 2008. "Could the Laws of Nature Change?" *Philosophy of Science* 75 (1): 69–92.

Lean, Collin H. 2020. "Why Wake the Dead? Identity and De-Extinction." *Journal of Agricultural and Environmental Ethics* 33 (3–6): 571–589.

Le Bihan, Soazig. 2016. "Enlightening Falsehoods: A Modal View of Understanding." In *Explaining Understanding: New Perspectives from Epistemology and*

Philosophy of Science, edited by Stephen R. Grimm, Christoph Baumberger, and Sabine Ammon, 111–135. Routledge.

Leonelli, Sabina. 2008. "Performing Abstraction: Two Ways of Modelling Arabidopsis thaliana." *Biology & Philosophy* 23: 509–528.

———. 2009. "Understanding in Biology: The Impure Nature of Biological Knowledge." In *Scientific Understanding: Philosophical Perspectives*, edited by Henk W. de Regt, Sabina Leonelli, and Kai Eigner, 189–209. University of Pittsburgh Press.

———. 2018. "The Time of Data: Timescales of Data Use in the Life Sciences." *Philosophy of Science* 85 (5): 741–754.

———. 2019. *Data-Centric Biology: A Philosophical Study*. University of Chicago Press.

Lerner, Gerda. 1998. *Why History Matters: Life and Thought*. Oxford University Press.

Levy, Arnon. 2015. "Modeling without Models." *Philosophical Studies* 172: 781–798.

———. 2024. "Models, Fiction and the Imagination." In *The Routledge Handbook of Philosophy of Scientific Modeling*, edited by Tarja Knuuttila, Natalia Carrillo, and Rami Koskinen. Routledge.

Levy, Aron, and Adrian Currie. 2015. "Model Organisms are not (Theoretical) Models." *British Journal for the Philosophy of Science* 66 (2): 327–348.

Levy, Arnon, and Peter Godfrey-Smith, eds. 2020. *The Scientific Imagination*. Oxford University Press.

Lewis, David K. 1976. "The Paradoxes of Time Travel." *American Philosophical Quarterly* 13 (2): 145–152.

———. 1978. "Truth in Fiction." *American Philosophical Quarterly* 15 (1): 37–46.

———. 1979. "Counterfactual Dependence and Time's Arrow." *Noûs* 13 (4): 455–476.

———. 1986. *On the Plurality of Worlds*. Blackwell.

Lewontin, Richard C. 2001. "Genes, Organisms and Environments." In *Cycles of Contingency: Developmental Systems and Evolution*, edited by Susan Oyama, Paul E. Griffiths, and Russell D. Gray, 59–66. MIT Press.

Lidgard, Scott, and Alan C. Love. 2018. "Rethinking Living Fossils." *BioScience* 68 (10): 760–770.

Lidgard, Scott, and Alan C. Love. 2021. "The Living Fossil Concept: Reply to Turner." *Biology & Philosophy* 36 (2): 13.

Ligrone, Roberto. 2019. "The Great Oxygenation Event." In *Biological Innovations That Built the World: A Four-Billion-Year Journey through Life and Earth History*, 129–154. Springer International Publishing.

Lindquist, Michael A. 2020. "Aesthetics at the Intersection of the Species Problem and De-Extinction Technology." *Environmental Values* 29 (5): 605–624.

Lipton, Peter. 2000. "Inference to the Best Explanation." In *A Companion to the Philosophy of Science*, edited by W. H. Newton-Smith, 184–193. Blackwell.

Liu, Alexander G. 2011. "Understanding the Ediacaran Assemblages of Avalonia: A Palaeoenvironmental, Taphonomic, and Ontogenetic Study." PhD diss., University of Oxford.

Liu, Alexander G., Charlotte G. Kenchington, and Emily G. Mitchell. 2015. "Remarkable Insights into the Paleoecology of the Avalonian Ediacaran Macrobiota." *Gondwana Research* 27 (4): 1355–1380.

Liu, Alexander G., Duncan McIlroy, and Martin D. Brasier. 2010. "First Evidence for Locomotion in the Ediacara Biota from the 565 Ma Mistaken Point Formation, Newfoundland." *Geology* 38 (2): 123–126.

Liu, Alexander G., Duncan McIlroy, Jonathan B. Antcliffe, and Martin D. Brasier. 2011. "Effaced Preservation in the Ediacara Biota and Its Implications for the Early Macrofossil Record." *Palaeontology* 54 (3): 607–630.

Liu, Hin-Yan, Kristian C. Lauta, and Matthijs M. Maas. 2018. "Governing Boring Apocalypses: A New Typology of Existential Vulnerabilities and Exposures for Existential Risk Research." *Futures* 102: 6–19.

Longino, Helen E. 2021. "Scaling Up; Scaling Down: What's Missing?" *Synthese* 198 (4): 2849–2863.

Lopes, Renato, Heinrich Theodor Frank, Francisco S. D. C. Buchmann, and Felipe Caron. 2017. "Megaichnus igen. nov.: Giant Paleoburrows Attributed to Extinct Cenozoic Mammals from South America." *Ichnos* 24 (2): 133–145.

Love, Alan C. 2008. "Explaining Evolutionary Innovations and Novelties: Criteria of Explanatory Adequacy and Epistemological Prerequisites." *Philosophy of Science* 75 (5): 874–886.

Lu, Wanyi, Andy Ridgwell, Ellen Thomas, Dalston S. Hardisty, Gemming Luo, Thomas J. Algeo, et al. 2018. "Late Inception of a Resiliently Oxygenated Upper Ocean." *Science* 361 (6398): 174–177.

Lucas, Gavin. 2015. "Evidence of What? On the Possibilities of Archaeological Interpretation." In *Material Evidence: Learning from Archaeological Practice*, edited by Robert Chapman and Alison Wylie, 311–323. Routledge.

Lyell, Charles. 1990 (1830). *Principles of Geology*. Vol. 1. University of Chicago Press.

Lyons, Timothy W., Charles W. Diamond, Noah J. Planavsky, Christopher T. Reinhard, and Chao Li. 2021. "Oxygenation, Life, and the Planetary System during Earth's Middle History: An Overview." *Astrobiology* 21 (8): 906–923.

Machamer, Peter, Lindley Darden, and Carl F. Craver. 2000. "Thinking about Mechanisms." *Philosophy of Science* 67 (1): 1–25.

Mackensen, Andreas, and Gerhard Schmiedl. 2019. "Stable Carbon Isotopes in Paleoceanography: Atmosphere, Oceans, and Sediments." *Earth-Science Reviews* 197: 102893.

Malaterre, Christophe. 2024. "Overdetermination, Underdetermination, and Epistemic Granularity in the Historical Sciences." *European Journal for Philosophy of Science* 14 (2): 23.

Manias, Chris. 2023. *The Age of Mammals: Nature, Development, and Paleontology in the Long Nineteenth Century.* University of Pittsburgh Press.

Mantell, Gideon Algernon. 1839. *The Wonders of Geology; or, A Familiar Exposition of Geological Phenomena: Being the Substance of a Course of Lectures Delivered at Brighton.* Vol. 1. A. H. Maltby.

Manucci, Fabio, and Marco Romano. 2023. "Reviewing the Iconography and the Central Role of 'Paleoart': Four Centuries of Geo-Palaeontological Art." *Historical Biology* 35 (1): 1–48.

Mariscal, Carlos. 2015. "Universal Biology: Assessing Universality from a Single Example." In *The Impact of Discovering Life beyond Earth*, edited by Steven J. Dick, 35–52. Cambridge University Press.

Massimi, Michela. 2019. "Two Kinds of Exploratory Models." *Philosophy of Science* 86 (5): 869–881.

———. 2021. "Realism, Perspectivism, and Disagreement in Science." *Synthese* 198 (Suppl 25): 6115–6141.

———. 2022. *Perspectival Realism.* Oxford University Press.

McAllister, James W. 2012. "Thought Experiment and the Exercise of Imagination in Science." In *Thought Experiments in Science, Philosophy, and the Arts*, edited by Melanie Frappier, Letitia Meynell, and James Robert Brown, 11–29. Routledge.

McConwell, Alison K. 2017. "Contingency and Individuality: A Plurality of Evolutionary Individuality Types." *Philosophy of Science* 84 (5): 1104–1116.

———. 2019. "Contingency's Causality and Structural Diversity." *Biology & Philosophy* 34 (2): 26.

McConwell, Alison K., and Adrian Currie. 2017. "Gouldian Arguments and the Sources of Contingency." *Biology & Philosophy* 32: 243–261.

McDermott, Alex. 2020. "Dinosaur Art Evolves with New Discoveries in Paleontology." *Proceedings of the National Academy of Sciences* 117 (6): 2728–2731.

McGhee, George R. 2011. *Convergent Evolution: Limited Forms Most Beautiful.* MIT Press.

McKenzie, Kerry. 2025. "Methods in the Metaphysics of Science." In *Methods in the Philosophy of Science: A User's Guide*, edited by Sophie Veigl and Adrian Currie. MIT Press.

McLaughlin, Alex. 2025. "Visionaries and Crackpots, Maniacs and Saints: Existential Risk and the Politics of Longtermism." *Ratio.* https://doi.org/10.1111/rati.12449.

McMahon, Sean. 2021. "Astrobiology (Overview)." *Oxford Research Encyclopedia of Planetary Science*. Oxford University Press.

Meadows, Victoria S., Christopher T. Reinhard, Giada N. Arney, Mary N. Parenteau, Edward W. Schwieterman, Shawn D. Domagal-Goldman, et al. 2018. "Exoplanet Biosignatures: Understanding Oxygen as a Biosignature in the Context of Its Environment." *Astrobiology* 18 (6): 630–662.

Meincke, Anne. 2023. "The Metaphysics of Development and Evolution: From Thing Ontology to Process Ontology." *Human Development* 67 (5–6): 233–256.

Meneganzin, Andra, and Adrian Currie. 2022. "Behavioural Modernity, Investigative Disintegration, and Rubicon Expectation." *Synthese* 200 (1): 47.

Meneganzin, Andra, and Adrian Currie. 2025. "Not Wasted on the Young: Childhood, Trait Complexes, and Human Behavioral Ecology." *Studies in History and Philosophy of Science* 109: 12–20.

Meneganzin, Andra, and Anton Killin. 2024. "Beyond Reasonable Doubt: Reconsidering Neanderthal Aesthetic Capacity." *Phenomenology and the Cognitive Sciences*: 1–33.

Meneganzin, Andra, Grant Ramsey, and James DiFrisco. 2024. "What Is a Trait? Lessons from the Human Chin." *Journal of Experimental Zoology Part B: Molecular and Developmental Evolution* 342 (2): 65–75.

Menon, Latha R., Duncan McIlroy, and Martin D. Brasier. 2013. "Evidence for Cnidaria-Like Behavior in ca. 560 Ma Ediacaran Aspidella." *Geology* 41: 895–898.

Meynell, Letitia. 2014. "Imagination and Insight: A New Account of the Content of Thought Experiments." *Synthese* 191: 4149–4168.

Miller, Boaz. 2016. "What Is Hacking's Argument for Entity Realism?" *Synthese* 193 (3): 991–1006.

Millikan, Ruth Garrett. 1999. "Historical Kinds and the 'Special Sciences.'" *Philosophical Studies* 95 (1–2): 45–65.

Millstein, Roberta L. 2011. "Chances and Causes in Evolutionary Biology: How Many Chances Become One Chance." In *Causality in the Sciences*, edited by Phyllis McKay Illari, Federica Russo, and Jon Williamson, 425–445. Oxford University Press.

Mitchell, Emily G., and Nicholas J. Butterfield. 2018. "Spatial Analyses of Ediacaran Communities at Mistaken Point." *Paleobiology* 44 (1): 40–57.

Mitchell, Emily G., Simon Harris, Charlotte G. Kenchington, Philip Vixseboxse, Lucy Roberts, Catherine Clark, et al. 2019. "The Importance of Neutral over Niche Processes in Structuring Ediacaran Early Animal Communities." *Ecology Letters* 22 (12): 2028–2038.

Mitchell, Ross N., Nan Zhang, Johanna Salminen, Yebo Liu, Christopher J. Spencer, Bernhard Steinberger, et al. 2021. "The Supercontinent Cycle." *Nature Reviews Earth & Environment* 2 (5): 358–374.

Mitchell, Sandra D. 1997. "Pragmatic Laws." *Philosophy of Science* 64 (S4): S468–S479.

———. 2002. "Integrative Pluralism." *Biology & Philosophy* 17: 55–70.

Monnin, Victor. 2023. "The Dinosaur Renaissance 1960s–80s: A Foundational Episode for the Historiography of Paleoart." *HoST: Journal of History of Science and Technology* 17 (1): 4–16.

Moyal, Ann, and Robert E. Marks. 2019. "The Scientists and Darwin's The Origin of Species in Nineteenth-Century Australia: A Re-Evaluation." *Journal and Proceedings of the Royal Society of New South Wales* 152 (471–472): 5–26.

Mukherjee, Indrani, Ross R. Large, Ross Corkrey, and Leonid V. Danyushevsky. 2018. "The Boring Billion, a Slingshot for Complex Life on Earth." *Scientific Reports* 8 (1): 4432.

Murphy, Alice. 2020. "The Aesthetic and Literary Qualities of Scientific Thought Experiments." In *The Aesthetics of Science*, edited by Milena Ivanova and Steven French, 146–166. Routledge.

Murphy, Alice, Adrian Currie, and Kirsten Walsh. Forthcoming. "Aesthetics and Agency in Experiments." *Australasian Journal of Philosophy.*

Myers, Natasha. 2008. "Molecular Embodiments and the Body-Work of Modeling in Protein Crystallography." *Social Studies of Science* 38 (2): 163–199.

Naish, Darren. 2017. "Introduction." In *All Your Yesterdays*, edited by C.M. Kosemen, 6–8. Irregular Press.

Nance, R. Damian, J. Brendan Murphy, and M. Santosh. 2014. "The Supercontinent Cycle: A Retrospective Essay." *Gondwana Research* 25 (1): 4–29.

Narbonne, Guy M. 2004. "Modular Construction of Early Ediacaran Complex Life Forms." *Science* 305 (5687): 1141–1144.

———. 2005. "The Ediacara Biota: Neoproterozoic Origin of Animals and Their Ecosystems." *Annual Review of Earth and Planetary Sciences* 33: 421–442.

Narbonne, Guy M., Shuhai Xiao, Graham A. Shields, and James G. Gehling. 2012. "The Ediacaran Period." In *The Geologic Time Scale 2012*, edited by Felix M. Gradstein, James G. Ogg, Mark D. Schmitz, and Gabi M. Ogg, 413–435. Elsevier.

Nersessian, Nancy J. 2007. "Thought Experimenting as Mental Modeling: Empiricism without Logic." *Croatian Journal of Philosophy* 7 (20): 125–161.

Neto, Celso. 2019. "What Is a Lineage?" *Philosophy of Science* 86 (5): 1099–1110.

Nicholson, Alice E., Stuart J. Daines, Nick J. Mayne, Jake K. Eager-Nash, Tim M. Lenton, and Krisztian Kohary. 2022. "Predicting Biosignatures for Nutrient-Limited Biospheres." *Monthly Notices of the Royal Astronomical Society* 517 (1): 222–239.

Nicholson, Daniel J., and John Dupré. 2018. *Everything Flows: Towards a Processual Philosophy of Biology.* Oxford University Press.

Nolan, Daniel. 2015. *David Lewis.* Routledge.

Northcott, Robert. 2025. *Science for a Fragile World.* Oxford University Press.

Norton, John D. 1996. "Are Thought Experiments Just What You Thought?" *Canadian Journal of Philosophy* 26 (3): 333–366.

Novick, Aaron. 2017. "Metaphysics and the Vera Causa Ideal: The Nun's Priest's Tale." *Erkenntnis* 82 (5): 1161–1176.

Novick, Aaron, Adrian M. Currie, Eden W. McQueen, and Nathan L. Brouwer. 2020. "Kon-Tiki Experiments." *Philosophy of Science* 87 (2): 213–236.

O'Connor, Ralph. 2008. *The Earth on Show: Fossils and the Poetics of Popular Science, 1802–1856.* University of Chicago Press.

———. 2015. "Hyena-Hunting and Byron-Bashing in the Old North: William Buckland, Geological Verse and the Radical Threat." In *Uncommon Contexts: Encounters Between Science and Literature*, edited by Ralph O'Connor, Ben Marzden, and Hazel Hutchinson, 55–82. Routledge.

Odenbaugh, Jay. 2023. "Philosophy and Ethics of De-Extinction." *Cambridge Prisms: Extinction* 1: e7.

Ohnesorge, Miguel, and Aja Watkins. 2024. "What Is Philosophy of the Geosciences?" *Philosophy Compass* 19 (2): e12962.

O'Malley, Maureen A. 2016. "Histories of Molecules: Reconciling the Past." *Studies in History and Philosophy of Science Part A* 55: 69–83.

Page, Meghan D. 2021. "The Role of Historical Science in Methodological Actualism." *Philosophy of Science* 88 (3): 461–482.

Palin, Richard M., and M. Santosh. 2021. "Plate Tectonics: What, Where, Why, and When?" *Gondwana Research* 100: 3–24.

Patton, Lydia. 2014. "Introduction to 'Catastrophism and Uniformitarianism.'" In *Philosophy, Science, and History: A Guide and Reader*, edited by Lydia Patton, 377–385. Routledge.

Paul, Gregory S. 1987. "The Science and Art of Restoring the Life Appearance of Dinosaurs and Their Relatives: A Rigorous How-To Guide." In *Dinosaurs Past and Present*, vol. 2, edited by Sylvia J. Czerkas and Everett C. Olson, 4–49. University of Washington Press.

Paul, L. A. 2012. "Metaphysics as Modeling: The Handmaiden's Tale." *Philosophical Studies* 160: 1–29.

Pemberton, S. George. 2010. "History of Ichnology: The Reverend William Buckland (1784–1856) and the Fugitive Poets." *Ichnos* 17 (4): 246–263.

Pinti, Jérôme, Tim DeVries, Tommy Norin, Camilla Serra-Pompei, Roland Proud, David A. Siegel, et al. 2023. "Model Estimates of Metazoans' Contributions to the Biological Carbon Pump." *Biogeosciences* 20 (5): 997–1009.

Potochnik, Angela. 2010. "Explanatory Independence and Epistemic Interdependence: A Case Study of the Optimality Approach." *The British Journal for the Philosophy of Science* 61 (1): 213–233.

———. 2017. *Idealization and the Aims of Science*. University of Chicago Press.

Powell, Russell. 2020. *Contingency and Convergence: Toward a Cosmic Biology of Body and Mind*. MIT Press.

Powell, Russell, and Carlos Mariscal. 2015. "Convergent Evolution as Natural Experiment: The Tape of Life Reconsidered." *Interface Focus* 5 (6): 20150040.

Preston, Beth. 2009. "Philosophical Theories of Artifact Function." In *Philosophy of Technology and Engineering Sciences*, edited by Anthonie W. M. Meijers, 213–233. North-Holland.

Raup, David M. 1992. *Extinction: Bad Genes or Bad Luck?* W. W. Norton.

Reinhard, Franziska. Forthcoming. "Direct and Circumstantial Traces." *Philosophy of Science*.

Reutlinger, Alexander, Dominik Hangleiter, and Stephan Hartmann. 2018. "Understanding (with) Toy Models." *The British Journal for the Philosophy of Science* 69 (4): 1069–1099.

Reydon, Thomas A. C. 2022. "Deflating the De-Extinction Debates: Domination and Artifactuality Are Not the Problem." *Ethics, Policy & Environment* 25 (2): 113–115.

———. 2023. "The Proper Role of History in Evolutionary Explanations." *Noûs* 57 (1): 162–187.

Rheinberger, Hans-Jörg. 1997. *Toward a History of Epistemic Things: Synthesizing Proteins in the Test Tube*. Stanford University Press.

———. 2010. *An Epistemology of the Concrete: Twentieth-Century Histories of Life*. Duke University Press.

———. 2023. *Split and Splice: A Phenomenology of Experimentation*. University of Chicago Press.

Rhodes, Catherine. 2024. "Introduction to Managing Extreme Technological Risk." In *Managing Extreme Technological Risk*, edited by Catherine Rhodes, 1–5. Routledge.

Richards, Robert J. 2016. *Biological Classification: A Philosophical Introduction*. Cambridge University Press.

Rieppel, Lukas. 2012. "Bringing Dinosaurs Back to Life: Exhibiting Prehistory at the American Museum of Natural History." *Isis* 103 (3): 460–490.

———. 2019. *Assembling the Dinosaur: Fossil Hunters, Tycoons, and the Making of a Spectacle*. Harvard University Press.

———. 2020. "How Dinosaurs Became Tyrants of the Prehistoric." *Environmental History* 25 (2): 215–241.

Rieppel, Olivier. 2023. "'Living Fossils' and the Mosaic Evolution of Characters." *Frontiers in Ecology and Evolution* 11: 1119418.

Rice, Collin. 2018. "Idealized Models, Holistic Distortions, and Universality." *Synthese* 195 (6): 2795–2819.

Rini, Adrian A., and Max J. Cresswell. 2012. *The World-Time Parallel: Tense and Modality in Logic and Metaphysics*. Cambridge University Press.

Rockström, Johan, Will Steffen, Kevin J. Noone, Åsa Persson, F. Stuart Chapin III, E. F. Lambin, et al. 2009. "Planetary Boundaries: Exploring the Safe Operating Space for Humanity." *Ecology and Society* 14 (2): 32.

Roberts, Tom, and Joel Krueger. 2022. "Musical Agency and Collaboration in the Digital Age." In *Collaborative Embodied Performance: Ecologies of Skill*, edited by Kath Bicknell and John Sutton. Bloomsbury.

Romano, Marco, and Richard L. Cifelli. 2015. "100 Years of Continental Drift." *Science* 350 (6263): 915–916.

Romano, Marco, Fabiana Console, Marco Pantaloni, and Jörg Fröbisch. 2017. "100 Years of Continental Drift: The Early Italian Reaction to Wegener's 'Visionary' Theory." *Historical Biology* 29: 266–287.

Roth, Paul A. 2019. *The Philosophical Structure of Historical Explanation*. Northwestern University Press.

Routledge, Bruce. 2023. "An Archaeology of Traces." *Cambridge Archaeological Journal*: 1–14.

Rucińska, Zuzanna, and Shaun Gallagher. 2021. "Making Imagination Even More Embodied: Imagination, Constraint and Epistemic Relevance." *Synthese* 199 (3): 8143–8170.

Rudwick, Martin J. S. 1970. "The Strategy of Lyell's Principles of Geology." *Isis* 61: 4–33.

———. 1975. "Caricature as a Source for the History of Science: De la Beche's Anti-Lyellian Sketches of 1831." *Isis* 66 (4): 534–560.

———. 1992. *Scenes from Deep Time: Early Pictorial Representations of the Prehistoric World*. University of Chicago Press.

———. 2005. *Bursting the Limits of Time: The Reconstruction of Geohistory in the Age of Revolution*. University of Chicago Press.

———. 2014. *Earth's Deep History: How It Was Discovered and Why It Matters*. University of Chicago Press.

Rupke, Nicolaas A. 1983. *The Great Chain of History: William Buckland and the English School of Geology (1814–1849).* Oxford University Press.

Salis, Fiora. 2020. "Of Predators and Prey: Imagination in Scientific Modeling." In *Imagination and Art: Explorations in Contemporary Theory*, edited by Keith Moser and Ananta Ch. Sukla, 451–474. Brill.

———. 2021. "Bridging the Gap: The Artifactual View Meets the Fiction View of Models." In *Models and Idealizations in Science: Artifactual and Fictional Approaches*, edited by Alejandro Cassini and Juan Redmond, 159–177. Springer International Publishing.

Salis, Fiora, and Roman Frigg. 2020. "Capturing the Scientific Imagination." In *The Scientific Imagination: Philosophical and Psychological Perspectives*, edited by Arnon Levy and Peter Godfrey-Smith, 17–50. Oxford University Press.

Sánchez-Dorado, Julia. 2024a. "The Novel Naturalness of Abstract Space." In *Abstraction in Science and Art: Philosophical Perspectives*, edited by Chiara Ambrosio and Julia Sánchez-Dorado, 143–169. Routledge.

———. 2024b. "The Earth in the Model: The Nomothetic, Idiographic, and Plural Epistemic Aims of Planetary Modelling." *The Anthropocene Review*. https://doi.org/10.1177/20530196241302084.

Savojardo, Valentina. 2024. "Imagination and Creativity in Science: An 'Embodied' Perspective." *Global Philosophy* 34 (1): 7.

Schuster, Joshua, and Derek Woods. 2021. *Calamity Theory: Three Critiques of Existential Risk*. University of Minnesota Press.

Scotese, Christopher R. 2018. "Atlas of Future Plate Tectonic Reconstructions: Modern World to Pangea Proxima." *CR Scotese PALEOMAP Project*.

Secord, James A. 2014. *Visions of Science: Books and Readers at the Dawn of the Victorian Age*. University of Chicago Press.

Seddon, Philip J. 2017. "The Ecology of De-Extinction." *Functional Ecology* 31 (5): 992–995.

Seilacher, Adolf. 1999. "Biomat-Related Lifestyles in the Precambrian." *Palaios* 14 (1): 86–93.

Senter, Phil, and Sara L. Juengst. 2016. "Record-Breaking Pain: The Largest Number and Variety of Forelimb Bone Maladies in a Theropod Dinosaur." *PLoS ONE* 11 (2): e0149140.

Sepkoski, David. 2019. *Rereading the Fossil Record: The Growth of Paleobiology as an Evolutionary Discipline*. University of Chicago Press.

———. 2020. *Catastrophic Thinking: Extinction and the Value of Diversity from Darwin to the Anthropocene*. University of Chicago Press.

Shapin, Steven. 1989. "The Invisible Technician." *American Scientist* 77 (6): 554–563.

Shapin, Steven, and Simon Schaffer. 1986. *Leviathan and the Air-Pump: Hobbes, Boyle, and the Experimental Life*. Princeton University Press.

Shapiro, Beth. 2015. *How to Clone a Mammoth: The Science of De-Extinction.* Princeton University Press.

Shen, Bing, Lin Dong, Shushai Xiao, and Michał Kowalewski. 2008. "The Avalon Explosion: Evolution of Ediacara Morphospace." *Science* 319 (5859): 81-84.

Sheredos, Benjamin, and William Bechtel. 2020. "Imagining Mechanisms with Diagrams." In *The Scientific Imagination: Philosophical and Psychological Perspectives*, edited by Arnon Levy and Peter Godfrey-Smith, 178–201. Oxford University Press.

Shoshani, Jeheskel. 1998. "Understanding Proboscidean Evolution: A Formidable Task." *Trends in Ecology & Evolution* 13 (12): 480–487.

Siipi, Helena, and Leonard Finkelman. 2017. "The Extinction and De-Extinction of Species." *Philosophy & Technology* 30: 427–441.

Smith, John Maynard, and Eörs Szathmáry. 1997. *The Major Transitions in Evolution.* Oxford University Press.

Sober, Elliot. 2002. "Reconstructing the Character States of Ancestors: A Likelihood Perspective on Cladistic Parsimony." *The Monist* 85 (1): 156–176.

Sober, Elliott, and Mike Steel. 2014. "Time and Knowability in Evolutionary Processes." *Philosophy of Science* 81 (4): 558–579.

Soler, Léna, Emiliano Trizio, and Andrew Pickering, eds. 2016. *Science as It Could Have Been: Discussing the Contingency/Inevitability Problem.* University of Pittsburgh Press.

Sommer, Marianne. 2003. "The Romantic Cave? The Scientific and Poetic Quests for Subterranean Spaces in Britain." *Earth Sciences History* 22 (2): 172–208.

Steel, Daniel. 2008. *Across the Boundaries: Extrapolation in Biology and Social Science.* Oxford University Press.

Steffen, Will, Katherine Richardson, Johan Rockström, Sarah E. Cornell, Ingo Fetzer, Elena M. Bennett, et al. 2015. "Planetary Boundaries: Guiding Human Development on a Changing Planet." *Science* 347 (6223): 1259855.

Sterelny, Kim. 1996. "Explanatory Pluralism in Evolutionary Biology." *Biology & Philosophy* 11: 193–214.

———. 2005. "Another View of Life." *Studies in History and Philosophy of Science Part C: Studies in History and Philosophy of Biological and Biomedical Sciences* 36 (3): 585–593.

———. 2010. "Minds: Extended or Scaffolded?" *Phenomenology and the Cognitive Sciences* 9 (4): 465–481.

———. 2016. "Contingency and History." *Philosophy of Science* 83 (4): 521–539.

Stern, Robert J. 2018. "The Evolution of Plate Tectonics." *Philosophical Transactions of the Royal Society A: Mathematical, Physical and Engineering Sciences* 376 (2132): 20170406.

Stiner, Mary C. 2017. "Love and Death in the Stone Age: What Constitutes First Evidence of Mortuary Treatment of the Human Body?" *Biological Theory* 12: 248–261.

Stokstad, Erik. 2015. "Bringing Back the Aurochs." *Science* 350: 1144–1147.

Strevens, Michael. 2009. *Bigger than Chaos: Understanding Complexity through Probability*. Harvard University Press.

Stuart, Michael T. 2016. "Taming Theory with Thought Experiments: Understanding and Scientific Progress." *Studies in History and Philosophy of Science Part A* 58: 24–33.

———. 2020. "The Productive Anarchy of Scientific Imagination." *Philosophy of Science* 87 (5): 968–978.

———. 2021. "Towards a Dual Process Epistemology of Imagination." *Synthese* 198 (2): 1329–1350.

———. 2022. "Sharpening the Tools of Imagination." *Synthese* 200 (6): 451.

———. 2023. "The Future Won't Be Pretty: The Nature and Value of Ugly, AI-Designed Experiments." In *The Aesthetics of Scientific Experiments*, edited by Milena Ivanova and Alice Murphy, 215–233. Routledge.

———. 2024. "The Qualitative Study of Scientific Imagination." *Qualitative Psychology* 11 (2): 277–292.

———. 2025. "A New Account of Pragmatic Understanding, Applied to the Case of AI-Assisted Science." *Philosophical Studies.*

Stuart, Michael T., and Anatoli Kozlov. 2024. "Moving Targets and Models of Nothing: A New Sense of Abstraction for Philosophy of Science." In *Abstraction in Science and Art: Philosophical Perspectives*, edited by Chiara Ambrosio and Julia Sánchez-Dorado, 118–142. Routledge.

Sundaram, Lalitha S., and Adrian Currie. 2024. "Extreme Risk and the Culture of Science: Two Challenges." In *Managing Extreme Technological Risk*, edited by Catherine Rhodes, 7–23. Routledge.

Sundaram, Lalitha S., Matthijs M. Maas, and S. J. Beard. 2024. "Seven Questions for Existential Risk Studies." In *Managing Extreme Technological Risk*, edited by Catherine Rhodes. Routledge.

Swaim, Daniel G. 2021. "What Is Narrative Possibility?" *Studies in History and Philosophy of Science Part A* 89: 257–266.

———. 2024. "Getting from Here to There: The Contingency of Historical Evidence and the Value of Speculation." *Studies in History and Philosophy of Science* 106: 118–125.

Swoyer, Chris. 1999. "How Ontology Might Be Possible: Explanation and Inference in Metaphysics." *Midwest Studies in Philosophy* 23: 100–131.

Tal, Eran. 2016. *Making Time: A Study in the Epistemology of Measurement*. Oxford University Press.

———. 2017. "Calibration: Modelling the Measurement Process." *Studies in History and Philosophy of Science Part A* 65: 33–45.

Tamborini, Marco. 2020. "Technoscientific Approaches to Deep Time." *Studies in History and Philosophy of Science Part A* 79: 57–67.

———. 2024. "Data, Aesthetics, and Visualizations of Deep Time." In *Handbook of the Historiography of the Earth and Environmental Sciences*, edited by Elena Aronova, David Sepkoski, and Marco Tamborini, 1–20. Springer Nature Switzerland.

Taylor, Rod S., Robert Nicholls, Jenna M. Neville, and Duncan Mcilroy. 2023. "Morphological Variation in the Rangeomorph Organism Fractofusus misrai from the Ediacaran of Newfoundland, Canada." *Geological Magazine* 160 (1): 146–166.

Teal, Lorna R., Mark Bulling, Eleanor R. Parker, and Martin Solan. 2008. "Global Patterns of Bioturbation Intensity and Mixed Depth of Marine Soft Sediments." *Aquatic Biology* 2 (3): 207–218.

Thomasson, Amie L. 2020. "If Models Were Fictions, Then What Would They Be?" In *The Scientific Imagination: Philosophical and Psychological Perspectives*, edited by Arnon Levy and Peter Godfrey-Smith, 51–74. Oxford University Press.

Tilly, Charles. 2006. "Why and How History Matters." In *The Oxford Handbook of Contextual Political Analysis*, edited by Robert E. Goodin and Charles Tilly, 417–437. Oxford University Press.

Tonn, Bruce, and Dorian Stiefel. 2013. "Evaluating Methods for Estimating Existential Risks." *Risk Analysis* 33 (10): 1772–1787.

Toon, Adam. 2012. *Models as Make-Believe: Imagination, Fiction and Scientific Representation*. Springer.

———. 2014. "Empiricism for Cyborgs." *Philosophical Issues* 24 (1): 409–425.

———. 2016. "Imagination in Scientific Modeling." In *The Routledge Handbook of Philosophy of Imagination*, edited by Amy Kind, 451–462. Routledge.

Torres, Phil. 2023. "Existential Risks: A Philosophical Analysis." *Inquiry* 66 (4): 614–639.

Tosh, John. 2019. *Why History Matters*. Bloomsbury Publishing.

Tucker, Aviezer. 1998. "Unique Events: The Underdetermination of Explanation." *Erkenntnis* 48: 61–83.

———. 2011. "Historical Science, Over- and Underdetermined: A Study of Darwin's Inference of Origins." *The British Journal for the Philosophy of Science* 62 (4): 805–829.

Turner, Derek D. 2004. "The Past vs. the Tiny: Historical Science and the Abductive Arguments for Realism." *Studies in History and Philosophy of Science Part A* 35 (1): 1–17.

———. 2005. "Local Underdetermination in Historical Science." *Philosophy of Science* 72 (1): 209–230.

———. 2007. *Making Prehistory: Historical Science and the Scientific Realism Debate.* Cambridge University Press.

———. 2009. "Beyond Detective Work: Empirical Testing in Paleontology." In *The Paleobiological Revolution: Essays on the Growth of Modern Paleontology*, edited by David Sepkoski and Michael Ruse, 201–214. University of Chicago Press.

———. 2016. "A Second Look at the Colors of the Dinosaurs." *Studies in History and Philosophy of Science Part A* 55: 60–68.

———. 2017. "De-Extinction as Artificial Species Selection." *Philosophy & Technology* 30: 395–411.

———. 2019a. *Paleoaesthetics and the Practice of Paleontology.* Cambridge University Press.

———. 2019b. "In Defense of Living Fossils." *Biology & Philosophy* 34 (2): 23.

Turner, Derek, and Joyce C. Havstad. 2025. "Philosophy of Macroevolution." In *The Stanford Encyclopedia of Philosophy*, Spring 2025 Edition, edited by Edward N. Zalta and Uri Nodelman. https://plato.stanford.edu/archives/spr2025/entries/macroevolution/.

Turner, Derek, and Junhyung Han. 2023. "Living Fossils and Conservation Values." *Frontiers in Earth Science* 11: 1086066.

Turner, Derek D., and Michelle I. Turner. 2021. "'I'm Not Saying It Was Aliens': An Archaeological and Philosophical Analysis of a Conspiracy Theory." In *Explorations in Archaeology and Philosophy*, edited by Anton Killin and Sean Allen-Hermanson, 7–24. Springer International Publishing.

Valde, Katherine. 2024. "From Metaphysics to Methods? Pluralism in Cancer Research." *Acta Analytica*: 1–19.

Valde, Katherine, and Eric Scarffe. 2025. *A Jurisprudential Philosophy of Science: Beyond the Value-Free Ideal.* Springer Nature.

van Fraassen, Bas C. 2008. *Scientific Representation: Paradoxes of Perspective.* Oxford University Press.

Varricchio, David J., Anthony J. Martin, and Yoshihiro Katsura. 2007. "First Trace and Body Fossil Evidence of a Burrowing, Denning Dinosaur." *Proceedings of the Royal Society B: Biological Sciences* 274 (1616), 1361–1368.

Walsh, Denis M. 2015. *Organisms, Agency, and Evolution.* Cambridge University Press.

Walsh, Kirsten. 2019. "Newton's Scaffolding: The Instrumental Roles of His Optical Hypotheses." In *Experiment, Speculation and Religion in Early Modern Philosophy*, edited by Alberto Vanzo and Peter R. Anstey, 125–157. Routledge.

Walsh, Kirsten, Adrian M. Currie, and Tom Roberts. 2025. "Situated Cognition in Early Modern Experimentation: The Case of Compelled Assent." *The British Journal for the Philosophy of Science*. https://doi.org/10.1086/738238.

Walton, Kendall L. 1993. *Mimesis as Make-Believe: On the Foundations of the Representational Arts*. Harvard University Press.

Waters, C. Kenneth. 2017. "No General Structure." In *Metaphysics and the Philosophy of Science: New Essays*, edited by Matthew H. Slater and Zanja Yudell, 81–108. Oxford University Press.

Watkins, Aja. 2021. "The Epistemic Value of the Living Fossils Concept." *Philosophy of Science* 88 (5): 1221–1233.

———. 2023. "Scaling Procedures in Climate Science: Using Temporal Scaling to Identify a Paleoclimate Analogue." *Studies in History and Philosophy of Science* 102: 31–44.

———. 2024a. "When Are Fossils Data?" *Philosophy, Theory, and Practice in Biology* 16 (2).

———. 2024b. "Paleoclimate Proxies and the Benefits of Disunity." *Philosophy of Science*: 1–18.

———. 2024c. "Using Paleoclimate Analogues to Inform Climate Projections." *Perspectives on Science*: 1–45.

Weber, Bruce H., and David J. Depew. 2004. "Darwinism, Design, and Complex Systems Dynamics." In *Debating Design: From Darwin to DNA*, edited by William A. Dembski and Michael Ruse, 173–190. Cambridge University Press.

Weber, Marcel. 2004. *Philosophy of Experimental Biology*. Cambridge University Press.

Weisberg, Michael. 2007a. "Who Is a Modeler?" *The British Journal for the Philosophy of Science* 58 (2): 207–233.

———. 2007b. "Three Kinds of Idealization." *The Journal of Philosophy* 104 (12): 639–659.

Werth, Alexander J., and W.A. Shear. 2014. "The Evolutionary Truth about Living Fossils." *American Scientist* 102 (6): 434–443.

Wilby, Philip R., John N. Carney, and Michael P. A. Howe. 2011. "A Rich Ediacaran Assemblage from Eastern Avalonia: Evidence of Early Widespread Diversity in the Deep Ocean." *Geology* 39 (7): 655–658.

Wilson, Joseph. 2023. "Paleoclimate Analogues and the Threshold Problem." *Synthese* 202 (1): 17.

———. 2024. "Is Model-Based Science a Kind of Historical Science?" *Perspectives on Science* 32 (4): 460–487.

Wilson, Joseph, and Frank G. Boudinot. 2022. "Proxy Measurement in Paleoclimatology." *European Journal for Philosophy of Science* 12 (1): 14.

Wilson, Robert A. 2004. *Boundaries of the Mind: The Individual in the Fragile Sciences: Cognition*. Cambridge University Press.

———. 2005. *Genes and the Agents of Life: The Individual in the Fragile Sciences: Biology*. Cambridge University Press.

Wimsatt, William C. 1987. "False Models as a Means to Truer Theories." In *Neutral Models in Biology*, edited by Matthew H. Nitecki and Antoni Hoffman, 23–55. Oxford University Press.

———. 2007. *Re-Engineering Philosophy for Limited Beings: Piecewise Approximations to Reality*. Harvard University Press.

Wirling, Ylwa Sjölin, and Till Grüne-Yanoff. 2021. "The Epistemology of Modal Modeling." *Philosophy Compass* 16 (10): e12775.

Wirling, Ylwa Sjölin, and Till Grüne-Yanoff. 2024. "Epistemic and Objective Possibility in Science." *The British Journal for the Philosophy of Science* 75 (4): 821–841.

Witteveen, Joeri. 2024. "Golden Spikes, Scientific Types, and the Ma(r)king of Deep Time." *Studies in History and Philosophy of Science* 106: 70–85.

Witton, Mark P. 2018. *The Palaeoartist's Handbook: Recreating Prehistoric Animals in Art*. The Crowood Press.

Witton, Mark P., Darren Naish, and John Conway. 2014. "State of the Palaeoart." *Palaeontologia Electronica* 17 (3): 17.3.3E.

Wood, Jamie R., Sarah J. Richardson, Matt S. McGlone, and Janet M. Wilmshurst. 2020. "The Diets of Moa (Aves: Dinornithiformes)." *New Zealand Journal of Ecology* 44 (1): 1–21.

Woodruff, D. Carey, and David J. Varricchio. 2011. "Experimental Modeling of a Possible Oryctodromeus Cubicularis (Dinosauria) Burrow." *Palaios* 26 (3), 140–151.

Wylie, Alison. 1985. "The Reaction Against Analogy." In *Advances in Archaeological Method and Theory*, edited by Michael B. Schiffer, 63–111. Academic Press.

———. 1999. "Rethinking Unity as a 'Working Hypothesis' for Philosophy of Science: How Archaeologists Exploit the Disunities of Science." *Perspectives on Science* 7 (3): 293–317.

———. 2017. "How Archaeological Evidence Bites Back: Strategies for Putting Old Data to Work in New Ways." *Science, Technology, & Human Values* 42 (2): 203–225.

———. 2019. "Rock, Bone, and Ruin: A Trace-Centric Appreciation." *Philosophy, Theory, and Practice in Biology* 11.

Wylie, Caitlin D. 2009. "Preparation in Action: Paleontological Skill and the Role of the Fossil Preparator." In *Methods in Fossil Preparation: Proceedings of the First Annual Fossil Preparation and Collections Symposium*, edited by M. A. Brown, John F. Kane, and W. G. Parker, 3–12.

———. 2015. "The Artist's Piece Is Already in the Stone: Constructing Creativity in Paleontology Laboratories." *Social Studies of Science* 45 (1): 31–55.

———. 2021. *Preparing Dinosaurs: The Work Behind the Scenes*. MIT Press.

———. 2024. "Timing Science: The Temporal Role of Scientists in the Construction of Data." *Philosophy, Theory, and Practice in Biology* 16 (2).

Xhignesse, Michelle-Antoine. 2024. "Distant Dinosaurs and the Aesthetics of Remote Art." *The British Journal of Aesthetics* 64 (3): 361–380.

Yoshida, Masaki. 2016. "Formation of a Future Supercontinent through Plate Motion–Driven Flow Coupled with Mantle Downwelling Flow." *Geology* 44 (9): 755–758.

Index

D

E

F

G

H

P

Q

R

S

T

U

V

W

Z

Kolibri 360 GmbH
Ferdinand-Jühlke-Straße 7
99095 Erfurt, Deutschland
produktsicherheit@kolibri360.de